Biotechnology from A to Z

DATE DUE

DEMCO 38-296

Biotechnology from A to Z

second edition

by William Bains

OXFORD
UNIVERSITY PRESS

OXFORD

UNIVERSITY PRESS

Great Clarendon Street, Oxford OX2 6DP

Oxford University Press is a department of the University of Oxford.
It furthers the University's objective of excellence in research, scholarship,
and education by publishing worldwide in

Oxford New York

Athens Auckland Bangkok Bogotá Buenos Aires Calcutta
Cape Town Chennai Dar es Salaam Delhi Florence Hong Kong Istanbul
Karachi Kuala Lumpur Madrid Melbourne Mexico City Mumbai
Nairobi Paris São Paulo Singapore Taipei Tokyo Toronto Warsaw

with associated companies in Berlin Ibadan

Oxford is a registered trade mark of Oxford University Press
in the UK and in certain other countries

Published in the United States
by Oxford University Press Inc., New York

First published 1998
Reprinted 2000

A catalogue record for this book is available from the British Library

Library of Congress Cataloging in Publication Data
Bains, William, 1955-   .
Biotechnology from A to Z / William Bains.—2nd ed.
Includes index.
1. Biotechnology—Dictionaries. I. Title.
TP248.16.B33 1998   660.6.'03—dc21   97–36132
ISBN 0 19 963693 1

Printed in Great Britain
on acid-free paper by
Bookcraft (Bath) Ltd., Midsomer Norton, Avon

Foreword

by G. Kirk Raab
Chairman of the Board
Connetics Corporation

In the foreword to the first edition of this book in 1993, I suggested that the increased knowledge it could bring about would help usher in 'the day when we cannot imagine life without biotechnology any more than we can imagine life without computers'. The many new entries in this revised and expanded edition suggest that this day is fast approaching, if not already here.

New entries for **artificial tissues**, **prosthetics**, and **transplants** suggest broad new medical applications. These have only begun to be realized primarily with skin and bone marrow tissue. The appearance of **fish farming, food**, and **flavour chemicals** implies how prevalent biotechnology has become in the food industry, which had long before relied on breeding techniques for genetic manipulation. New entries for **brewing** and for **cows** also point to biotechnology's utility in food production. Genetically engineered yeasts and proteins have boosted beer and milk production, respectively. (Of course these two latter entries also reflect on the impact biotechnology has had on large-scale production of genetically engineered proteins.)

The introduction of entries like **textiles, ornamental plants, fingerprinting**, and **waste disposal** makes clear that biotechnology will fill unique niche markets. The addition of the word **zoonosis** (which brings us AIDS, the ebola virus, and the hantavirus) reminds us that as new medical problems arise, biotechnology may be our best hope to tackle them.

Speeding biotechnology's impact on our lives is the advent of new tools. Long-time industry analyst G. Steven Burrill recently observed, 'Early in the industry we were defining [biotechnology] around a very narrow set of sciences that included such things as genetic engineering and recombinant technology. Today it's a whole toolbox of enabling technologies'.

Many of the shiny new tools appear as new entries in this book: **combinatorial chemistry, functional genomics, high throughput screening, LIMS, panning**, and **xenograft**, for example. Some of the new tools/entries — such as **bioprospecting, extremophile**, and **marine biotechnology** — simply reflect the natural resources with practical potential still to be tapped by biotechnology.

And don't worry. The older, trusty tools of **genetic engineering, recombinant DNA technology, DNA probes, PCR**, etc. are all still here, just as they are all still being used in laboratories around the world every day. Enhanced by the new tools, basic approaches of biotechnology have even more to offer. While small synthetic molecules and custom proteins are increasingly the future targets of pharmaceutical biotechnology, many natural proteins — and certainly the genes that code for them — have still much to contribute. Since the publication of the first edition, this has been evidenced by the anxious homesteading of the human genome (*see* **DNA sequence patents**) and the marketing of new medicines based on natural proteins or induced antibodies. Two examples are beta-interferon for treating multiple sclerosis and ReaproR for enhancing the success of angioplasty in the treatment of cardiovascular diseases.

But a review of the list of approved new biotechnology products versus failed products shows that the latter list is much longer than the former. Success in biotechnology continually proves to be extremely difficult to achieve. For biopharmaceutical companies, clinical success is often based as much on luck as it is on scientific prowess. A glance at some new entries in this book, such as **capital** and **health care reform**, reminds us that the industry's obstacles to success go far beyond the science.

But many established companies and dozens of products on the market prove that success is indeed possible. And I am convinced that the potential for this industry is greater than ever. In health care alone, in the next few years we can look for several new treatments for cancer with *minimal* side effects. We can expect advances in treating a variety of neurological diseases, an area that has largely remained a medical frontier. (One notable pioneering success so far was the recent approval of t-Pa as the first emergency treatment available for stroke.) We can expect new, stronger antibiotics to tackle the new, more resistant strains of bacteria now sneaking by our immune barriers. And we can expect some developments to help millions of people, and others to help dozens — with the advent of some new treatments for common, chronic disorders, such as diabetes, atopic dermatitis and asthma, and others for rare, inherited disorders, such as scleroderma.

If my allusion to the many diverse new entries has stimulated your curiosity enough to cause you to turn to some of them (who can resist ornamental plants or extremophile?) then I have done my job. If it has not yet, here's another curious point: symbiosis is rampant in this book. In many cases definitions discuss both how biotechnology impacts the topic and how the topic impacts biotechnology. I already mentioned breeding and cows. Consider also **antibiotics**, **artificial intelligence**, and **transplants** to name just a few.

I recommend you read and use this book. Not only because it is fascinating reading, but because, as in 1993 — whether you are a student, an educator, a legislator, a reporter, a health care provider, a patient, a consumer, an industry insider, or even a scientist — the understanding it can help bring about is essential to the continued success of biotechnology and the opportunity to improve the quality and length of human lives. Only with such understanding can we read about, for example, cloned sheep and do better than respond with the **yuk factor** (go ahead, look it up!). Instead, as enlightened U.S. legislators recently demonstrated, we can intelligently address the concerns it raises then move hopefully forward with a focus on the tremendous potential it offers.

Acknowledgements

The second edition of this book would have been less complete without the generous comments of Professor Tony Atkinson and Dr Nick Ashley. I also appreciated the comments of John Hodgson (Nature Biotechnology), Andy Richards (Chiroscience), and Keith Redpath and Lyn Scott (PA Consulting). My thanks also to Drs Jane Devereaux, Chris Lowe, Peter Hammond, Brian Richards for their contributions to the first issue and, surprising as it may seem, to many people who reviewed the first edition with the words 'but why does it not include . . .'. Now it does. Needless to say, none of the above bear any responsibility for what I put in, or what I left out.

How to use this book

This is not a textbook. It is an extended glossary of terms which will give you quick insight into a term or concept in one of the most exciting areas of applied science. The entries give a quick description of the concept, mention any related terms or ideas, and where relevant describe what practical application that idea has. It is a book for dipping into, for quick reference when puzzled or momentarily lost. You do not have to read half of it to get to the explanation you want.

The book is for the non-expert, and so it does not assume that you have a Ph.D. in biochemistry. Some familiarity with the basic ideas of modern biology is useful: if you are floored by terms such as 'bacterium' or 'DNA', then this book is definitely not for you. If you can get that far, then we can work together.

The book consists of around 380 entries for key terms or concepts in biotechnology. They are arranged alphabetically, so you can thumb through them to find what you want or just browse and see what catches your eye. However, if you want an answer fast then turn to the index at the back. Any term that is described in the book is listed there.

Explaining each term completely in a self-contained entry would mean a lot of redundancy, so many of the entries cut short discussions of topics which are covered elsewhere in the book, and simply refer you to that entry. You can pick your way through these connections until you feel that you have found enough of the background to the term you started with.

ADEPT (antibody-directed enzyme prodrug therapy)

This is a way to target a drug to a specific tissue. The targeting mechanism and the drug are administered separately. The drug is administered as an inactive prodrug (*see* **Drug delivery**) that does not itself have any effect. The prodrug can be converted into an active drug by an enzyme. In ADEPT, the converting enzyme can be, and indeed preferably is, one that does not occur normally in humans. Instead, it is administered with a second injection, coupled to an antibody that concentrates it in the target tissue. When the enzyme has arrived at the target tissue, the prodrug is activated there to form the active drug, while elsewhere it remains inactive.

This is being developed for tumour treatment. The prodrugs being considered are prodrugs of highly toxic antitumour compounds, which in their normal form have severe side-effects because they kill many cells other than the tumour cells. Using ADEPT these drugs can be targeted to the tumour by using an antibody that binds specifically to the tumour, thus sparing the rest of the body.

A new, closely related idea is GDEPT—gene-directed enzyme prodrug therapy. Here a cell is transfected with a gene that makes an enzyme, which itself converts a prodrug to a drug. A favourite is the thymidine kinase gene (TK) from herpes simplex virus, which converts gancyclovir into its active form. This is a version of using a 'suicide gene' in gene therapy.

Affinity chromatography

This is a method of separating molecules by using their ability to bind specifically to other molecules. This is of particular use in the separation of biological molecules because so many biomolecules bind very strongly and specifically to other molecules—their substrates, inhibitors, regulators, etc. Collectively, the molecule that binds to a larger molecule is called a **ligand**.

There are two types of biological affinity chromatography. Either a biological molecule can be immobilized and a smaller molecule that it is to bind to can be stuck to it, or the smaller ligand can be immobilized and the macromolecule stuck to it. (Of course, both 'sticker' and 'stickee' can be biological, too). A variant is to use an antibody as the immobilized molecule and use it to 'capture' its antigen.

Biological molecules used to separate smaller molecules include:

- Enzymes, to isolate substrates (only works if at least one substrate is missing from the mix, otherwise the enzyme just destroys what you are trying to separate).

- Antibodies (which when immobilized on a column, called an 'immunoaffinity column', can separate their antigen from a complex mixture).

- Cyclodextrins (for separating lipophilic materials in particular).

- Lectins (proteins that bind specific sugars very strongly and specifically, and which are therefore used for separating carbohydrates and anything with carbohydrates attached).

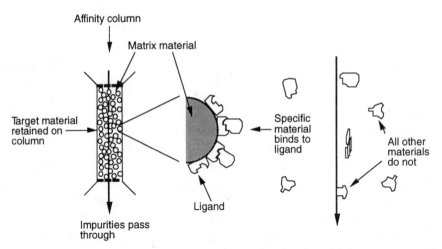

A variation is pseudoaffinity chromatography, in which a compound that is like a biological ligand is immobilized on a solid material and enzymes or other proteins are bound to it. A range of complex organic dyes are very good at binding some types of enzymes (especially dehydrogenases) because of their similarity to the enzyme's 'real' substrate NAD or NADP (nicotinamide–adenine dinucleotide or its phosphate). This is also called dye-ligand affinity chromatography. Other methods include metal affinity chromatography, where a metal ion is immobilized on a solid support: metal ions bind tightly and specifically to many biomolecules. The metal ion is bound to a chelator or chelating group, a chemical group that binds the metal extremely tightly. Such metal columns are used to purify proteins with His_6 **affinity tags**.

There are a wide range of support materials used in affinity chromatography. They are discussed further in the entry on **chromatography**.

To produce an affinity separation material, the solid support material to which the binder is going to be attached must be chemically activated. This is a process that takes the rather inert chemical material and adds a reactive chemical group to it, so that when the affinity-binding molecule is added to the support it reacts with it, forming a covalent link. Otherwise the affinity material would simply be washed off.

Affinity chromatography is widely used in research. It is also used in production, although the materials are usually too expensive for it to be useful for large-scale purification. It is used where a valuable product is to be isolated from a complex mixture of similar chemicals, of which the product is a minor component. Thus Armour Pharmaceuticals and Baxter Healthcare both isolate Factor VIII (used to treat haemophilia A) from blood using affinity chromatography. An antibody is linked on to a 'column' of solid material, and the plasma passed over it: the Factor VIII sticks and all the other proteins do not, resulting in a purification of roughly 200 000-fold.

Affinity tag

This is also sometimes called a purification tag, an affinity tail, or an affinity handle. An affinity tag is a bit of the amino acid sequence of a protein that has been engineered into the protein to make its purification easier. These can work in a number of ways.

If the protein is being produced as a fusion protein (i.e. several proteins made as one single polypeptide by the cell, which need to be chopped up afterwards by the biotechnologist), then an affinity tag can be a short amino acid sequence between the 'units' of the fusion protein that allows the protein to be chopped up easily. It could be the specific sequence recognized by a peptidase or protease, for example the Leu–Val–Pro–Arg–Gly–Ser sequence recognized by the enzyme thrombin (it cleaves between the Arg and the Gly).

The tag could be another protein, for example an enzyme (making the new protein easier to detect) or a protein that binds to some other material very avidly (such as avidin, which binds to the vitamin biotin very strongly), which would allow the protein to be purified by affinity chromatography. Enzymes sometimes fulfil both roles, since they catalyse the reaction of substrates and bind to inhibitors very strongly. Short segments of cellulase (an enzyme that breaks down cellulose) have been used to make fusion proteins that stick to a cellulose affinity matrix. Other examples are using the last 26 amino acids of the calcium-binding protein calmodulin, which binds to muscle myosin kinase, a relatively cheap protein to produce

The tag could be a short amino acid sequence, either random or selected from some other protein, that is recognized by an antibody. The antibody would then bind to the protein when it would not have done so before. One such short peptide, called FLAG, has been designed so that it is particularly easy to make antibodies against it.

A short string of amino acids can also be used as a chemical tag on the protein. A string of positively charged amino acids, for example, will bind very strongly to a negatively charged filter: this could be used as the basis of a separation system. Histidine binds strongly to some metals, and short runs of six histidines (His_6 tag) can be added to the end of a protein to make it bind extremely tightly to nickel metal columns.

Ageing

Biotechnology and molecular biology have claimed to have discovered the 'secret' of ageing several times. All vertebrates age, because it is more efficient from an evolutionary standpoint to have lots of children and then die than to expend all your physiological energy on staying alive. As well as developing treatments for specific disease, biotechnology has sought to understand and treat ageing itself.

Approaches to treatment include:

Organ replacements. This suggests that, as each organ wears out, we replace it with a new one. *See* **Transplant** and **Artificial tissues**. Obviously, this is a stop-gap measure.

Human growth hormone. Touted as a cure for ageing, this has been found in clinical trials to reduce some of the symptoms of ageing, such as the replacement of muscle by fat, but not the underlying process.

Telomeres. Telomeres are unusual genetic sequences at the end of chromosomes. The telomeres gradually shorten as cells age. This is probably a defence against cancer; when they have shortened below a certain critical point the chromosome is destroyed and the cell dies. The telomere theory suggests that ageing is an effect of shortening the telomeres in all cells of the body, so they can no longer divide and repair themselves. One approach to 'fundamentally' curing ageing is therefore to stimulate an enzyme (telomerase) that rebuilds the telomeres. It is still too early to have any idea whether this is a sensible approach, let alone practical. Telomerase is also a target for anticancer therapy: by blocking its action, you could, in principle, make cancer cells mortal again, so that their growth stops after a few cell divisions.

AGE. Another 'fundamental' change that goes with ageing is the accumulation of proteins linked together by sugars, called advanced glycosylation end-products (AGE). These accumulate and clog up cells and the intracellular matrix. However, it is unclear what anyone can do to remove them.

Genes. Some very rare genetic diseases result in what looks like greatly accelerated ageing, so patients die 'of old age' in their late teens. The group of diseases is called the progerias. Small animals can be selected that are longer lived than their cousins, and they have identifiable genetic changes. Both lines of argument suggest that single genes can have a substantial effect on how fast we age, and imply that we could slow the rate at which we age if we understood how those genes work. Several genome companies are seeking ageing-related genes.

Starvation. It is a well-documented fact that small mammals who are semi-starved from birth live up to 40% longer than their well-fed brothers and sisters. There is no convincing explanation for this, but it opens the way to modifying metabolism in a way that mimics the effect of starvation but without the need to reduce caloric intake.

Agrobacterium tumefaciens

Agrobacterium tumefaciens is a bacterium that causes crown gall disease in some plants. The bacterium infects a wound, and injects a short stretch of DNA into some of the cells around the wound. The DNA comes from a large plasmid—the Ti (tumour induction) plasmid. A short region of the Ti plasmid (called T-DNA) is transferred to the plant cell, where it causes the cell to grow into a tumour-like structure. The T-DNA contains genes that, among other things, allow the plants cells to make two unusual compounds, nopaline and octopine, which are characteristic of transformed cells. The transformed cells form the gall, a hard lumpy growth on the plant, which is a benign home for the bacterium.

This DNA transfer mechanism has been harnessed as a way of genetically engineering plants. The Ti plasmid is engineered so that a foreign gene is transferred into the plant cell along with the nopaline synthesis genes. When the bacterium is cultured with isolated plant cells or with wounded plant tissue, the 'new' gene is injected into the cells and ends up integrated into the chromosomes of the plant.

Agrobacterium tumefaciens usually only infects some dicotyledons, because their response to wounding is compatible with *A. tumefaciens'* DNA transfer mechanism. When they are wounded, dicotyledons make specific phenolic chemicals that are part of their wound-protection mechanism. *Agrobacterium tumefaciens* uses these compounds both as chemotactic agents (i.e. they swim towards the source of the compound, and so 'find' the wound) and to stimulate DNA transfer.

Monocotyledons do not respond in this way, and so are resistant to A.tumefaciens. This has been a problem in the past for biotechnologists since grain crops are monocotyledons. However, manipulation of the plasmid and the conditions under which it transfers its DNA in culture have allowed grain crops (including rice and maize) to be transformed with T-DNA. However, this is still a difficult route for getting DNA inside some plants. The difficulties arise from the limited range of plants that *A. tumefaciens* will attack, and the problems of manipulating the large Ti plasmid, which is too big to cut up and reassemble using conventional recombinant DNA techniques.

Agrobacterium tumefaciens has been used particularly to get DNA into trees. Trees are difficult plants to breed because of their size and long life cycles, and so genetic engineering techniques offer unusual advantages of speed and the ability to engineer millions of clones. Walnut, poplar, apple, and plum trees have all had DNA transferred to them using *A.tumefaciens*.

AIDS

AIDS (acquired immune deficiency syndrome) is the final stage of infection of humans by the human immunodeficiency virus (HIV). Infection is currently believed to be irreversible and fatal, although how long it takes to die varies enormously between patients. Once thought to be caused solely by HIV, there is growing evidence that HIV alone cannot cause AIDS. In particular, it is believed that if someone has an infection of mycoplasma (a type of bacterium), then they are very much more susceptible to getting an HIV infection if exposed to the virus, and that cytomegalovirus (CMV), which many people carry around all the time, may trigger the change from an apparently harmless HIV infection to full AIDS. There is also a theory, the HIVER theory, that says that most of the damage caused by the disease is actually caused by an autoimmune problem. AIDS is the immune system destroying itself, triggered by the virus, rather than the virus being destructive. However, the effectiveness of anti-HIV drugs shows that HIV must have an important role to play.

There are several areas in which biotechnology has made progress in analysing the disease, in developing methods for diagnosis and treatment, and in working towards cures and prevention of AIDS.

Fundamental research. The complete characterization of HIV was essentially complete within six years of recognizing that the disease existed, something of a record in medical history. That this could take place so fast was the result of the techniques of molecular biology, and the ready availability of reagents to serve those techniques.

Diagnosis. AIDS is a very slow disease, and HIV-positive people may be infectious but show no symptoms for years. Consequently, there is a lot of interest in diagnosing HIV infection as soon as possible. A large number of tests based on monoclonal antibodies have been developed and marketed. Others based on **DNA probes**, and especially **PCR** (polymerase chain reaction), have been launched as services performed by reference laboratories (since they are too complicated for use in doctors' offices).

Therapy. The first accepted AIDS therapy was treatment with AZT (Retrovir), a drug that blocks the retrovirus-specific enzyme reverse transcriptase (RT). A range of other drugs are in development, some based on the conventional pharmaceutical research done in the last few years. It is becoming apparent that a combination of drugs acting by different routes is the best approach, and especially a combination of an RT inhibitor with an inhibitor of another viral enzyme, the protease.

Biotechnological products such as the CD4-based proteins, which aim to stop the virus from ever linking on to a cell, and so stop new infection, have also been tried. CD4 is the cell protein which the virus binds to. The protein gp120 (and its parent protein gp160) is the virus protein that does the binding. Covering either with some other protein will, in theory, stop the virus locking on to the cell. Genentech, Biogen, Chiron, and many of the other 'big names' in biotechnology were pursuing variations on this type of AIDS therapy, but clinical trials have not been very promising for most of them, and now researchers are concentrating on more conventional drugs.

Vaccines. Developing a therapeutic vaccine for something that destroys the immune system is difficult. Current development targets include the gp120 and p24 and p55 proteins of the virus' core, but usually several proteins or peptides are included in one vaccine in order in increase the chance that the immune system will 'see' the vaccine. A prophylactic vaccine (one that protects people who do not have an HIV infection from catching the virus) should be easier to develop, but is much harder to test: who are you going to give it to, and what are you going to do to show that they are protected?.

A substantial effect that the AIDS epidemic has had on the biotechnology industry is to speed up the regulatory process for some drugs. People with AIDS became fed up with how slow the official regulatory process was and started to try out drugs with potential effects on AIDS informally on themselves. A range of possible antiviral compounds, including interferon (which is not for sale as an anti-AIDS drug in the USA), have been tried out by people with AIDS. This has in turn caused politicians to develop 'fast track' drug approval processes for drugs for AIDS, and potentially for other terminal disease states.

AIDS has a high political profile. The 1992 'AIDS Awareness' concert in memory of Queen lead singer Freddy Mercury attracted a billion viewers (compared with 250 million for the 'Live Aid' concert for African famine relief). Funding for research into AIDS therapeutics has been easier to obtain than for many other diseases, and this is one reason that the biotechnology industry has worked extensively on AIDS. Another reason is that this is a 'new' disease with no existing treatment, and the sheer scale of the potential infected population, with around three million people infected in the Western world, means that the demand for any effective treatment is huge.

Airlift fermentor

Airlift fermentors, or airlift reactors (ALRs), are a type of loop fermentor (*see* **Loop bioreactors**) and are very popular in many applications. The fermentor consists of two parts, a riser and a downcomer. The liquid fermentation medium circulates between the two, driven up the riser by air (or other gas, sometimes pure oxygen) pumped into the bottom through a sparger. Thus, there is no mechanical stirring mechanism in the fermentor. There is usually a gas separator at the top of the riser. This separates ('disengages') the gas from the liquid, so the bubbles of gas are not sucked back down the downcomer where they would try to rise, bringing the circulation of liquid to a halt.

The popularity of this type of fermentor is mainly because of the fact that the air drives the liquid round the fermentor quite gently, so reducing shear forces that can otherwise occur as stirring plates churn through the medium and which might break open delicate mammalian cultured cells, or damage long fungal hyphae. Airlift fermentors were very popular for making monoclonal antibodies in bulk, although the trend has now swung rather towards using **hollow fibre** reactors instead for all except very large-scale production.

A variation on the airlift fermentor is the bubble column fermentor. Here, the cells are kept suspended in solution by the bubbles rising through a tube. The difference is that in an ALR the bubble flow drives the liquid round the fermentor circuit (sometimes helped by pumps). In a bubble column fermentor the bubbles are there to keep individual cells suspended, but not to move the bulk liquid around.

Algae: commercial uses

Algae have been used for centuries in Asia to produce food ingredients and other materials, but their use elsewhere to produce biotechnology products is relatively new. Algae comprise the seaweeds and a host of smaller plants, including single-celled organisms. Among their uses are:

Production of gums. Many of the gums used in food, medicine (e.g. in wound dressings), and for a variety of other things, such as research reagents and components of printing inks, come from seaweed. Among the more common compounds prepared commercially from seaweed are: alginate (an acetylated polymer of mannuronic acid and glucuronic acid), carageenan, agar, and agarose.

Production of chemicals. Some chemicals are produced commercially from algae, such as the food dyes beta-carotene and astaxanthin: the latter is added to salmon food to give farmed fish pinker flesh (which wild fish get from eating plankton). A variety of human food ingredients are made from cultured algae, such as Omega-3 fatty acids DHA (docosahexaenoic acid) and ARA (arachidonic acid) from dinoflagellates by Merck.

Single-celled algae have also been used as food—the algae *Chlorella*, *Scenedesmus* (in Japan), *Spirulina* (in Israel, USA, and Mexico) are used to make algal 'biomass' for single-cell protein (**SCP**).

Amino acids

The amino acids, key components of all living things, are produced by biotechnology in bulk using **fermentation** and **biotransformation**. Several Japanese companies dominate the world production of amino acids. They use fermentation systems in which they grow bacteria or fungi that have been selected to over-produce specific amino acids, which they excrete into the fermentation medium. Harvesting the medium and removing the other components yields amino acids in total amounts of hundreds or thousands of tonnes a year.

Commercially produced amino acids include:

Glutamic acid. This amino acid is produced in larger amounts than any other, because it is widely used in the food industry in the form of monosodium glutamate (MSG) as a flavour enhancer, and in the Far East as a table condiment. *Corynebacterium glutamicum* is used commercially to make glutamate, and can convert up to 50% of the carbon it is supplied with into glutamate.

Lysine. This is the second most abundantly produced amino acid, and is used as a supplement for animal feeds (which are often deficient in some amino acids, and particularly in lysine which is an essential amino acid in mammals).

Cysteine and methionine. These are the sulfur-containing amino acids, and again are used as animal feed supplements.

Phenylalanine. As well as being used to a small extent as a feed supplement, phenylalanine is the most expensive chemical ingredient in the manufacture of Aspartame.

Tryptophan. Tryptophan hit the headlines in 1990 when the tryptophan produced by a new genetically engineered *Bacillus amyloliquefaciens* made by Showa Denko KK was linked to a rare degenerative disorder called eosinophila-myalgia syndrome (EMS). Despite loud and prolonged claims that this was proof that genetic engineering was dangerous, the problem was eventually traced to a chemical generated during the (perfectly conventional) purification procedure, and had nothing to do with recombinant DNA.

Several other essential amino acids, i.e. amino acids that our bodies cannot make themselves and hence have to be eaten in our diet, are also made in substantial amounts for human or, more usually, animal consumption. Indeed, most of the other 15 'natural' amino acids (those found in proteins) are produced by fermentation in thousand tonne amounts. Other amino acids, not found in proteins, and especially the D-isomers, are made by biotransformations for use as chemical intermediates. Biotransformations are used for these materials because they are not found in nature, or are only found in tiny amounts. The D-amino acids, for example, are used in antibiotic manufacture. (D-amino acids are ones with the opposite 'handedness' to natural amino acids—*see* **Chirality**).

Animal cell immobilization

Animal cells are used widely in biotechnology to produce natural products or genetically engineered proteins. Animal cells have the advantage that they already produce many proteins of pharmacological interest, and that genetically engineered proteins are produced by them with the post-translational modifications normal to animals. However, animal cells are much more fragile than bacterial ones, and so cannot be exposed to the high shear forces of repeated stirring, which bacterial cells can put up with in a commercial fermentation process.

Any cell (indeed, any small particle) can be immobilized by entrapping it in some solid material, either by having it grow there or by forming the material around it after it has grown. Entrapment in some form is common and popular, from microencapsulation to growing cells inside a hollow fibre in a bioreactor (*see* **Hollow fibre**). As well as these general approaches, there are some methods and materials that are specific to animal cells.

Surface adherent cells. The simplest is to use the natural adherence of animal cells to some materials. Many animal cells stick down flat on a suitable surface, hugging it as they would hug other cells or connective matrices in the body. If grown on suitable plastic surfaces, on glass, or on many ceramics these cells will stick to them. This makes them much easier to keep in one place. In general, between 10 000 and 100 000 mammalian cells will grow on 1 cm^2 of surface (the number depends on the surface and the type of cell).

This is a bulky way to grow cells unless the surfaces are folded in some way. Hollow fibre or membrane bioreactors can provide a way of doing just that, but one of the favoured ways is to use porous carriers. These can be polysaccharide, protein (especially collagen), plastic, or ceramic materials with microscopic holes in them a few tens or hundreds of micrometres across. Such materials are called microcarriers or microbeads. The cells grow into these holes, greatly increasing the surface area available to them while not increasing the bulk of the culture: the Opticore ceramic culture matrix, for example, has 8 cm^2 of surface per cm^3 of volume of solid material. The carriers can be formulated into small particles or into sheets or tubes. As well as ceramics, they can be made of polysaccharides (dextrans, alginates, agar) with various chemical modifications to give them a surface charge: these are popular because they mimic some aspects of the membranes on which cells grow in the body, and so the cells stick to them firmly.

Animal cloning

A clone is a group of genetically identical organisms. As animals nearly all reproduce sexually, normal offspring are not genetically identical to either parent. So we must have another way of making many copies of an animal that has a useful genotype. This is called 'cloning'.

Unlike plants, we cannot grow new animals from 'cuttings'. The only cells known to be capable of growing naturally into a new mammal are fertilized egg cells or cells from very early embryos. So we can split an early embryo into eight cells and allow them all to grow into genetically identical fetuses, and hence adults. In mammals, if we split the embryo into more than eight, very few or none of the resulting clumps of cells develop into an embryo. We can also seek to inject the nucleus from another type of cell into a fertilized egg, from which we have removed the nucleus (enucleated egg), and grow up the result. This was done successfully in sheep in 1996, when Scottish researchers injected the nuclei from a cell line made from Welsh mountain sheep into eggs from Scottish blackface sheep mothers. Of five lambs born, only two survived more than 10 days.

Fish and frogs are more amenable to cloning, and frogs were cloned by nuclear injection in the early 1970s. However, even this is a very difficult endeavour.

Cloning animals is useful because it allows us to make carefully one genetically very desirable animal, by transgenic technology or decades of breeding, and then make many more with exactly the same genes. This is important when the genetic character we are seeking is determined by many alleles, which would be separated into different animals by sexual reproduction. Thus, highly productive strains of farm animals or extremely fast racehorses could be cloned to provide the core of a breeding population rapidly. This shows why cloning humans is unlikely to be done much; unlike cows, which are genetically selected for a few simple traits, humans are not genetically selected for anything. So sexual reproduction will do as good a job as cloning, and is easier, cheaper, and more fun.

See also **Embryo technology**.

Antibiotics

Much of biotechnology is directed towards discovering new drugs. One class of drugs is the antibiotics. There is increasing pressure to discover new antibiotics as resistance to conventional antibiotics spreads among the bacterial populations, producing 'killer bugs' such as the MRSA (methicillin-resistant *Staphylococcus aureus*).

The 'established' antibiotics, especially penicillins, were the first products of the industry that is now called biotechnology. Produced from fungi in fermentation systems, the penicillins, streptomycins, and a host of other antibiotics were breakthrough drugs in the 1940s and 1950s, and are still major products of fermentation industries. Since then, biotechnology has built on this base to develop a range of new antibiotics. There are four routes to developing new antibiotics (as opposed to improving the production of existing ones) with biotechnological components.

Hybrid antibiotics. The synthesis of an antibiotic is the result of many enzyme steps in a bacterium or fungus. Some work is aimed at producing hybrid antibiotics, i.e. molecules that have bits from two different antibiotics. This is done by putting selected enzymes from two different antibiotic-producing cells into one bacterium. This work has gone furthest using genetically engineered streptomycetes. This approach is believed to be especially valuable for generating new macrolide antibiotics (macrolides are large antibiotic molecules that are made by a complex series of reactions each of which adds a unique feature to the molecule). (Compare the synthesis of proteins, which are made by the sequential addition of amino acids on to a chain—the protein synthesis apparatus will assemble any sequence of amino acids, given the correct instructions, while the enzyme complex that makes a macrolide antibiotic can only assemble that one macrolide.)

Novel metabolites. It is probable that there are vastly more antibiotics produced by microorganisms and plants than humans have discovered yet. Biotechnology is using its ability to grow new bacteria and fungi on large scales to screen new bacteria for compounds that have useful drug activities. Xenova has specialized in this.

Animal antibacterials. Animals, especially invertebrates (which do not have the same sophisticated, adaptable immune system as mammals) produce a wide range of materials that kill bacteria, usually proteins or peptides, which could be developed into drugs. Target peptides are those produced by cells of the immune system, which normally destroy invading bacteria, and cells that produce proteins of the complement system, a set of proteins that knock holes in virus-infected cells. Some of these peptides do not destroy the cells themselves, but increase the chance that a white blood cell wil destroy them (a process called opsonization). Others, such as the defensin peptides and the bacterial permeability increasing protein (BPI), bactenecin peptides, azurocidin, and the enzyme lysozyme actually kill bacterial cells. A third group, typified by lactoferrin, inhibit bacterial growth, in this case by removing the free iron that the bacteria need from their surroundings, and binding it in a tight, inaccessible complex.

Pathogenesis-targeted antibiotics. These approaches attempt to find out what makes a bacterium a pathogen, and then block that molecular mechanism, rather than killing the bacterium outright. It is hoped that this will get round the problems of bacterial resistance in two ways. First, the drugs will be completely different from those currently in use, so resistant bacteria will be sensitive to them. Secondly, they will not put as much selection pressure on the bacteria to acquire resistance; providing the bacteria do not cause disease, they will not be affected. This second point is still speculative. The bacterial genome projects (to find the DNA sequence of every gene in the genomes of several bacteria) are aimed at finding out the genes that determine pathogenesis.

Many antibiotics are semi-synthetic, as are some other drugs like steroids. Biotechnological routes are used to make the core structure, such as the beta-lactam ring structure of the penicillins, and the organic chemistry modifies it to produce the end-product.

Antibodies

Antibodies are proteins made by the immune system to combat infection. Each antibody is made to recognize one target antigen molecule. If the antigen is a small molecule, then the antibody will recognize all of it. If the antigen is a large molecule, then the antibody will recognize only a part of the antigen, which will be called that antibody's 'epitope'. The binding site of the antibody latches on to this antigen very specifically and very strongly. This latching-on allows the body to recognize the antigen as part of something that should not be there—a virus, a bacterium, a toxin—and so start the process of removing it.

Mammals make antibodies to almost anything that is a 'non-self' molecule, i.e. which is not normally part of the body. Thus, you can get a mammal to make an antibody against almost any molecule by injecting it into the bloodstream. The immune system recognizes it as foreign material and makes a suitable antibody. In fact, it makes a whole range of slightly different antibodies: the blood of most people usually contains a vast host of different antibody molecules targeted at the various disease agents and other foreign molecules that have got into their blood in the past. For this reason the antibodies that are prepared from mammalian blood are called polyclonal antibodies, because they come from a large number of 'clones' (i.e. identical sets) of cells. This contrasts with the synthetic **monoclonal antibodies**.

Antibodies have been enormously useful to biotechnology because of their ability to latch tightly on to only one antigen, ignoring all others. For example, they would accurately distinguish sucrose from glucose, right-hand from left-hand amino acids (*see* **Chirality**), human blood proteins from ape proteins, etc. Thus they are the basis of many processes where great discrimination is needed (*see also* **Immunotoxins**, **Immunodiagnostics**, **Affinity chromatography**.

Technically, the antibody proteins are called immunoglobulins. There are four types usually mentioned:

- IgM—the first type made by the body when it encounters a foreign material.

- IgG—the most common type, made after a prolonged encounter (as during a disease).

- IgE—the type responsible for allergic reactions.

- IgA—a rarer type which is present in saliva and some other non-blood fluids.

Antibodies are made by lymphocytes—they are actually made by B lymphocytes (B cells) in a process helped by T cells.

See also **Antibody structure**.

Antibody structure

Antibodies have a well-defined structure. Antibodies are made of equal numbers of 'light' chains and 'heavy' chains: IgG antibodies (the sort most commonly used as reagents in biology and biotechnology) contain two of each. The antigen-binding region or binding site ('complementarity determining region') lies at the end of the light and heavy chains; it is therefore formed from both chains. The chains fold up into discrete

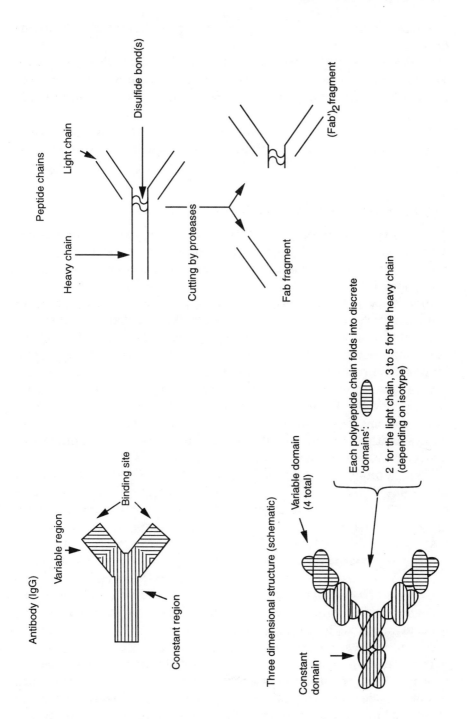

Antibody (IgG)

Peptide chains

Light chain

Disulfide bond(s)

Heavy chain

Cutting by proteases

Fab fragment

(Fab')₂ fragment

Variable region

Binding site

Constant region

Three dimensional structure (schematic)

Variable domain
(4 total)

Constant
domain

Each polypeptide chain folds into discrete
'domains':

2 for the light chain, 3 to 5 for the heavy chain
(depending on isotype)

blobs called domains: a 'single domain antibody' (Dab) is just one domain of an antibody.

The amino-terminal domains of both heavy and light chains are called the variable domains because they vary between antibodies. The other domains are constant domains, i.e. they are the same between antibodies of the same class and subclass.

The antibody can be cut by proteases into several fragments known as Fab, Fab', and Fc (for historical reasons). These also feature in some biotechnological literature.

Anti-idiotype antibodies

Anti-idiotype antibodies are antibodies that recognize the binding sites of other antibodies. Their binding sites are complementary to the binding site of another immunoglobulin. They are important to biotechnology in three ways.

First, they occur in normal blood. When we become immune to something, we not only acquire antibodies against that something, we also acquire antibodies against those antibodies (and antibodies against *those* antibodies, etc.). This forms a network of antibodies that can all bind to each other to various degrees, a network that helps to regulate the immune response. It is possible that allergic responses are in part a result of the breakdown of this sort of regulation. Thus anti-idiotype antibodies are important to the regulation of the immune system, and understanding how and why they are produced is an important part of understanding how the immune system works.

If an antibody is a 'key' exactly selected to fit the 'lock' of a virus or a bacterium, then an anti-idiotype antibody is a 'lock' exactly selected to fit that 'key'; in other words, it must have some similarity to the original antigen, the material that the original antibody reacted to. This means that making anti-idiotype antibodies could be a way of replicating the functional properties of such proteins as hormones or hormone receptor molecules. By raising an antibody against that molecule, and then raising an anti-idiotype antibody against the antibody, you will create an immunoglobulin with some of the functional characteristics of the original hormone or hormone receptor molecule, but which can be produced more easily and is chemically quite distinct.

In practice, antibodies only recognize a small region of the surface of a protein. Thus, an anti-idiotype antibody can only mimic the properties or functions of that region of the protein, and these are likely to be rather limited. Thus, for example, an anti-idiotype antibody that binds to an antibody against insulin (and hence will have a binding site looking like some of the insulin molecule) will sometimes bind to the insulin receptor molecule. However, it does not necessarily make a cell respond in the same way as it would to insulin, because the receptor is not bound in the same way as when the insulin molecule itself binds. These relatively subtle differences have limited the use of anti-idiotype antibodies.

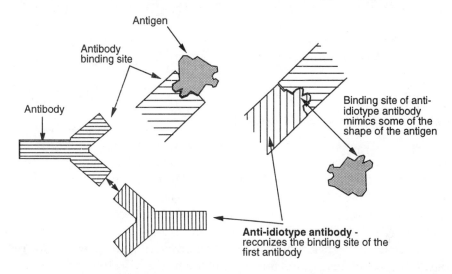

Antigen

Antibody binding site

Antibody

Binding site of anti-idiotype antibody mimics some of the shape of the antigen

Anti-idiotype antibody - reconizes the binding site of the first antibody

Anti-idiotype antibodies also have potential as vaccines. Here again, they are used to mimic a protein, in this case part of the surface of a virus or a bacterium. However, it is not so critical in this case to mimic the whole viral coat protein. Provided the anti-idiotype antibody mimics a part of the virus' surface that is readily accessible to the immune system (so that it is easily recognized in the final virus), then it could be used to stimulate the immune system to make a suitable antibody. This is a good idea because it allows a vaccine to be developed without ever using the live virus to make it, so there is no risk of contaminating the vaccine with live virus. However, the link of the virus being used to make antibody, which is used to make anti-idiotype antibody, which is injected to get the body to make antibody, seems to be too tenuous in the experiments done so far for the resulting antibody to recognize the virus properly.

Antisense

Antisense RNA or DNA is a single-stranded nucleic acid that is complementary to the coding, or 'sense', strand of a gene, and hence is also complementary to the mRNA produced from that gene. If the antisense RNA is present in the cell at the same time as the mRNA, it hybridizes to it, forming a double helix. This double helical RNA cannot then be translated by ribosomes to make protein. Thus, antisense RNA can be used to block the expression of genes that make proteins.

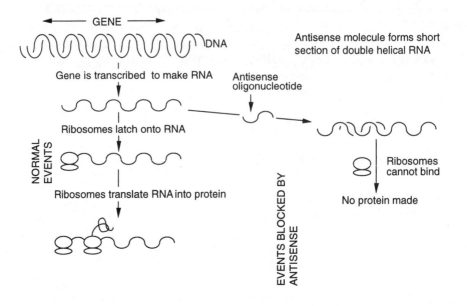

Antisense RNA is a powerful way of modifying gene activity because it is a positive genetic engineering step, not a selection of negative mutants of a gene. Thus, rather than try to knock out all copies of a gene in, say, a plant, the genetic engineer only has to put in one gene that produces antisense RNA, and the antisense will prevent any mRNA from any copy of that gene being used by the cell.

Although it has a 'code-blocker' function (stopping ribosomes from making protein), antisense also causes the cell to break down its target RNA, through the action of an enzyme called RNAase H. RNAase H recognizes double-stranded RNA and cuts it (probably as part of a defence mechanism against viruses, which sometimes have double-stranded RNA).

Antisense RNA has been suggested as a potential drug, because it can block the effect of one gene without affecting any others. In particular, it could be used to block the effect of **oncogenes**, and so slow or prevent the development of cancer. It could also block the effect of viral genes, and so be used as an antiviral drug (*see* **Antiviral compounds**). Preliminary experiments show that antisense holds great promise in these areas, and Genta, Hybridon, and ISIS Pharmaceuticals have antisense pharmaceuticals in clinical trials.

A major problem in taking that promise from experimental models involving cultured cells into more realistic animal models is how to get the antisense into the affected cells. Since genetically engineering humans is not practical, the pharmaceutical chemist has to be able to deliver intact antisense RNA or DNA as a chemical to all affected cells. This is doubly difficult because RNA is quite unstable, and is very easily broken down by RNAases, enzymes found in very many tissues and very hard to destroy. One way round this is to use antisense DNA, or modified DNA (such as phosphorothioate DNA, which has one of the oxygen atoms in the phosphate groups replaced with a sulfur atom), which are more resistant to enzyme attack.

In theory, an antisense oligonucleotide has to be at least 15 bases long to make sure that it targets only its target mRNA. However, it now seems that much shorter DNAs, of only seven or eight bases will work, probably because the target needs to be 'available' for binding (i.e. not folded into secondary structure or hidden in chromatin), and few potential targets are available.

A more immediate application of antisense is in genetic engineering of plants and animals. Plant genetic engineering in particular has benefited from antisense technology, where several groups have blocked specific enzyme genes. Most famously, the genes for polygalacturonidase have been blocked in tomatoes by several groups in industry and academia. Polygalacturonidase is one of the key enzymes in breaking down the walls of cells in the flesh of ripening tomatoes, and so making them get squashy. If a gene that makes antisense to the polygalacturonidase mRNA is spliced into the tomato plant, then its antisense product blocks the manufacture of that enzyme in the tomato, and the tomato stays firm much longer as it ripens. Calgene have marketed a tomato strain based on these lines as the 'Flavr Savr' tomato.

Antisense oligonucleotides often have substantial '**aptamer**' effects, which confuses their effect *in vivo*.

Antiviral compounds

One of the areas in which biotechnology is playing a substantial role in developing new drugs is in producing antiviral compounds. The work has focused on a range of technical approaches.

One of the more established approaches is via a range of immune system boosters. Some, such as the interferons, are specifically antiviral. They stimulate cellular defences against viruses on many levels, from reducing cell DNA synthesis, and so making cells more resistant to hijack by viral genes, to enhancing cellular immune responses. Originally, it was hoped that interferons, some of the first products of recombinant DNA technology, would be genuine broad-spectrum antivirals, but their value has been limited to use in combination with other drugs as immune boosters, and in a few specialist applications (*see* **Biological response modifiers**).

Biotechnology has been very active in preparing complex chemicals with antiviral properties. The most popular line of approach is to make compounds that look like the nucleotides in DNA. These then block up the enzymes that allow the virus to make its own DNA without damaging the cell. Wellcome's AZT (Retrovir, the anti-AIDS drug) is one such nucleotide 'analogue'. Since these are complex compounds and must be synthesized in their correct stereoisomer if they are to work, using enzymatic synthesis for at least part of their production can be useful. A range of enzymes that make part of the nucleotide molecules have been purified (phosphoryl transferases, glycosyl transferases, and enzymes that modify the bases) and are so efficient that they can work usefully fast on the nucleotide analogues even though these analogues are not their normal substrates. A range of nucleotide analogues, especially the 'carbocyclic' analogues (compounds in which the oxygen in the sugar ring is replaced by carbon), are being very actively investigated as potential antivirals for treating long-term viral diseases.

The second approach is to use genetic engineering to create proteins that block viral replication. The approach here depends on the virus concerned, but generally depends on making a protein that binds to the protein on the cells that is the 'docking' protein for the virus, or to the virus' protein that is the 'docking probe'. In the former case a segment of the viral protein can do the trick, in the latter case it is a fragment of the cell receptor protein. (See the discussion of CD4 in the entry on **AIDS.**) Many other strategies have been suggested, but have not resulted in products near to clinical trials.

The third approach is to use **antisense** RNA or **ribozymes**.

Apoptosis

Apoptosis, or 'programmed cell death', is the process by which cells deliberately destroy themselves. Mammalian cells are not meant to live forever, and some of them are meant to self-destruct as part of the normal programme of development. Examples are unwanted immune cells (e.g. ones that might react with the body's own tissues) and the surplus nerve cells that populate the brain before birth, but which must be removed to generate the functioning networks that enable us to see and remember things in postnatal life.

Apoptosis is a deliberate process—the cell dismantles itself systematically. It is unlike necrosis, where the cell is killed, often by chemicals or toxic attack. Necrotic cells burst and release their contents, causing inflammatory responses. Apoptotic cells shrink and are engulfed by cells around them. A characteristic marker is the 'nucleosomal ladder'; their DNA is broken up into short segments which show up as a 'ladder' on agarose gels.

Apoptosis is controlled by specific signals and genes, and mutations in those genes can be important in disease. Signals include receptors for the nerve growth factor (NGF) family of proteins, and for tissue necrosis factor (TNF). Absence of NGF can cause cells to self-destruct; NGF is not really a growth factor, but rather a 'don't die' or survival factor. (Other apparent growth factors may also act like this, rather than really promoting growth.) Cells undergoing apoptosis also send 'eat me' signals to neighbouring cells.

Mutations in apoptotic genes are important in cell immortalization and cancer. *bcl-2* is a gene that blocks apoptosis—some cancers overproduce the gene's protein and so become resistant to signals from the body that would otherwise cause it to self-destruct. Other genes, like *bax*, are active apoptosis triggers. *bax* is a tumour suppressor gene—destroying its function allows the tumour to grow, in this case, because mutating it stops apoptosis. An approach to cancer therapy is to give cancer cells back the ability to destroy themselves by apoptosis.

Apoptosis is also important in other diseases. In the normal immune system, white blood cells are generated that could 'recognize' every molecule in existence as foreign. Those that are targeted against the body's own molecules are then removed. Apoptosis may be critical in this removal. Apoptosis may even lie at the heart of the ageing process itself, as cells systematically destroy themselves after their normal life, so as to limit the growth of tumours (see **Immortalization**). This hypothesis suggests that controlling apoptosis may enable primary cells to be cultured indefinitely.

Apoptosis is pronounced as it is spelt: 'a-pop-toe-sis'. Only Greek scholars and pedants insist that it should be 'ay-po-toe-sis'.

Aptamer

Aptamers are nucleic acid molecules, usually short lengths of DNA, that have been selected to bind to a target other than another nucleic acid. Just as a chain of amino acids can fold to form a shape that exactly binds to some target molecule, so a nucleic acids can fold to form a specific binding shape. Nucleic acids that bind to specific proteins (such as thrombin, an experiment performed by Gilead Sciences) or small molecules such as NADPH or ATP have been developed by **Darwinian cloning**.

Just as there are specific structural motifs in proteins such as alpha helices, so there are in aptamers. The most common is the G quartet—four guanosines in a row in a molecule make the molecule assemble into four-strand helices. Many other structural forms have been found.

The advantage of aptamers over proteins is that you can amplify a successful aptamer through **PCR**. The disadvantage is that nucleic acids are much more chemically limited than proteins.

Aptamer effects are a major problem for developers of antisense agents. Most oligonucleotides will have effects on living things, but this is often an unwanted, aptamer effect rather than the desired antisense one. Aptamers have been developed as products in their own right, with attempts made to harness their protein-binding effects for therapeutics (as Gilead is doing) or as diagnostic reagents analogous to monoclonal antibodies.

Aquaculture

Aquaculture is growing water plants and animals in 'farms', rather than harvesting them from wherever they happen to grow naturally in rivers or seas. A related and rather more specific term is pisciculture, the culture of fish (*see* **Fish farming**). Usually aquaculture uses fresh water. When it does not, using sea water instead, is can be called mariculture. It is considered a part of biotechnology because, as a new commercial development, it uses the latest technologies rather than traditional ones, and often involves growing organisms in large volumes of water, which has similarities to growing large volumes of yeast or bacteria, biotechnology's 'heartland'.

Aquaculture is a growing industry, and with a range of products:

- *Fish*, especially high-value fish such as salmon and rainbow trout. (*see* **Fish Farming**)

- *Crayfish, lobsters, oysters, shrimp, and other molluscs.* These are farmed even more intensively (i.e. with more animal mass per cubic metre of water) than fish.

Aquaculture has also been used to mass-produce macro- and micro-algae (see **Algae**, **Biomass**). These are grown for food in the Far East, but also for chemicals (agars and gums), vitamins, and pigments such as astaxanthin and beta-carotene.

Biotechnology comes into animal aquaculture in providing clean, well-aerated water for the animals to grow in. It also often provides food, as krill, powdered synthetic food, and food additives such as astaxanthins (a bright red-pink pigment) to ensure the fish and prawns have the right colour.

In both animal and plant areas, biotechnology has been using genetic methods on aquacultured species, in particular to produce triploid and tetraploid organisms, and hybrid algae through plant cell fusions. Triploid trout, for example, are sterile, and can be used for biocontrol of weeds without the threat of being able to breed themselves. Triploid oysters are considered to taste better than normal ones by the US market, and, being sterile, put more energy into muscle production and less into reproductive organs.

Artificial intelligence

Artificial intelligence (AI) is an area of computer programming techniques that attempts to capture aspects of human 'intelligence' in programmes. As no-one can decide what 'intelligence' is, useful AI products tend to use human intelligence as a model and then develop a specific computer version of that model. There are a number of applications in biotechnology.

Expert systems. There are lots of types of expert systems, which attempt to capture human expertise in a formal computer program, i.e. as a set of formal, explicit rules or formulae. They have been used in bioreactor control systems, where they can usefully combine quantitative and qualitative 'rules' to guide a fermentor.

Neural nets. These were originally computer simulations of how nerves work. They are programs that can be 'trained' to recognize patterns in large masses of data, by 'training' them on previously identified examples of the patterns. Neural nets have been used in many applications, including:

- bioreactor control (where the computer is 'trained' to identify the conditions in the bioreactor, derived from many sensors, under which the reactor performs optimally).

- DNA sequence analysis (where the computer is 'trained' with examples of genetic features, e.g. coding regions, and then used to identify the same type of region in newly sequenced DNA).

Fuzzy logic. Normally, logic decides between statements that are true or false. Fuzzy logic, and older variants such as Bayesian logic, quantify the degree of truth or falseness, or our confidence in truth, and so can handle situations where we cannot decide between absolutes.

Genetic algorithms (including 'artificial life'). This is an AI technique inspired by living systems. Possible solutions to a problem (e.g. equations to predict what a reactor will do) 'compete' within a computer. They 'reproduce' (i.e. are copied) and in doing so introduce new variants of the equations. If one of them is better than the others, it will be more successful. Thus, this uses the power of natural selection to find a solution to a problem. Unfortunately, it also has the inefficiency of natural selection. A related concept is 'artificial life' which is simulating very simple living things in a computer.

Cellular automata. These are computer simulations where each unit of memory or 'cell' behaves according to what is going on in other cells. They have been used to simulate immune system function at a fairly primitive level, as well as being of more abstract interest to computer scientists. The most popular cellular automaton is a game called 'Life', invented by Cambridge mathematician John Horton Conway in the 1960s.

Biotechnology is also skirting the idea of artificially enhancing human intelligence. Using drugs, especially biotech-based drugs, to boost intelligence in people is also a popular speculation, with a long history going back at least to the isolation of cocaine from the coca plant in 1855. A wide variety of drugs are claimed to be able to increase attention, memory, or general intelligence. For example, tobacco smokers claim that nicotine enhances their ability to concentrate (although it is more likely that it just alleviates the craving that stops them concentrating when they have not had a cigarette). Some compounds being developed for treating organic disease, such as Alzheimer's disease, have demonstrable effects on memory and attention in ill people, but increasing intelligence in people with no disease still seems to need luck, genetics. and 20 years of education.

Artificial sweeteners

Many substances are used to make food taste sweeter without increasing its 'calorie value'. Among those of interest to biotechnology are:

Thaumatin. A protein produced by *Thaumatococcus danielli* in its fruit. Thaumatins are 3000 times as sweet as sugar, and at low concentrations enhance other flavours as well. Because they are proteins they can and have been produced by genetic engineering in bacteria, so avoiding having to go to the tropics to harvest the fruit. Thaumatins have been produced in *Escherichia coli*, *Bacillus subtilis*, *Streptomyces lividans*, and *Saccharomyces cerevisiae*, and the genes have been put into higher plants too.

Aspartame. Known as Nutrasweet, this is one of the most commonly used commercial artificial sweeteners. It is a dipeptide (aspartate–phenylalanine–methyl). Because it is made from two amino acids, there are two parts of its manufacture that may be of interest to a biotechnologist. One amino acid, phenylalanine, is relatively expensive, so selection, genetic engineering, or other manipulation of a fermentation that produces phenylalanine more efficiently is a worthwhile goal as part of the production of aspartame. The other aspect is the synthesis of the dipeptide by enzymes, and particularly by using a protease to join the two amino acids together (rather than its more normal reaction of separating them). Both areas are undergoing continuing commercial development.

Artificial tissues

Scientists have attempted to build artificial versions of many tissues. All approaches need one to isolate relevant cells, 'expand' them (i.e. grow lots of them *in vitro*), and then introduce them back into the body in a form that will function and not be destroyed. The simplest example is bone marrow transplant, where bone marrow cells are taken out, grown up, and simply injected. However, for most other tissues a more complicated structure is needed to support the new cells.

Skin. One of the most successful artificial tissues is skin substitutes. Keratinocytes, the cells that build up the epidermis, can be cultured in sheets from a skin biopsy: a 2 cm^2 biopsy will produce enough keratinocyte sheet to cover an adult in a month. This sheet can itself be used to replace skin lost in ulceration or burns. The cells are usually derived from the patient themselves, to avoid rejection. More sophisticated grafts will be made of composites of several cell types, including the fibroblasts of the dermis, and of matrix materials to encourage the cells to grow. Skin-like structures form when keratinocytes are grown in contact with air.

More complicated tissues have also been developed at an experimental level. In late 1995 experiments moulding a synthetic matrix on which to grow cartilage hit the headlines with stories of the 'artificial ear' grown on a mouse (for toxicity tests). There are many developments attempting to get a semi-synthetic bone matrix to work as a substrate for growing new bone. A range of proteins under the general heading of bone morphogenic protein (BMP) have been tried in such matrices to enhance bone formation.

One of the most developed complex tissues is the 'artificial pancreas', giving diabetics new pancreatic islet cells to produce insulin. The cells, which would normally be destroyed by the patient's immune system, can be encapsulated in a semi-permeable membrane so that they get nutrients from the body but the immune system does not 'see' them. It is quite difficult to design a system that will allow cells to get enough nutrients, especially oxygen, but will keep them immunologically isolated. Debris from dead cells can get through most semi-permeable membranes and trigger the body to mount an antibody response. Despite this, encapsulation is one of the more promising approaches to an 'artificial pancreas'.

This overlaps with the studies of wound-healing technology, where materials are developed with biomimetic properties that can stimulate the growth of new tissue to cover a wound. Unfortunately, wound healing is stimulated by a concert of many growth regulatory factors, and so no single factor has proven to be very useful. Integrins are used widely in such product ideas, because they encourage cells to migrate and adhere to a substrate.

An alternative approach to artificially building a tissue is to grow it in another animal, a technology known as **xenografting**.

A related product is artificial blood, which is made chemically or biochemically. On its own haemoglobin is rapidly extracted from blood plasma by the kidneys, which it then poisons, so the red pigment has to be held in a cell-like form. This can be haemoglobin encapsulated in artificial 'red cells' (liposomes or some equivalent), cross-linked into large aggregates, or covalently linked to polyethylene glycol. Biotechnology produces both the haemoglobin and the encapsulation technology. It can also produce blood-bulking agents—polymers such as dextrans that can bulk out real blood to make one unit go further in an emergency. These substitutes are increasingly being used at a time when concerns about the safety of blood transfusion are high.

Aseptic processing

Aseptic processing is processing materials so that no microorganisms can get into them. This either means that the production process machinery must keep the material sterile, or the whole manufacturing environment must be aseptic, i.e. a clean-room environment. In many pharmaceutical production processes, both approaches are adopted. UHT (ultra heat treated) milk and pasteurized fruit juice is usually packaged aseptically.

Issues in aseptic processing are:

- the balance between the growth rate of bacteria and the processing. Pasteurized milk, for example, is not sterile, so it has to be processed fast enough so that the bacteria in it do not have time to grow. If the starting materials are sterile, then this is not a concern.

- monitoring the microbial contamination, which can be a slow process, slower than bacteria can grow in the production line.

- cleaning and sterilizing equipment, which has to be done by **cleaning-in-place** technology.

- validation and quality control (QC), testing to make sure that the process actually is aseptic. Some groups do this by 'media fill'. A filling line will be run with bacterial growth medium in it instead of the product, resulting in bottles or cartons full of sterile growth medium. If one of them is not sterile, the rapid growth of bacteria will make the medium murky, pointing to a failure.

Many aseptic processing lines used to rely on constant checking to prove that no contamination was getting through. Now most manufacturers, even of drug products, use parametric release (also called parametric control). Here, the parameters of the production line (such as processing times, flow rates, temperatures, pressures, etc.) are rigorously monitored. If these fall within the proper range, then the operator knows that the product must be OK. For example, if an autoclave really did 'cook' the product at 121°C for 30 minutes, then it is sterile. Batches of product can be released just on tracking of the process parameters, and do not need to be tested for microbial growth before they can be shipped.

Assay

An assay is a specific chemical test for something. In biotechnology this usually means a specific biomolecule in a mix of other ingredients. This means that an assay must be able to select the analyte molecule (the thing to be measured) from among a host of others, and measure its concentration. This in turn means that at least one of the reagents involved must be very specific for the analyte.

Lots of assays are developed by and used in biotechnology. Among the more common technologies are **immunoassays** and **enzymes**, discussed in separate entries.

Key parameters in an assay are its sensitivity and specificity. There are two meanings for each of these words. Chemical sensitivity is a simple measure of how little of the analyte the assay can respond to. Some PCR assays, for example, can detect a single molecule of their target DNA. Their sensitivity is one molecule. Usually sensitivities are given as concentrations (micrograms per millilitre) or as molar concentrations (moles/litre). There is no standard way of measuring the chemical specificity of an assay, so usually it is tested against all the chemicals that might give a false result, and thus show that those chemicals have no effect.

Diagnostic sensitivity and specificity are slightly different. These ask how good the assay is at telling me whether or not something has happened. Assays are usually quantitative, giving a quantitative answer. They can be converted to a qualitative, yes/no test by setting a threshold above which a response is taken to be a 'yes' (positive), and below which it is a 'no' (negative). But how reliable is this dichotomy? Any test will give true positives (i.e. tests that show up positive when the sample really is positive—P), false positives (showing up positive when the sample is negative—F), true negatives (not showing positive when there is nothing there—N), and false negatives (not showing up positive when there is something there—G). The sensitivity of a test is defined as:

$$\frac{P}{P+G}$$

i.e. what fraction of the test that should be positive is positive. The specificity of a test is defined as:

$$\frac{N}{N+F}$$

i.e. what fraction of the tests that should be negative are negative. The sensitivity and selectivity of a test can be traded against each other by adjusting the threshold; for a test for explosives at an airport, for example, the threshold is set very low so the slightest hint of a bomb gives a positive. F (the number of false positive alarms) is very high, but G (the number missed) is (hopefully) very low.

Note that sensitivity and specificity can be analytical or clinical. The analytical sensitivity is how well the test measures the chemical analyte. The clinical (or functional, if the test is not a medical test) sensitivity measures how well it detects the disease. A thermometer has 100% sensitivity for measuring elevated temperature, but only 50% sensitivity for detecting the common cold, because this is only weakly linked to the thing you are measuring.

Automation

Automation means getting machines to do something, rather than having a human do it. In biotechnology, this usually means getting machines to perform repetitive laboratory procedures. It is particularly prevalent in drug discovery, where hundreds of thousands of tests have to be run to discover one candidate drug. However, other areas of biotechnology, such as process monitoring in fermentation, environmental sample processing, even testing meat for quality, are also highly automated.

Automation is an engineering discipline, not a biotechnological one, but despite this many biotechnologists think that they can build automated systems. Strong engineers can be reduced to tears by the systems that some biotechnologists call 'automation'. Common problems they encounter are:

● automating the scientist. How a scientist does something is convenient for them, using human limbs to perform a small number of tests. It may not be the best way for a machine to do a very much larger number of tests.

- not having the machines talk to each other. This means both physically (so the output of one machine fits into the input of the next) and in handling information.

- not building robustly. Having a robot that needs a PhD standing over it all day to keep it working is no better than having the PhD doing the experiments. However, building a robot that will work for 24 hours without even one tiny fault is a major challenge.

- failing to think 'factory'. A related idea, this says that if you are going to turn science into a factory-like process, then you must organize it like a factory, not like a scientist's normally messy and unstructured workplace. Again, this applies to both hardware layout and work flow.

- reinventing the wheel. Humble industries such as food manufacture and brewing have been automated for decades, but biotechnologists often do not realize this, and try to invent their systems from scratch.

Autoradiograph

Radioactive labels are often detected by autoradiography (autorad for short), a technique where an X-ray film is laid against something in the dark, and the film darkens where the radiation struck it, making an image of the radioactive regions of the object. (This was how Antoine Becquerel discovered radioactivity to start with, in 1896). Autorads were a standard way of detecting DNA or protein samples in a wide variety of gels from electrophoresis, and are still common, although they have been supplanted to a degree by non-radioactive methods. Typically, autorads produce dark patches of exposure on transparent film, and so are presented as black spots or bands on a white background.

In some experiments, the film emulsion is actually laid on to the experiment directly, so that, for example, the radiation trapped in specific sections of cells can be detected.

The same approach can be used to visualize the very weak light given out by luminescent labels; an X-ray film is laid directly on the experiment, and the blackened areas show where the label is. Confusingly, this technique is also often called 'autoradiography' and the product an 'autorad'.

Patterns of radioactivity can also be captured by the electronic equivalent of an X-ray film, called an imager or phosphorimager.

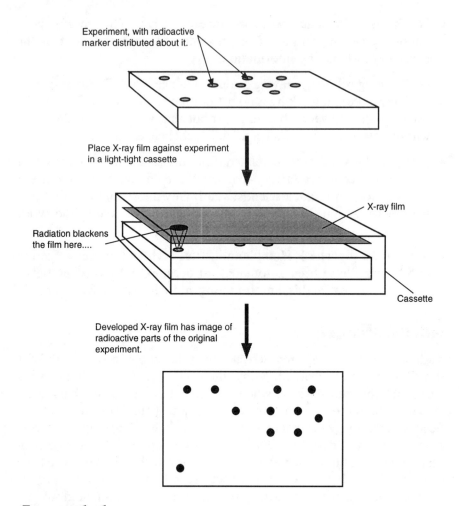

Experiment, with radioactive marker distributed about it.

Place X-ray film against experiment in a light-tight cassette

X-ray film

Radiation blackens the film here....

Cassette

Developed X-ray film has image of radioactive parts of the original experiment.

Auxostat

An auxostat is a chemostat in which the dilution rate can vary. Usually a chemostat, a closed culture vessel into which new medium is continuously fed in and old medium and organisms continuously removed, has a fixed rate of dilution, i.e. a fixed rate at which the new material is added and the old removed. This determines how fast the organism in the chemostat will grow. In an auxostat the rate at which new material is added and old material removed is determined by some feature of the culture. For example, the amount of bacteria may be measured by measuring the cloudiness (turbidity) of the culture, and the amount of material added adjusted to keep the turbidity constant. Alternatively, if the bacteria reduce the pH of the culture as they grow (as bacteria often do), the pH may be used to control the rate of dilution. The former is called a turbidostat, the latter a pHauxostat.

Auxostats have the advantage that the maximum growth rate or yield can be obtained much more easily than with a chemostat. If the dilution rate is not high enough in a chemostat, the culture will grow at less than its maximum growth rate. If it is too high, the organisms will not be able to keep up with the addition of new medium and so will be diluted out— you will end up with an empty chemostat. An auxostat can be adjusted to keep up automatically with bacterial growth, and so maximize the growth rate. At such high growth rates bacteria that can grow fast are selected over ones that grow slowly. Thus, natural selection acts on the population, selecting fast-growing variants of the bacteria. Depending on what the auxostat is to be used for, this can be a good or a bad thing.

In practice, all large, continuous industrial fermentation systems are auxostats rather than chemostats, since they have many feedback controls, which allow the operator to adjust what materials the fermentation receives as it proceeds.

Bacteriophage

A bacteriophage is a virus that attacks bacteria. They have been used extensively in DNA cloning work, where they are the basis of convenient vector molecules. The bacteriophage (or 'phage', for short) used most are derived from two 'wild' phages called m13 and lambda: P1 is another, less commonly used, phage.

Lambda phages are used to clone large pieces of DNA or RNA. They are a 'lytic' phage, i.e. they replicate by breaking open (lysing) their host cell. If a few phage are spread on a huge number of bacterial cells, they lyse open the cells they hit, releasing more phage, which then lyse open the neighbouring cells, releasing more phage, and so on. On a bacteriological plate, this results in a small clear zone, a plaque, where each original phage landed. In a bulk liquid culture it results in an enormous number of phage particles, 10^{14} per litre in some cases. Both plaques and bulk culture are useful sources of large amounts of the bacteriophage DNA, for analysis. Some lambda vectors have also been developed that are expression vectors, producing foreign protein from a cloned gene.

The other principal bacteriophage vector is the m13 system. This phage can grow inside a bacterium as a plasmid, so that it does not destroy the cell it infects but causes it to make new phage continuously. It is a single-stranded DNA phage, and is used for the Sanger dideoxy DNA sequencing method (which requires single-stranded DNA as a starting material). Messing has developed a well-known series of m13 vectors for cloning pieces of DNA into m13 for sequencing.

Both these phage grow on *Escherichia coli* as a host bacterium. Many other phages, both of *E. coli* and of other bacteria, are used in more specialist research applications.

Baculovirus

Baculovirus is a class of insect virus that has been used to make DNA cloning vectors for gene expression in eukaryotic cells. The vector system was derived from *Autographa californica* nuclear polyhedrosis virus to enable biotechnologists to make large quantities of proteins from cloned genes in insect cells (the cells usually used are a cell line derived from the fall armyworm). Baculoviruses have a gene that is expressed late on during their infection cycle at very high levels, filling the nucleus of the cell with many-sided bodies full of a protein that is not necessary to produce more viruses, but is necessary for the virus' spread in the wild. In the vector cloning system this gene is replaced by the one that we want expressed.

Production of the protein can be up to 50% of the cells' protein content, and several proteins can be made at once so that multi-subunit enzymes could (in principle) be made by this system. However, the main strength of the baculovirus system is that, as a genuine animal expression system, it produces proteins that are glycosylated like the proteins in animals. This, in combination with the relatively high levels of expression, can make this an attractive option for proteins that are to be used as biopharmaceuticals, outweighing the problems of growing a virus in mammalian cell culture. In addition, baculovirus is non-infective and non-pathogenic to vertebrates.

The baculovirus DNA is very large (100–150 kb), and so conventional recombinant DNA methods cannot be used to engineer it. Instead, plasmids containing the desired gene are recombined with the virus *in vivo* through homologous recombination

A novel use for baculovirus systems is as viral insecticides. A gene is inserted into the virus that is lethal to an insect (e.g. the endotoxin gene from *Bacillus thuringiensis*), but does not affect isolated cells. This is then used to produce infective virus, which can, in principle, infect insects and kill them. There are technical problems with this, however (such as whether the virus is still infective in a real organism), as well as regulatory ones.

Binding

Much of biochemistry and molecular biology is about molecules binding to each other. This means that they stick to each other because the exact shape and chemical nature of parts of their surfaces mean that they are 'complementary': a common model is 'lock-and-key', much used to describe how enzymes fit around their substrate. Biology only works because many biological molecules bind extremely tightly and specifically to other molecules—enzymes to their substrates, antibodies to their antigens, DNA strands to their complementary strands, and so on. This binding is entirely spontaneous, and depends on the chemical nature of the molecules concerned.

Binding can be characterized by a binding constant or association constant (K_a), or its inverse the dissociation constant (K_d). Mathematically, if two molecules (Mol-1 and Mol-2) link up to form a complex, then

$$K_a = \frac{[\text{Complex}]}{[\text{Mol-1}] \times [\text{Mol-2}]} \qquad K_d = \frac{[\text{Mol-1}] \times [\text{Mol-2}]}{[\text{Complex}]}$$

where [something] is the concentration of that something. For any given concentration of Mol-1 and Mol-2, the higher the K_a (or the lower the K_d) value the more complex and the less free Mol-1 and Mol-2 there will be around. In general, in biotechnology when someone talks about K_a or K_d, they want a very 'tight' binding, so the bigger the K_a or the smaller the K_d the better. Antibodies generally have K_a values between 10^7 (bad) and 10^{10} (good). Hormones bind to receptors with K_a values from 10^4 to 10^8. Proteins such as cytokines or growth factors can bind to their receptors much more tightly, with K_a values of 10^{10}–10^{12}. The prize goes to streptavidin, the protein that binds biotin (*see* **Biotin**). The K_a for the biotin–streptavidin binding is around 10^{16}, i.e. enough for streptavidin to be able to suck just 3 micrograms of biotin out of a small aircraft hanger full of water.

K_a and K_d are thermodynamic measures. They describe what would happen if we had enough time. Often a system is kinetically limited, i.e. the reactions do not happen fast enough. The key rates are 'on rate' (how fast the molecules stick together) and 'off rate' (how fast the come apart again). In the biotin example, although in theory a few milligrams of streptavidin in a tube could suck all the biotin out of a large swimming pool, in practice it would take thousands of years unless we pumped the pool past the streptavidin.

Bioaccumulation

This is the accumulation of materials that are not critical components of an organism by that organism. Usually it refers to the accumulation of metal. Many organisms (e.g. plants, fungi, protists, bacteria) accumulate metals when grown in a solution of them. Sometimes this is part of their defence mechanism against the poisonous effect of those metals, sometimes it is a side-effect of the chemistry of their cell walls.

In a few cases, bioaccumulation is economically important as part of the microbial mining cycle. Using this biosorption process, metals present in very low concentrations in water can be accumulated in the cell walls of living organisms and subsequently harvested. Bioaccumulation and the use of bacteria to remove toxic metals from waste water as a purification ('bioremediation') process are clearly closely related (*see* **Biosorption, Microbial mining**).

Bioassay

A bioassay is a way of measuring something that has a living system as a key component. Usually it means a way of measuring the concentration of a chemical, although bioassays for magnetic fields (using homing pigeons or magnetic bacteria), ionizing radiation (measuring mutation), or other physical effects are possible.

Many bioassays have been in traditional use—the proverbial canary in the coal mine was a bioassay for poisonous gas, the canary being the biological element. Animals have been used extensively in drug research in bioassays for the pharmacological activity of drugs. However, new bioassays are usually developed using bacteria or animal or plant cells, as these are normally much easier to handle than whole animals or plants and are cheaper to make and keep. Thus, bacterial bioassays for biological oxygen demand (BOD) (*see* **Sewage treatment**) and poisons in general are in use in the water industry. Here, bacteria are mixed in with a sample of water, and how well they can metabolize (and hence use up oxygen, produce carbon dioxide, or, in one case, give out light) is measured by an instrument.

Many of the cytokines and growth factors that biotechnologists are now producing using recombinant DNA methods were originally identified using bioassays in which mammalian cells were used to detect minuscule amounts of the compounds concerned through their potent effects on the cell's behaviour. They are often still measured by bioassay, because this is the only reliable way to distinguish the functional, active protein from similar proteins or degraded material. Bioassays also reflect the mode of action of the material, i.e. what it meant to do, rather than an arbitrary definition of chemical purity. Many very active biological substances are still measured in 'international units', which are defined as the amount required to have a specific effect in a bioassay.

On the borderline between bioassays and chemical assays are immunoassays (*see* **immunodiagnostics/ immunoassays**) and enzyme assays. These use proteins, made from a biological system, in an otherwise entirely standard chemical-type measurement.

Bioassays can be turned into sensor systems (*see* **Immobilized cell biosensors**), although as bioassays are no more convenient to use than any other chemical reaction, this is quite a difficult task.

Bioconversion

Bioconversion is the conversion of one chemical into another by living organisms, as opposed to their conversion by enzymes (which is biotransformation) or chemical processes. Synonyms are biological transformation or microbial transformation. Bioconversion has been used for a long time to make a few chemicals such as alcohol (from sugar) and, more recently, the drug ephedrine. However, it is only since the Second World War that bioconversion methods have become commonplace.

The usefulness of bioconversion is much the same as that of biotransformation, especially its extreme specificity and ability to work in moderate conditions. However, bioconversion has several different properties. The positive points are that bioconversions can involve several chemical steps, and the enzymes involved can be quite unstable, because the cell continuously remakes them as they break down. The problem with bioconversion is that most bacteria either convert chemicals very inefficiently, in which case they are not much use to the biotechnologist, or they convert them very efficiently into more bacterium, which is also not much use. Thus, to make an effective bioconversion process the bacterial strain has to be optimized so that it converts substrate to product efficiently but does not convert the product into something else.

Bioconversions of a number of chemicals have been reported, and some are used commercially. A major commercial application is in the manufacture of steroids. The 'basic' steroid molecule, often isolated from plants, is itself a very complicated molecule, and not one that it is easy to modify by normal chemical means in order to produce the very specific molecules needed for drug use. However, a variety of bioconversions that attack only specific bits of the molecule can be used. Bioconversion is particularly useful for introducing chemical changes at specific points in large, complex molecules, such as steroids. In many cases bioconversion is used together with more traditional organic chemistry to complete a complex synthesis.

Among the applications of bioconversion are bioremediation and bacterial mining (or biohydrometallurgy), which are discussed elsewhere. Other applications centre around the degradation of compounds that are difficult to attack chemically. A major class of such compounds are the hydrocarbons in oil, which bioconversion seeks to transform into more reactive alcohols and aldehydes. This can be done chemically, but requires extreme conditions and metallic catalysts, and usually results in a complex mixture of products. Bioconversion takes place under much milder conditions, and produces primarily a single product. Bacterial oxidation systems that can convert hydrocarbons to alcohols, aldehydes, or acids are known in many bacteria, such as *Pseudomonas oleovorans*, which soil bacterium has been the subject of a lot of work to make it more industrially efficient. *Pseudomonas* species contain a wide variety of plasmids that allow them to break down many organic chemicals, and so they could be of use in bioconversion processes.

Bioconversion in organic solvents

Many chemical reactions that are to be targeted for bioconversion or biotransformation are conventionally carried out in organic solvents, not in water, because the reagents are insoluble in water or because water causes undesirable side reactions. Enzymes can also be used in organic solvents, but interest is growing in the use of bacteria in solvents other than water.

Some bacterial bioconversions can be carried out in mixed phases, where the bacterium is tough enough to survive living next to droplets of the solvent. This has the advantage that a large number of enzymes, or very unstable enzymes that would not survive on their own in a bioreactor, can be used for a bioconversion. The disadvantage is that the bacterium must be kept alive, and that the bacteria produce all sorts of other metabolites in addition to the one you are after.

See also **Organic phase catalysis**.

Biocosmetics

A term for cosmetics with a biotechnological ingredient, or an activity or action based on biological knowledge (rather than cosmetic industry experience or belief, or marketing ploys).

Biocosmetics can be divided into three areas: biomaterials, biologically based ingredients, and medically rationalized products. The last class includes hypoallergenic products and UV-blocking agents, whose mode of action is supported by medical research but that are not 'biotechnological' as such. It also includes products such as liposome-based preparations, which may or may not have the effects claimed for them, but whose use of biotechnology products or buzzwords undoubtedly gives them a marketing edge.

Biomaterials include the use in cosmetics of collagen and collagen hydrolysate, a wide range of lipids as 'moisturizers' (including liposomes, which are claimed to carry active ingredients into the skin), fibronectin, and hyaluronic acid. These, and especially the last, are water-retention agents, which are meant to prevent the skin drying and wrinkling. Lipids such as gamma-linolenic acid are also claimed to have anti-inflammatory effects.

Biological ingredients include biotin, cyclodextrins, sphingosine, and a range of pigments. These are all 'natural' products, i.e. are made in a living organism rather than by chemical synthesis, and so can be produced biotechnologically: however, their actual effect is debated by medical professionals.

Cosmetics cannot be too effective since any cosmetic that had a substantial effect on skin physiology would be classed as a drug, and so have to undergo all the rigorous proof of efficacy and safety that drugs must undergo. Cosmetics that are deliberately being developed to have drug-like effects are called cosmeceutics (*see* **Nutraceuticals**). An example is suntan lotion which blocks UV light (rather than just replacing skin lipids). More conventional drug-like activity includes the use of retinoids for reducing skin wrinkling, and hydroxy-acid compounds, called exfoliation agents, which increase the rate at which the skin cells are replaced and so give a more youthful appearance.

Biodegradable materials

Biotechnology has preceded the 'green' bandwagon by some years in developing biodegradable materials. These endeavours generally fall into three areas.

1. Developing organisms that can break down normal materials, especially plastics (*see* **Bioremediation**).
2. Developing composite materials. Most 'biodegradable' plastics are composites of a conventional plastic laced with a biodegradable material such as starch. They break down when soil bacteria digest the starch, leaving small granules of the plastic behind. There is controversy over whether this is an improvement, especially since these materials are much weaker than unadulterated plastic, so you need to use much more of them to make bottles and containers of the required strength.
3. Biopolymers. Most living things produce polymers to make cell walls or other structural materials. Some of these can be used to make things: however, most are easily wetted, and tend to fall apart if left in the rain for a long time. A few exceptions have been found, the most well-developed of which are polyhydroxybutyrate (PHB), developed by ICI, and polycaprolactone (PCL) (*see* **Biopolymers**).

One strong, flexible, water-resistant, biodegradable polymeric material not usually talked about is wood. Quite a lot of plant biotechnology is aimed at trees, and biotechnology is indeed working on genetic engineering of trees (*see Agrobacterium tumefaciens*). However, growing wood deliberately as an 'industrial' production process is itself controversial.

Biodiversity

This means the diversity of life. Usually, the measure of biodiversity is genetic diversity, a measure of how many distinct genetic species there are in an area or population. These are sometimes called genetic resources, because they can be used as the starting point for new products or processes. Biodiversity is used in two contexts: the diversity of natural organisms, and the diversity of farmed crops or animals.

In the larger world, biodiversity concerns the vast range of plants (and animals, although these are considered less important from a biotechnological point of view) currently alive. Many of them may produce something useful to humans: a new drug, a new foodstuff, a new material. If they are allowed to be wiped out (and most species of plants live in the tropics, which are now under substantial threat) that potential will be lost for ever. There is substantial concern that the biodiversity of 'Third World' (i.e. not rich) countries is being exploited by rich ones (*see* **Bioprospecting**).

On the farm, biodiversity is considered to be a 'good thing'. If a country plants only one type of crop (for example), then a pathogen can sweep the entire country's crop from the fields. This happened to US wheat farmers in a wave of epidemics in the 1960s. Thus, not having a country plant just a single type (or cultivar) of a crop is protection against pandemics.

The role of biotechnology in this is double-edged. If biotechnology produces a new, wonder-wheat, then that will be planted in place of all other cultivars, and the wheat-growing world will end up being a monoculture, biodiversity will have been reduced. On the other hand, biotechnological methods are such that if you can transform one cereal with a gene you can transform a lot more, so biotechnology could actually increase biodiversity by increasing the number of crops with desirable genes in them. And it has been argued that the 'green revolution', including biotechnology, has been so successful that farmers are no longer under pressure to grow monocultures of the most productive crop: many farmers in Europe are paid to let fields lie fallow to reduce production, and so would be under pressure to diversify.

On the rain forest question biotechnology is less vociferous, but one of the key technologies in plant biotechnology is plant cloning, storage, and micropropagation, which could be used to store and propagate rare or endangered species.

Bioethics

Bioethics is the branch of ethics, philosophy, and social commentary that deals with the life sciences and their potential impact on society. At one extreme it can be enormously useful in focusing attention on problems that need to be solved. At the other extreme it can become a name-calling exercise between the 'pro-biotechnology' and 'anti-biotechnology' schools of thought which, as it reduces complex arguments to clichés and abuse, makes better sound bites. The Human Genome Project has set aside around 3% of its budget solely to consider ethical issues. The medical genetics and pharmaceutical communities have employed ethical experts for years. Thus the industry and regulators of biotechnology have a substantial concern in bioethics. The more serious side of bioethics is sometimes called ELSI—ethical, legal, and social issues.

Bioethics is not strictly confined to classical ethics, but rather spills over into social policy and even politics. Issues of current concern are almost all to do with how far biotechnology should be encouraged or (as is usually believed to be the case) prevented from doing something. These issues include:

- The validity of making animal models for human diseases; for example, transgenic models of cancer. (Animal rights as a whole is also a major issue for some biotechnology companies, but is not really a biotechnology-specific issue.)

- The use and abuse of information about people's genetic make-up (*see* **Genetic information**).

- The problem of trading off a new drug's potential side-effects with a need to have patients benefit from it as quickly as possible (*see* **Treatment protocol programme**).

- How people should be informed for trials of experimental procedures and products. (This goes beyond drug trials into food safety and environmental issues).

- The conditions under which recombinant organisms can be released into the world.

- The role of biotechnology in embryo and fetal research.

- The justification for patenting life forms and pieces of living material (especially genes).

- The use of genetic resources in biotechnology exploitation (*see* **Bioprospecting**).

Bioethicists have identified a number of common themes among the arguments that are brought to bear on biotechnological issues. The most controversial is probably the '**yuk factor**'. Others include the need for individuals to determine their own fate, the need to protect the vulnerable from the unscrupulous, and so on, which are common to many wider ethical arguments.

There is also a strong component of public opinion in bioethics, although often the reason that people feel some particular way about the technology is not examined (*see* **Mythogenesis**).

Biofilms

A biofilm is a layer of microorganisms growing on a surface, in a bed of polymer material which they themselves have made. Biofilms tend to form wherever a surface on which bacteria can grow is exposed to some suitable medium and a supply of bacteria. Thus biofilms form on sites as diverse as domestic plumbing, hyperbolic cooling towers in power stations, sewage plants, and teeth.

The bacteria stick to the surface by a combination of corrosion and glue. The bacterial film is rarely a single type of organism—rather they are communities (or 'consortia') of different organisms. Some corrode the surface. This process, called biocorrosion, can go on on its own. It leaves the surface rougher and more chemically 'sticky'. Other bacteria synthesize extensive networks of sticky mucopolysaccharide polymers to stick themselves, and any other bacteria nearby, to the surface. The resulting films can be amazingly hard to get off. They also increase the surface's roughness (and so the amount of pumping needed in a pipe, for example), and block the diffusion of oxygen through membranes.

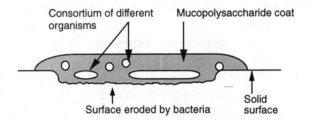

Consortium of different organisms • Mucopolysaccharide coat • Surface eroded by bacteria • Solid surface

The process of covering surfaces with biofilm in this way is called biofouling. It is a particular problem where liquid is recirculated round a closed loop of pipework (so any bacteria washed off the film can get the chance to stick back on next time round), or when filter membranes are exposed to bacteria. Unlike normal fouling of membranes by solids or large molecules, biofouling is an active process which, once under way, cannot easily be reversed by cross-filtering or reversing the flow through the membrane. The biocorrosion can also start to break down the membrane, making it leaky. Thus there is a lot of interest in the use of biocides (both in the solution and impregnated into the surface) to stop biofilm production.

Biofouling and biocorrosion can affect nearly all materials known. Bob Tatnell of DuPont estimates that about 50% of all metal corrosion world-wide is caused, at least in part, by biocorrosion. However, biofilms can also be used; some biosensors use a biofilm of cells to detect when the water flowing past them contains a poison, and biofilms growing on permeable membranes have been used to break down organic wastes.

Biofilms form rapidly where non-sterile water containing nutrients flows over a surface (the slime forming on stones on the bottom of streams is an example and also shows that, if the water is flowing fast enough, film cannot form). However, biofilm has been seen even when no obvious nutrient was available in ultra-pure water.

Biofilter

A biofilter is a large-scale filtering system where the surface area of the filter provides a support for organisms. Waste fluid flowing through the filter is broken down into harmless products by the organisms. There are three variants.

Biofilters for treatment of waste gases. Gases from food and chemical factories, especially for processes that produce a lot of organic volatiles such as acetone or hydrocarbons, can be adsorbed efficiently by a large filter mesh of microorganisms. These have to be kept moist and supplied with nitrogen and trace elements, but otherwise the filter is self-sustaining, with the organisms getting carbon from the gases and oxygen from air. This is often cheaper than conventional chemical scrubbers.

Trickle bed sewage treatment. This uses the same principle, but for treating liquid waste (see **Sewage treatment**).

Reed beds. A variant on the biofilter idea, reed beds are becoming increasingly used for bioremediation of contaminated water. The reeds, planted in gravel or loose soil, grow from hollow, air-containing roots (rhizomes) and the hollow-stemmed shoots grow from these. Oxygen passes down the plant and out through the root hairs, so the water immediately around the root is highly oxygenated. This supports the growth of lots of bacteria, which can in turn use up the carbon pollutants in high BOD (biological oxygen demand) water. The bacteria can also remove nitrogen and phosphorus, pathogenic bacteria, and some metals and other toxins. The region around the roots is called the rhizosphere (in reed beds and other plant ecosystems too).

Two types of system are used. In horizontal flow systems water flows through a dense reed bed as in a stream, and usually flows just below the surface (because the surface is full of reed stems). This is used to treat relatively lightly contaminated water. More heavily contaminated water is passed through a vertical bed system. Here, water is pumped on to the top of a reed bed based in sand or gravel and air is pumped into the base of the bed. Water is taken out of the base. After a few days such a bed becomes clogged, so it is used in 'shifts' of a few days each to give it time to recover. This is similar to the trickle bed system for sewage treatment, but with reeds as well as bacteria in the bed.

Reed beds are usually used for secondary treatment of farm and municipal waste; they cannot handle raw sewage. They have the advantage of being relatively cheap to build and very 'green' in the literal and political senses.

Biofuels

Biofuels are fuels made from bulk biological materials such as cane sugar or wood pulp. There are a range of ways of converting these rather bulky, inconvenient fuel materials into fuels that are useful for industrial or transport use, or as starting materials for the chemical industry. Converting biomass into replacements for gasoline attracted a lot of interest in the 1980s after the 'oil crisis' of the late 1970s.

Wet biomass materials, such as starch, sugar, bagasse (the solid residue left after the juice has been squeezed from cane sugar), sewage, waste waters, etc., can be used. They are broken down by fermentation, sometimes assisted by enzyme digestion, to make a variety of simple molecules, usually ethanol or methane. Ethanol for use as a fuel is made from cane sugar by fermentation and distillation in commercial quantities in Brazil, where the economics are unusually favourable, and 'Proalcool' is a major fuel there: 14 billion litres were made in 1989. In the USA various initiatives to promote 'gasohol', various gasoline–ethanol mixes, have had mixed responses in the past, owing to changing political support and general discouragement from the oil industry. Most fuel alcohol made in the USA is made by fermentation of corn (maize) starch. Ethanol made in this way is sometimes called 'agricultural ethanol'. The use of methanol has also been suggested, but is harder to make and is more corrosive. Methane is widely used as a heating fuel, and some biofuel methane has been tried out for electric power generation (*see* **Biogas**).

Some crops are grown specifically to provide they raw material for fuel production in this way; they are called energy crops. Sugar cane is used in this way in Brazil (for fermentation into ethanol) as is sweet sorghum (a tall, grass-like plant) in other parts of the world.

The other gaseous biofuel is hydrogen, made primarily by photolysis of water. This is what photosynthesis does, but in normal living systems the hydrogen is not created as gas, but rather is used to make sugars. One aim of this area of biofuel research is to get organisms, probably single-celled algae, to release hydrogen gas when exposed to sunlight. In 1995 scientists showed that the enzymes hydrogenase and glucose dehydrogenase could use glucose to generate hydrogen. They speculate that wood pulp could be the source of the glucose; in theory, an enzymatic process could use the recycled newspapers generated in the USA to power a city of one million inhabitants by hydrogen alone (*see also* **Solar energy**). Hydrogen made in this way is inevitably called biohydrogen, although it is exactly the same as any other hydrogen.

The other route to making biofuels is chemical. If any dry biological matter is heated up slowly, it undergoes 'pyrolysis' and produces a complex mixture of oily materials and charred polymers. These oils can be distilled in the same way as conventional mineral oil, to give fractions with similar properties to gasoline, diesel, lubricating oil, etc. The charred remains can themselves be burned, possibly to heat the pyrolysis reactors and stills. The chemical properties of the resulting oils can be rather different from conventional oils, and so far no one has succeeded in making this sort of process competitive with mineral oil production. Pyrolysis can be used to make other chemicals as well, but is no more effective a route than conventional oil chemistry.

Biogas

Biogas is the name given to methane ('natural gas') produced by fermenting waste, and particularly sewage. This is an alternative method of waste disposal to landfill or conventional sewage plants. Waste is incubated with suitable bacteria in a digestor in the complete absence of air (an anaerobic fermentor). The organic matter in the waste is converted mainly into methane and carbon dioxide, and the methane can be burned to provide power, heat, etc. In sewage treatment plants using anaerobic fermentation, methane is often used as a power source for the plant itself. The process is also called anaerobic digestion.

Anaerobic sewage disposal has some advantages over conventional systems (such as the activated sludge system). It produces less microbial mass to get rid of, and does not require aeration (which is expensive because it requires power). However, it only works well on concentrated wastes: solid food waste or sewage sludge. Anaerobic fermentation is rarely a practical option for treatment of raw sewage, which is rather dilute.

The bacteria responsible for generating methane from waste are the methanogenic bacteria, an unusual group which can turn a limited number of carbon substrates into carbon dioxide and methane. To break the waste down into things that the methanogens can 'eat' requires other bacteria. Thus an anaerobic digestor needs a specialized population of bacteria to work well. In practice, waste digestion processes tend to use whatever organisms are on the waste, with consequently lower efficiency.

Biohydrometallurgy

This is the use of bacteria to perform processes involving metals. It encompasses a wide range of industrial processes, including microbial mining, oil recovery, desulfurization, and a range of physiological processes, including biosorption and the redox metabolism of bacteria. It is also the study of how bacteria corrode metal and metal-containing surfaces, a process known as biocorrosion (*see* **Biofilms**).

Biohydrometallurgy includes two broad areas of bacterial activity.

Biosorption. This is the selective absorption of metal ions by bacteria and bacterial materials (such as their isolated cell walls) (*see* **Biosorption**).

Redox reactions. These are reactions where the bacterium uses the metal ion, or a mineral in which the metal is immobilized, for its metabolism. A major use is the oxidation of sulfide to sufate, which reaction some bacteria use as a source of energy (the reaction releases substantial chemical energy when carried out in air). As sulfides are frequently insoluble, and sulfates are often soluble, this is a fairly general method of releasing the metals in sulfide ores. The same reactions can be used to oxidize sulfides in one compound, releasing sulfuric acid, which then dissolves another compound, or to pre-oxidize a metal ore to make it more amenable to further processing. These topics are covered in separate entries on **microbial mining** and its subcomponents.

Bacteria can also oxidize or reduce metals themselves. The manganese nodules on the sea floor and the Banded Iron Formation rock strata (laid down 1000 million years ago) are probably the result of bacterial reduction of manganese and oxidation of iron, respectively.

Bioinformatics

This is the use and organization of information of biological, almost always molecular biological, interest. In particular, it is concerned with organizing biomolecular databases, in getting useful information out of such databases, and in integrating information from disparate sources.

Bioinformatics covers four distinct types of work.

Data acquisition and analysis. This is the realm of **LIMS** systems.

Database development. The rapid growth of DNA sequencing drove the initial requirement for databases that could store lots of biological data. There is a lot of computer science expertise out there concerning databases, but it needs adaptation to work on the analogue, 'woolly' concepts of biology. In the 1990s DNA sequence databases have been equalled in size by databases of 3D molecular structure, and surpassed by databases from high throughput screening work (HTS) in drug discovery (*see* **High throughput screening**).

There is a debate about the 'curation' of public databases, the extent to which a curator manages and prunes the data. Some databases, such as dbEST (which collects expressed sequence tag sequences), have very little curation—providing the data is entered in the right format, the database managers will include it. Some, such as SwissProt (protein sequence information), have substantial curation, with the database 'owner' performing many checks on the data, sending it to experts for additional commentary, and adding their own comments and links to other data. This makes the data entries much more valuable to the user, but slows down the process of releasing the data and means that some data just does not appear. Several databases have got into public trouble with the scientific community for too much or too little curation (often the same databases, sometimes at the same time).

By mid-1996 there were well over 100 biotechnology-related databases available free over the Internet, ranging from DNA sequence, to biomedical literature, to product safety data, to fish farming tips. Many pharmaceutical and biotechnological pharmaceutical companies also have very substantial in-house, private databases of genomic, HTS, or other data.

There is also much debase about the 'correct' structure for databases for specific data, which means the format best suited to: (a) efficient storage, (b) efficient search, or (c) efficient analysis. There are dozens of different formats for many types of data. This makes searching the data automatically very tiresome: in practice, scientists use just a small number of databases that they have become familiar with. Most databases are relational (i.e. are based on tables of data and associated tables of links that relate pieces of data to each other) or object-oriented (i.e. treat data as separate 'objects' that have a set of predefined properties). However these are approaches to constructing a database, not prescriptions for them; saying that something is a relational database does not tell you very much about what is stored in it or how it could be used.

Integration and links. Scientists need links between databases to point to where related information is stored. This is difficult to do, especially if the database you are pointing to changes its numbering scheme (or even disappears completely). Building stable links and 'pointers' (pieces of information that tell you unambiguously that this entry in my database refers to the same gene or protein as that entry in your database) is a major task of the database curators. A related term is interoperability, which means that data from one database can make sense in the context of another (they both define 'gene' the same way, for example).

Analysis. Once you have got all that data, you want to analyse it. This usually means:

- searching for similar sequence or structural data and comparing the items with each other (*see also* **Phylogeny**). The programs BLAST, FASTA, and BLITZ are fast methods for searching DNA or protein sequence databases for things that are similar to a 'query' sequence that you type in.

- predicting structure from sequence This is a major area of research, and is edging closer to being realistic. Generally, a combination of many methods will give quite good results unless your sequence is like nothing seen before. One expert system-based predictor is even accessible over the World-Wide Web.

Nearly all bioinformatics resources were developed as public domain 'freeware' until the early 1990s, and much is still available free over the Internet. Some companies have developed proprietary databases or analytical software. In some cases, they have done so on the back of public domain work, which has raised scientific hackles: for example, in 1996, the database company MDL started to sell a commercial mole-cular modelling programme based around RasMol, a popular 'freeware' modelling tool that had been sponsored by drug giant Glaxo. MDL claimed that they had added a lot of functionality to the program. Scientists were not convinced.

Biolistics

This is a method, developed at Cornell University and commercialized by DuPont, to get DNA into cells. The DNA is mixed with small metal particles, usually tungsten, a fraction of a micrometre across. These are then fired into a cell at very high speed. They puncture it and carry the DNA into the cell. In the original system a .22 cartridge was used to drive the particles, hence this is also called a 'particle gun' system.

Biolistics has an advantage over transfection, transduction, etc. (*see* **Transfection, transduction, transformation**) because it can be applied to *any* cell, or indeed to parts of a cell. Thus, biolistics has put DNA into animal, plant, and fungal cells, and into the mitochondria inside cells.

The force to drive the particles into the cells can also be electrical. A spark is used to vaporize a water droplet, which explodes like a small cartridge. This has the advantage that the current, and hence the energy of the explosion, can be varied at will. However, it is more complicated to set up.

As well as getting DNA into isolated cells, biolistics have been used to transfect DNA into animal tissues. Mouse skin and ears have been transfected by a suitably modified biolistic gun in whole, live mice, suggesting that this could be a route to somatic cell gene therapy in humans. The key to getting this to work is to limit the damage the gun-like propulsion causes: curiously, this is not caused by the particles themselves, but by the blast of air or gas that accompanies them. The DNA was only active for a few days, however, before the cells broke it up.

Biological containment

This is restricting how far an organism can travel by arranging for their biology to prevent them from growing outside the laboratory.

Biological containment can take two forms: making the organism unable to survive in the outside environment, or making the outside environment inhospitable to the organism. The latter is rarely suitable for bacteria which, in principle, could survive almost anywhere. Thus, for bacteria and yeasts the favoured approach is to mutate the genes in the bug so that they need to have a supply of a nutrient that is usually only found in the laboratory. If they get out, they then cannot grow. Other mutants may have weakened cell walls so that they fall apart outside the laboratory, or may even have 'destructor genes' in them which destroy the cells if the temperature becomes lower or higher than the laboratory optimum.

Making the environment unfriendly to the organism is partly a biological control, partly a physical one. For example, some of the first genetically engineered rice strains were developed in England (which is too cold to grow rice) and tried in the field in Arizona (where it is too dry). Thus there was no rice growing nearby to cross-pollinate with the genetically engineered rice, and if any rice 'escaped' it would die. This is containment based on the biology of the plant, but without altering the plant specially.

Biological control

Also called biocontrol, this is the control of one species by another which has been deliberately introduced for that purpose. The most famous example is the introduction of myxamatosis into Australia to control rabbits, although biological control is much more ancient, dating back at least to the ancient Chinese, who used pharaohs ants to combat destructive insects in their grain stores.

Biotechnology has looked at many potential biological control agents: sometimes these overlap with **biopesticides**. For example, *Bacillus thuringiensis* produces an anti-lepidopteric (caterpillar-killer) protein. *B.thuringiensis* has been used as a biocontrol agent for many years, but recently biotechnology has isolated the protein responsible to make it into a biopesticide. Half-way house versions have been tested, such as the baculovirus (*Autographica californica*) which contained a scorpion toxin gene-tested on cabbages in the UK in 1993. The virus is still a living entity, but has been engineered to be a pathogen to the caterpillars infesting the cabbages. (This particular experiment raised substantial public opposition, as it was a field trial of a genetically engineered virus, i.e. a deliberate release experiment.)

Biotechnologists approach biological control in several ways. Fungi, viruses, or bacteria that are known to attack a pest can be cultured in large amounts and sprayed on to a crop, to kill that particular pest. Entamophagous fungi (fungi that infect insects) are favourites here, as they can infect insects through the cuticle, and so do not have to be eaten to be effective. Such fungi are termed mycoinsecticides, and about a dozen are under scale-up development.

Some mycoinsecticide fungi produce short epidemics (called epizooics) among the target pest population without creating a stable presence in the ecology: they can only continue to spread while there is a high density of the insect pathogen around after this they die out. Bacteria, such as *Bacillus popillae* (which attacks the Japanese beetle *Popilla japonicum*) have been used. Insects can also be controlled with nematodes, as is done against sciarid fly and phorid fly, which attack commercial mushroom farms.

In essence culturing fungal pathogens is the same as culturing any other fungus, with the additional constraint that the fungi usually require very specific and unusual culture media.

Fungi, bacteria, and insects are also considered as weed-control agents: microorganisms to attack northern jointvetch and milkweed vine (weeds of rice and citrus trees, respectively) are in use, and others are under development. Such agents are called mycoherbicides.

Biocontrol can also be aimed at pathogenic fungi: Gary Strobel gained some notoriety in 1987 when he inoculated elm trees with a genetically engineered bacterium designed to protect them from Dutch elm disease without obtaining proper Federal approval. Monsanto performed field trials of a bacterial biocontrol agent against the fungus that causes the wheat disease 'take-all' in 1988.

Biotechnology has been more visible in producing viral biocontrol agents. Here, genetic engineering and advances in culturing viruses in insect cells (*see* **Baculovirus**) have enabled biotechnologists to manipulate insect viruses to be, potentially, more effective biocontrol agents. The aim is to increase or alter the host range of a virus by altering the specificity of the viral proteins that bind to the cell surface, or to increase the virulence of a normally benign but very infectious virus by engineering in a toxin gene or the 'pathogenesis' genes from another virus. In practice these aims are rather hard to achieve, since viral infection is a very complex process. Some trials have marked viruses with marker genes so that their spread can be monitored: this gives a measure of how well a simpler form of viral control—growing large amounts of the virus and then spraying it on to the crop—is working. Such field trials have been carried out, most famously in Scotland where pine trees were sprayed with an anti-moth viral biocontrol agent without (as it turned out subsequently) permission being given for release of this genetically modified organism (GMO).

The key to any biocontrol programme is to isolate an effective organism, one that can spread rapidly and effectively through the target pest population but that will not spread to other species (and hence become a pest in its own right). As pests are often foreign organisms introduced into an area where they have no natural enemies (e.g. water hyacinths in much of Africa, tumbleweed in the US, Dutch elm disease in many temperate climates), the best source of a potential biocontrol agent is often the original home of the pest. Biotechnology helps in this by providing tools for cultivating strange organisms in the laboratory, and for following them in the environment, even if the organisms themselves are not altered in any way.

Biological response modifiers

This is a very general term, usually meaning proteins that affect how the immune system works. In this meaning it is almost synonymous with '**cytokine**'. The term is widely used because of the existence of the FDA Biological Response Modifiers Advisory Committee, which oversees biopharmaceuticals that modify biological response mechanisms (i.e. all of them to date). Normally, biological response modifiers act in concert, not as isolated chemical entities. Thus, there has been much agonizing about how the components of biological response modifier drugs, cloned as pure proteins but used in combinations, can be regulated by the drug regulatory agencies, and particularly by the FDA (Federal Drug Administration). Cetus had well-publicized problems trying to get its IL-2 (interleukin-2) approved as an anticancer drug, because IL-2 was not effective on its own. Cetus wanted to use it in combination with other biopharmaceuticals, but as it was not effective enough as an isolated entity it was refused approval. (The case was not helped by Cetus' inexperience at presenting data to the FDA.)

Biomass

Biomass means the mass of organic material in any bulk biological material, and, by extension, any large mass of biological matter. Fermentations generate biomass as well as the fermentation product, and it is important to control the mount of biomass to maintain the rate of fermentation. The most important thing to measure is the active biomass, which is the amount of living cells in a fermentation. However, this can be hard to measure, as a dead cell looks very like a living cell.

Biomass products are ones made from the biomass of your process. Usually, in a fermentation system, the biomass is a side product, or even a waste. In brewing, for example, we want the alcohol, not the yeast. Biomass products are made from the mass of cells themselves, and some processes grow plants, fungi, or bacteria specifically for their biomass. Such biomass production is split into several areas of interest.

Single-cell protein (SCP). See **SCP**.

Algal biomass. Single-celled plants such as *Chlorella* and *Spirulina* are grown commercially in ponds to make food materials. *Spirulina* enjoyed a vogue as a health food a few years ago, because of an unfounded belief that it was extraordinarily nutritious. Like most algae (including some seaweeds) it is quite a good food, but *Spirulina* is not outstanding. *Chorella* is grown commercially to make into fish food: it is fed to zooplankton (microscopic animals), and these in turn are harvested to feed the fish in fish farms. This is a way of converting sunlight into food in a more convenient and controllable way than normal farming.

Plant biomass. Crop plants such as sugar cane have also been grown for biomass. This is usually used as the start of a chemical production process (growing plants for food is usually called 'farming'). Brazil spent a considerable amount of effort and money growing sugar to make ethanol by fermentation—relatively unprocessed sugar cane was used as the substrate, and the quite disgusting product used to run cars. This is using biomass as a way of converting sunlight into useful chemicals.

See also **Biofuel**.

Biomaterials

'Biomaterial' is a general term for any biologically derived material that is used for the sake of its material property, rather than because it is a catalyst or a pharmaceutical. Thus, DNA could be a biomaterial if you used it to make paper-clips or cranes, rather than using it to store information. 'Biomaterial' is also sometimes used to mean the material that is compatible with biology, and particularly with medical devices such as inhalers or catheters. This is not the usual use in a biotechnological context (*see* **Prosthetics**).

The most common biomaterials are some proteins, many carbohydrates, and some specialized polymers (*see* **Biopolymers**). The proteins considered for biomaterials applications are usually proteins used as structural elements in animals or, occasionally, plants. Collagen, the protein in bone and connective tissues in a wide variety of animals, is a common candidate, and has been used (controversially) as a cosmetic biomaterial, being used as a 'natural filler' for plastic surgery operations. Fibroin, the protein in silk, has been suggested as a protein with sufficient strength to rival nylon, and spider silk could challenge even Kevlar as a structural material. Most of these structural materials have fairly simple amino acid sequences, as they are made of short blocks of amino acids repeated many times. Thus the rigid, central sections of the collagen molecule, which give it its elastic strength, are made mostly of repeats of the three-amino acid unit glycine–X–proline (where X can be one of several amino acids). Biotechnologists have therefore made synthetic proteins with simple repeating patterns in the search for new biomaterials. All these proteins can be produced as amorphous materials by conventional cloning methods, and then spun into ordered fibres.

Carbohydrates have been used as structural materials for millennia: the strength of paper and papyrus derives from the properties of their carbohydrate, principally cellulose, components. Biotechnology has produced a wide range of carbohydrates with modified properties, which act as lubricants for biomedical applications or as texture modifying or bulking agents in food manufacture. These include rare but natural materials made from bacteria, such as polydextrose, carbohydrates modified by enzymes to have improved properties, and entirely artificial polymers (*see also* **Wood)**.

Other polymers include 'natural' plastics, such as polyhydroxybutyrate (*see* **Biopolymers**), or rubbers produced by bacteria or fungi. The company Enzymol has developed the enzyme soybean peroxidase to make polyphenol resins, a common industrial plastic.

The properties of a polymer that are crucial in determining whether it will make a 'good' biomaterial for a specific application include:

- tensile strength (both elasticity and breaking strength),

- hydration (How much water does it bind? How much does it need to bind to keep its properties?),

- visco-elastic properties, and

- viscosity.

Biomimetic

Literally meaning 'imitating life', this is the area of chemistry that seeks to develop reagents that perform some of the functions of biological molecules. The reason for doing this is that many biological molecules are chemically inconvenient to produce, handle, or apply in large amounts and using cheap processes. By using chemical mimics of such molecules the biotechnologist hopes to achieve a more flexible and more commercially useful way of achieving the same ends. Thus Arris Pharmaceuticals is developing a small molecule 'mimetic' of erythropoietin, and DuPont Merck are developing a neurotensin mimetic: both programmes aim to mimic the molecular effect of a large protein (EPO and neurotensin, respectively) with a small, stable, cheap organic molecule.

Areas of chemical research in the general field of biomimetics include the following.

Cofactor substitutes. Many enzyme cofactors are complex and labile molecules: in particular NAD and NADP are difficult to work with on a large scale. Two lines of research seek to replace them with other molecules. Triazine dyes have been used as replacements for NAD in affinity chromatography applications. Here the dye is bound to a column, and a mix of materials containing a dehydrogenase enzyme is passed over the column. The triazine dye binds to the dehydrogenase as NAD would, and so holds it on to the column, all other materials pass through. This has been used very successfully for many purifications.

The other application of cofactor substitutes is as actual substitutes for the substrates, especially for NAD, NADP, and FAD (flavine–adenine dinucleotide) in reactions catalysed by dehydrogenases. Here again the aim is to find a small molecule that will do the chemical work of NAD, etc., with the enzyme.

Peptide biomimetics (peptidomimetics). These are very valuable as drug candidates. Peptides are easy targets for manipulation through recombinant DNA techniques, but make poor drugs. A peptidomimetic is a molecule that has the same effect as a peptide (usually because it has the same critical 'shape'), but is not itself a peptide, and hence is not broken down by proteases and is cheaper to make. About 20 peptidomimetics have been developed to the point where they are potentially interesting as drugs. These compounds have to be made as special organic chemicals, based around the known structure of a peptide.

DNA substitutes. DNA is also a very valuable reagent, but one that is attacked rapidly by the body and is expensive to make. Biotechnological chemists are working on altering the basic 'backbone' of nucleic acids so as to make them more stable and potentially easier to make. In early

1992, a 'DNA substitute' called polyamide nucleic acid (PNA) was reported that had no sugar–phosphate backbone at all: in its place was a polyamide chain looking more like a protein. This material bound tightly to single-stranded DNA in a way that suggested that it was forming the correct base pairs. This has applications in **antisense**, as such molecules could be much easier to get into cells and totally resistant to breakdown by nucleases or proteases.

Glycomimetics. These are organic molecules that have some of the properties of sugars and are for potential use in affecting interactions that are usually mediated by sugars.

Synzymes. These are low molecular weight molecules that act as artificial enzymes, i.e. as highly specific catalysts. Usually they are synthesized to copy deliberately the 3D structure of the 'active site' of an enzyme, but using non-peptide chemical building blocks. Unlike more usual organic chemistry catalysts, which catalyse a broad range of reactions, the aim is to make synzymes as specific as enzymes are.

Molecular imprinting. This is another approach to making an artificial, enzyme-like catalyst (*see* **Molecular imprinting**).

Biomineralization

This is the deposition of minerals by living organisms. In some applications it is related to microbial mining (the breakup of minerals by microorganisms) and hence is part of biohydrometallurgy. However, biomineralization extends beyond this. There are two general areas of interest to the biotechnologist.

Microbial biomineralization. This is the laying down of minerals by microorganisms. If the minerals are deposited inside the bacterial cell, then they are, of necessity, laid down as extremely small crystals or granules. The magnetite made by magnetic bacteria is of this sort; this magnetic mineral is made as tiny inclusion bodies inside some bacteria, which are thereby enabled to swim preferentially along lines of magnetic field. (This enables them to swim towards the bottom of ponds in temperate zones.) Many larger mineral forms are also made partly by bacteria, and this has been suggested as a way in which minerals could be isolated or purified using biotechnology.

Multicellular biomineralization. In many plants and animals minerals are used to give strength. Thus, vertebrate bone contains calcium phosphate, and many grasses have silica in their leaves to give them a hard cutting edge to dissuade animals from eating them. The control of biomineralization is of substantial interest in several human diseases, notably osteoporosis, a disease in which too much calcium and phosphorus is lost from the bones.

Biomineralization is of interest to the materials scientist as well. Biological systems manage to deposit minerals in unusually useful forms. Thus, bone and teeth are much stronger than 'raw' calcium phosphate. Additional strength and specific crystal forms are potentially useful as ways of extending the range of mineral materials available for construction, electronics, and chemical industries. Living things achieve these feats by incorporating specific proteins into the growing mineral, to force crystal growth into the form they require or to reduce the propagation of cracks through the mineral when it is stressed.

Biopesticide

A biopesticide is a pesticide, i.e. a compound that kills animal pests, that is based on specific biological effects, and not on broader chemical poisons. Specific types are also called bioinsecticides and biofungicides. Biopesticides are different from biocontrol agents (*see* **Biological control**) in that biopesticides are not living, and so cannot reproduce themselves in a target population, whereas biocontrol agents are active, living things that seek out the pest to be destroyed.

There are a wide range of materials that plants produce to foil pest attack—the caffeine in coffee beans is probably such a material. However, the most attractive for biotechnologists are protein anti-pest materials, such as the much-hyped *Bacillus thuringiensis* (B.t., sometimes B.t.k.) toxin, which specifically interferes with the absorption of food from the guts of some insects but is harmless to mammals. This protein (which has been used as a pesticide for some time as a bacterial suspension) has been 'cloned' into more amenable bacteria. The gene for the protein has also been inserted into petunia by Calgene, making a plant that is more resistant to pest attack.

The rationale behind developing biopesticides, in contrast with conventional pesticides, is threefold. First, they are more likely to be biodegradable than chemical entities that are not normally found in nature. Secondly, they are intended to be more specific (and sometimes, as a consequence, more potent), since they are targeted at specific elements of the pest's metabolism. This is an important aspect in Third World applications, where toxicity to people is a major problem. (It also means that it is not worth stealing them for use as broad-spectrum pesticides for crop protection, which is the fate of some of the more conventional pesticides used to control malaria and sleeping sickness in poor countries.) Thirdly, their targets are less likely to evolve resistance to them. This is speculative, and some resistance of *Culex* mosquitoes to B.k.t. toxin has been recorded.

Biological control agents are sometimes also known as biopesticides.

By the end of 1991 there were 45 biopesticides or biocontrol agents aimed at insects (mostly bacteria, bacterially derived proteins, or viruses), 10 at organisms that cause plant diseases, and two at weeds.

Biopesticides are sometimes grouped with probiotics, biocontrol agents, and other biologically derived products in being called agrobiologicals.

Biopolymers

This can mean two things. Biopolymers can be made synthetically, by polymerizing a simple biological monomer. A popular example is the hyaluronate polymers (hyaluronic acid is a sugar derivative), which are used in a range of artificial tissues as the base material on which to grow cells, and as a material in advanced wound dressings. Another is the range of dextrans, which have uses ranging from chromatography columns to the fillings in low calorie chocolate bars.

The other, related, meaning is a polymer made by a living organism. Most living things produce polymers to make cell walls or other structural materials, or to store energy or food. Cellulose is a biopolymer, but the more common meaning of the word is the plastic- or rubber-like materials produced by bacteria and fungi. The most well-developed examples are polyhydroxybutyrate (PHB), developed by ICI, and polycaprolactone (PCL). Both materials can be moulded like other plastics, and are water-resistant and water-tight. However, their structures can be attacked slowly by bacteria, and so after a period of months to years will be completely broken down. PHB is one of a class of polymers called polyhydroxyalkanoates (PHAs), which also includes polyhydrovaleric acid (PHV). A copolymer of PHB and PHV is produced by fermentation on *Alcaligenes eutrophus*, and sold by Zeneca as Biopol. (The Biopol product line was sold to Monsanto at the end of 1996.) The only problem remaining is what to make out of it. For demonstration purposes, Zeneca has made entirely biodegradable coffin handles; however, this will not alter the waste budget of the Western world significantly. Hundreds of tonnes of PHB are produced annually. It is adapted to a range of applications by mixing it with small amounts of other biodegradable polymers.

Biopreservation

This is usually the preservation of food using biological materials. The leading example is nicin, a bacterial protein that acts as a broad-spectrum antibiotic and which has been approved for food use in Japan. There are a range of other bacteriocins that could be developed, but the main barrier to their use is that they have to undergo severe regulatory tests before manufacturers are allowed to use them in food.

Other approaches to biopreservation are:

Probiotics. The growth of bacteria on food can be suppressed by other bacteria. Some experimental products have tried spraying 'good' bacteria on to food to block the growth of 'bad' bacteria. This has been marginally successful, but has not been accepted by regulators as a safe treatment method.

Enzymatic suppression of fermentation effects. Often, some bacterial growth is not harmful in itself, but generates undesirable side products. The smelly amines produced by rotting fish are an example. A biopreservation approach is to use enzymes to break down these products, so that the bacterial growth does not matter.

Bioprospecting

This is searching for new biological resources, usually new plant or microbial strains, that could be sources of **natural products** and phytopharmaceuticals. There is a substantial argument about who actually owns the resources, between the countries in which these genetic resources reside and the companies who turn them into valuable products. Shaman Pharmaceuticals and Phytogenetics are among the specialist biotechnology companies prospecting in this way. Countries resent having their resources taken and essentially getting nothing in return, and call such collection 'biopiracy'. This is illustrated by the 1995 row over the Neem tree. The US company W.R.Grace was granted a patent for a new process for manufacturing a pesticide extract from this tree, which has been used in India for pest control and traditional medicine for centuries. The Indian government and a range of pressure groups objected furiously, although, in fact, it was the manufacturing process, and not the tree, that was the subject of the patent.

Regulation of the use of native biodiversity (a fancy term for things that grow where you live) is the aim of the Convention on Biodiversity (CBD), signed, with various delays, after the Earth Summit conference in Rio de Janeiro in 1992 (as of April 1996, the USA had not ratified the treaty). It provides rules concerning access to biological material, the sustainable use of those resources, and fair distribution of the resulting benefit. It does not seek to say how that benefit should be protected, and so does not address patent or ownership issues directly.

The preferred approach is 'bilateralism' (which will no doubt be called 'biolateralism' in time), where the company taking the genetic resources has a long-term commitment to give back some of the value resulting to the originating country. Most companies in the field have agreements of this sort.

Bioreactor

A bioreactor is a vessel in which a biological reaction or change takes place, usually a fermentation or biotransformation. Bioreactors, and indeed fermentation and biotransformation, are central to much of biotechnology—everything from baking bread to producing genetically engineered interferon takes place in a fermentation, and hence uses a bioreactor.

Bioreactors are conventionally divided up into three size classes. Laboratory bioreactors cover bench-top fermentors (up to 3 litres volume) and larger, stand-alone units (up to about 50 litres). These are used for research, and are usually used to create the fermentation process. Pilot plant fermentors are used to **scale up** a fermentation process, and to optimize it. They typically have a capacity of 50–1000 litres. Pilot plants have to be quite flexible to allow for process optimization. Production units can have any capacity, but usually hold at least 1000 litres, and can go up to the 1 000 000 litres of the ICI Pruteen plant. Generally, production units are much more specialized than pilot plants, being designed to operate one process with maximum efficiency.

There are a number of separate entries about bioreactors. They cover different types of bioreactors:

- **tank bioreactors** (which is most of them)

- **immobilized cell bioreactors**

- **hollow fibre** and membrane bioreactors

- **digestors**

Other simpler types of reactor are not covered specifically. These include pond reactors and tower fermentors. Ponds are simply ponds: they are used mainly for growing algae. Often they are called oxidation ponds, because their large surface area allows for more rapid oxygen transfer, and so most of the degradation is by oxidation, rapidly reducing the BOD of waste. Tower reactors are relatively simple towers in which nutrient is injected at the base and product collected at the top. They are used typically for anaerobic fermentation, i.e. fermentation where no air is needed, as, for example, in brewing. Fermentation material is injected at the base of the tower, the flocculent organisms settle at the base of the tower, and the product is collected at the top through a series of baffles which separate foam from bulk liquid.

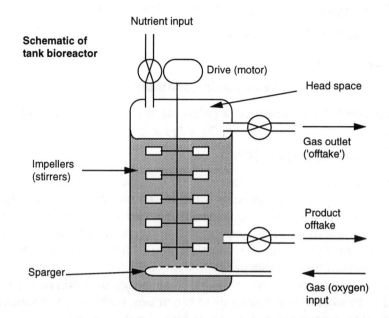

Nutrient input

**Schematic of
tank bioreactor**

Drive (motor)

Head space

Gas outlet
('offtake')

Impellers
(stirrers)

Product
offtake

Sparger

Gas (oxygen)
input

Many of the classifications of bioreactors and fermentors depend on how the materials in them are cycled and recycled. One extreme is the completely mixed bioreactor, in which the contents are mixed uniformly during their passage through the reactor, and then pass *en masse* to the next stage. Most bioreactors take some of the output, usually some of the biomass, and feed it back into the reactor again.

A general type of reactor is the plug flow reactor. Here a substrate flows past a plug of solid support material, emerging from the end having been changed by the plug. Often the reactor is essentially a pipe, although it is sometimes a flat bed. The solid support can contain an enzyme or an organism. This is, in fact, a bioreactor equivalent of column chromatography (*see* **Chromatography**). Strictly speaking, a plug flow reactor is an unmixed system. In practice, many plug flow reactors have some stirring in the flow mechanism to even out the reaction across the 'plug'.

In almost all bioreactors, mixing the contents is important to their efficiency. Often this is achieved through the use of stirring paddles or turbines. However, the inflow of liquid can be used to mix the content, as in the deep jet fermentor, where fluid is injected at high pressure at the base of the fermentor to stir the whole contents. The inflow of gas can be used to stir the fermentor contents too, as in the **airlift fermentor** or the deep jet fermentor, where an air jet, often starting above the liquid content of the reactor, plunges into the liquid and stirs it without the need for mixers.

Other topics covered under fermentation include:

- **fermentation processes**

- **gas transfer**

- sensor systems (*see* **Bioreactor control**)

- **fermentation substrates** (what the microorganism grows on)

Bioreactor control

Very small and simple bioreactors, such as garden composts or 'home brewing', can run successfully without any control on what happens. Any larger process requires control to make sure that the conditions (chemical additions, pH, gas content, temperature control, and so on) remain appropriate for the reaction. Large fermentation systems can have thousands of sensors linked to hundreds of control valves through a substantial computer. Since the exact efficiency of a bioreactor can be critical to whether it is economic to run or not, such control systems are a very important topic in a range of biotechnological areas.

Bioreactor control is complex for two reasons. Firstly, most biological systems are themselves complicated. A fermentor growing a microorganism must be monitored to keep the concentration of substrate chemicals, pH, gas levels, temperature, cell mass, and amount of product material within tight boundaries, otherwise, at best, the yield of product will decline, and, at worst, the organisms will die completely. In particular, in continuous bioreactors these conditions must be monitored in real time (i.e. as they happen, also called 'online'), and cannot be monitored offline (i.e. put on a laboratory bench and analysed tomorrow). Secondly, many of the key parameters are hard to measure. The method used to measure something must be sterilizable, because it is going to measure it inside the bioreactor. This means, usually, that it must stand up to autoclaving, live steam, corrosive chemicals, or all three. Thermometers and pH meters can be built to do this quite easily. Chemical and biomass measurements are very much harder to do. A great deal of ingenuity has gone into devising 'non-contact' or 'non-invasive' sensors for bioreactor control. Among the success stories are:

Capacitative and inductive sensors for biomass. Alternating electric fields are affected in a characteristic way by biomass in the liquid between two electrodes. This can be used to find out the amount of biomass there. This does not work with only small amounts of biomass.

Real-time offline sensors. These take a sample out of the bioreactor and measure its properties. Providing the test is fast, this works OK, and the test does not have to be sterilized. However, it is expensive and complicated to engineer, since the sample has to be taken out of the reactor in a way that guarantees that nothing gets in.

Many things that a scientist wants to measure about a fermentation, such as the concentration of a particular chemical, simply cannot be measured quickly using any existing technology. They can fall back on chemometrics, which is the use of sophisticated statistical techniques to identify the concentration of one chemical species from the spectral 'signature' of a complicated mixture, or they can use 'surrogate measures', things that behave in the same way as what you are trying to measure, but are not that thing. An example would be to measure cloudiness (turbidity) in a microorganism culture as a surrogate for biomass; this is not actually measuring biomass, but usually comes out with a pretty close approximation.

Bioreactor control also needs complex algorithms to make it work. These are among the few applications of **artificial intelligence** techniques to biotechnology. Expert systems (computer programs that capture human knowledge as a set of formal rules) and neural nets (programs that learn their own rules through 'training') are among the systems that have been used with some success in controlling the more complex bioreactors.

Bioremediation

Bioremediation is the use of biological systems, usually microorganisms, to clean up a contaminated site ('environment'). **Sewage treatment** plants perform this in a limited way. Bioremediation covers the use of microorganisms to destroy more toxic materials than are usually found in sewage, and to destroy them *in situ*, usually in soil or in waste tips. The same approach (although using different technologies and organisms) is used for 'bioscrubbing'—removing waste materials from flue gases.

The basic approach to most bioremediation projects is:

Selection of the microorganism. This is best done from soil that has been contaminated with the target chemical for some time. Often, such soil is found near a pipe junction or tank overflow valves in the plant that makes the chemical. Variants of an organism, which grow faster or degrade the target chemical more efficiently, are then created in the laboratory, by combinations of traditional microbial genetics, recombinant DNA methods, or selection. Typically, bioremediation methods use a consortium of organisms, rather than a single organism, that can catalyse the breakdown of different components of a pollutant or can perform different parts of the breakdown of a complex molecule. Even so, some molecules are quite hard to destroy: PCBs can be dechlorinated by obligate anaerobes (bacteria killed by oxygen), and the carbon skeleton broken down by aerobes (organisms needing oxygen); however, clearly these two cannot work together at the same site.

Development of organism physiology. Often, the right organisms are already present at the site, just not working fast enough. In this case, the scientist develops a cocktail of nutrients to boost the rate at which the soil organisms break down the target chemicals, a mixture often called an enhancer. Designs of enhancers are complex, since they must direct the bacterium's metabolism towards digesting the target chemicals, and not just feed the bug.

Inoculation of the environment. The microorganism is introduced into the site, usually with a nutrient mix to support its growth and encourage it to break down the target compound. Oxygen is usually a limiting factor, since most targets for bioremediation are complex hydrocarbon-based compounds which must be metabolized by oxidation: nitrogen and phosphorus are also commonly added so that the bacterium's growth is limited by the availability of carbon. Thus, the bacterium is under continued selective pressure to use all the carbon available in the soil for growth, including that present in the target compound. This phase of bioremediation is as critical as identifying a suitable microorganism, and requires a substantial knowledge of microbial physiology and ecology. The main cause of failure of practical bioremediation projects is that the organism selected cannot perform the breakdown at a useful rate at the site, despite performing well in the laboratory. Clays, for example, are particularly poorly suited to bioremediation: because they are very densely packed, water penetrates them very slowly, and air hardly at all.

Typical target compounds are chlorinated aromatics (although disposing of PCBs has met with only limited success), vinyl chloride, solvent residues, gasoline fractions, and crude oil. Alpha Environmental has hit the headlines on several occasions with its oil-eating bacterial preparation, used to digest oil spills at sea into soluble molecules that other bacteria can digest. Its most public application was in the Persian Gulf in 1991. This breakdown of compounds into biomass is a type of biodegradation. Other, non-organic, materials can also be metabolized if their end-product is non-toxic or volatile: selenium has been removed from soil by conversion to volatile compounds or elemental selenium, and nitrates have been removed from sewage waste by biological reduction to nitrogen gas for decades.

If a target site is very highly contaminated, or too cold or dry for bacteria to flourish, then the soil can be placed in a tank bioreactor and the bioremediation carried out there. These bioreactors are essentially large, insulated tanks into which soil or waste is placed with a bacterial inoculum. Air is blown through the mass to keep it oxygenated. Peter Wilderer at Hamburg has used a biofilm-based tank reactor to remove aromatic hydrocarbons (specifically benzene, toluene, and xylene, the BTX mixture) from landfill site leachate. A film of organisms growing on a permeable membrane was used in order to capture the volatile hydrocarbons from the water.

Biosensors

Biosensors are devices that use a biological element as an intimate part of a sensor. For example, an electrode could have an enzyme immobilized on its surface so that it generated a current or voltage whenever it encountered that enzyme's substrate. There are several classes of biosensor:

- **ISFET** (ion-sensitive field-effect transistor)-based devices

- physical sensors (including sensors for mass and **thermal sensors**)

- **enzyme electrodes**

- **immobilized cell biosensors**

- **immunosensors**

- **optical biosensors**

Other biosensors use DNA probes as the biological element, or even multicellular organisms such as *Daphnia* (a small freshwater shrimp) or trout.

Biosensors are also classified as potentiometric or ampometric. Ampometric sensors measure the current generated by a reaction, and are operated at as low a voltage as possible. Potentiometric sensors measure the voltage generated by the reaction, and are operated with as little current flowing through them as possible.

Biosensors have the potential for being extremely sensitive and specific ways of detecting something. However, their practical application has been hampered by the biological element being very prone to destruction by whatever is meant to be being analysed. Thus, while for commercial applications, a sensor system should be either very cheap and disposable or able to operate continuously for some time, nearly all biosensors are very difficult to make in bulk and last for only a few measurements. The major problems found have been:

Stability. The biological element 'went off' quite rapidly with use. Some went off in minutes when operational requirements were for days or weeks of operation.

Shelf life. Even when they were not operating, the electrode went off unless stored in a fridge or (in extreme cases) a freezer. Papers on biosensors often claim stability for weeks of operation, but this usually means that they are used once a day and kept in the fridge in between uses, a far cry from being used in a production line 24 hours a day.

Manufacturability. Most biosensors are very difficult to make, and constructing an assembly line to make them in commercial quantities requires a well-defined way of making them. Even commercially successful sensors are hard to make sufficiently reliable for such a method to be defined.

The most prominent exception is the glucose biosensor, an enzyme electrode based on glucose oxidase and commercialized by several companies, notably as the Exactech test for blood glucose levels. These work where others fail because the amounts of glucose being measured are large (so the electrode does not have to be very sensitive) and the enzyme glucose oxidase is unusually stable.

A related idea is the biochip. There are two interpretations of this word. The more speculative refers to the use of biological components in computer chips (*see* **Molecular electronics**). The second refers to immobilizing or building biological molecules on a very small 'chip' to use in an assay or a sensor. The most advanced biochips are ones with a huge number of DNA molecules immobilized on them to perform hybridization reactions (*see* **DNA probe**). Biochips, and many biosensors, require supporting microfluidic systems. These are any system that can deliver nanolitre amounts of liquid where you want it. This is quite hard. (At scales of a few micrometres, the behaviour of water is dominated by surface tension, not by gravity. It behaves like rubber covered with glue. Pumping it into and out of biosensor or biochip elements is very difficult.) Technologies for building such systems usually come from the semiconductor industry, and are collectively called microfabrication technologies. They typically involve etching away the structure you want chemically rather than putting it together from separate parts.

Biosorption

Biosorption is the sequestering (i.e. capture from solution) of chemicals (usually metals) by materials of biological origin. Biosorption is widely talked about and little used as a method for removing materials from waste or of purifying rare metals. Many organisms have components that bind metal ions: human bone matrix material, for example, binds strontium rather well. In some cases this is an active process (the organism uses energy to take the metal ions inside and trap them in an insoluble form) in others the process is passive (the metals stick of their own accord to a material that the organism makes). In both cases organisms can be selected that can accumulate more of the 'target' metal, or that accumulate one metal specifically. For industrial use, bacteria or yeast are almost always the organisms used, although many other organisms, such as protozoa, simple plants, even trees, can accumulate substantial amounts of metals.

Among the ways in which organisms actively accumulate metal ions is by precipitating them as phosphates or sulfides by 'pumping' them into special sections of the cell. 'Passive' systems include proteins that bind the metal specifically (metallothioneins, for example; sulfur-containing proteins found in many organisms), lignin (from wood), chitin, chitosan, and some cellulose derivatives.

Biosorption is a biological phenomenon, and is interesting for its insight into how organisms cope with metal poisoning, lack of essential nutrient, etc. It can also be adapted to direct industrial use as a purification system, by immobilizing the organisms on filters or in pellets, by using a recycling reactor system that passes the water to be treated through a bed of bacteria in a fermentor, or by extracting the biosorptive material from the organism and using that on its own. This latter option allows non-microbial biosorption systems: chitin, for example, absorbs a number of metal ions and is produced from waste prawn shells.

One of the most common waste removal targets is removal of heavy metals from industrial waste water, especially nuclear waste streams, where the metals are present at low concentrations but are the most hazardous element in the water. There is also substantial interest in using biosorption to purify precious metals such as silver and gold from very low grade ores by washing the metal out of the ore and then concentrating it from the leachate using biosorption.

To be useful, biosorption must be specific and efficient. For metal removal from waste streams, removal must be at least 90% efficient to be of any industrial use, and the organisms or polymers must be able to remove at least 15% of their own weight in metal. Any less efficient system costs more to use than conventional removal systems (such as 'ion-exchange' materials). The efficiencies for precious metal extraction can be lower, depending on how valuable the metal is, but must be very specific: there is no point purifying gold if you purify a lot of lead along with it. As well as being improved by breeding and selection systems, biosorption can be improved, in principle, by genetic manipulation, by altering the structure of metal-binding proteins such as the metallothioneins, or by altering the enzymes that make other materials such as the chitosans or lignins. However, although it is talked about a lot, biosorption is not understood well enough to make such genetic engineering feasible as yet.

Biostatistics

This is simply statistics applied to biology or biological research results. The term is often used in drug development, where it means using statistics to design a clinical trial properly and analyse the results. In particular, biostatistics addresses topics such as:

- how to select a 'random' sample of organisms for a trial (this is especially important in selecting people for clinical trials),

- how to decide the sample sizes you need, how to analyse qualitative measurements of success, like 'a bit better' or 'does not hurt so much',

- analysis of variance (where is the 'noise' in your experiment coming from, and is it actually useful information),

- how to decide whether you have a significant result, and if so what 'significant' actually means.

This is a specialized branch of applied statistics, and many clinical trials have been rendered useless by not paying proper attention to it.

Biotechnology

Biotechnology is the application of knowledge of living systems in order to use those systems or their components for industrial purposes. The word 'biotechnology' was first used by the Hungarian agricultural economist Kark Ereky in 1919, to mean 'all lines of work by which products are produced from raw materials with the aid of living organisms'. The definition has been broadened slightly to include producing things with the aid of materials from living organisms (such as enzymes) and some raw materials that are produced from living organisms themselves (such as alginates or biomass), and narrowed to focus on new technologies, rather than traditional production processes. The *Oxford English Dictionary* follows this, defining biotechnology as 'technology using modern forms of production utilizing organisms, especially microorganisms, and their biological processes.' The UK government in *Developments in Biotechnology* (1992) focused more on the product than the production process, with '. . . the production of innovative products, devices and organisms by exploitation of biological processes'. Either way, farming is not considered part of biotechnology.

So, biotechnology is the pragmatic combination of science and technology to make use of our knowledge of living systems for practical applications. This includes a wide variety of applied biological science, but also includes aspects of chemistry, chemical technology, engineering, and specialist disciplines in specific industries such as pharmaceuticals, environmental treatment, or agricultural industries. There are about 1300 biotechnology companies in the USA, with 260 being public companies (whose shares are traded on the stock markets), 485 in Europe, with only 30 public and those nearly all in Britain, which reflects the breadth of ground covered by the industry.

Some argue that biotechnology is just an extension of brewing and baking. Usually, this is a prelude to an argument about how safe biotechnology is. The record shows that biotechnology really is amazingly safe—Orsen Welles' famous radio broadcast of *The War of the Worlds* caused more deaths than biotechnology has done, suggesting that radio plays are more dangerous than genetic engineering. However, saying that it is an extension of traditional brewing is wrong. The difference is that biotechnology seeks to use rational approaches to developing its technologies and products, rather than traditional craft or trial-and-error. (Modern brewing, of course, is a highly technical and scientific branch of biotechnology.)

Note also that biotechnology is not the same as 'applied molecular biology'. Genes, DNA, and all the science associated with them get a lot of press nowadays, but equally important for the end-users of biotechnology (i.e. you) are the food, beer, vitamins, and other products manufactured by means of the wide range of technologies outlined in this book.

Biotin

Biotin, a natural coenzyme, turns up in some unexpected places in biotechnology as a 'label' system. Biotin can be linked on to many different macromolecules by chemical reaction, a process called biotinylation. The protein avidin (usually made from egg white) or its bacterial counterpart streptavidin binds to biotin extremely tightly (far tighter than an antibody binds to its antigen). The avidin can be labelled with an enzyme, a fluorescent group, a coloured bead, etc. This will then seek out and recognize the biotinylated molecules, and will not stick to any others. This can be preferable to trying to link the enzyme, fluorescent tag, or other label on to the target macromolecule directly because: (i) you can get more biotin on to a macromolecule than enzyme molecules, and (ii) the biotin is very stable, and so can be treated with extreme pH, boiled, or irradiated, whereas an enzyme would be destroyed by these conditions.

Biotransformation

Biotransformation is the conversion of one chemical or material into another using a biological catalyst: a near synonym is biocatalysis, and hence the catalyst used is called a biocatalyst. Usually the catalyst is an enzyme, or whole, dead microorganisms that contain an enzyme or several enzymes. The advent of catalytic antibodies and ribozymes will broaden the definition somewhat. Conversion of one material into another using whole living organisms is usually called **bioconversion**.

Biotransformation is one of the largest areas of applied biotechnology (as opposed to research technologies): around 5% by volume of the enzymes used industrially are used for biotransformation (nearly all the rest are used in the food industry, or in detergents). A wide range of materials are made by biotransformation, from commodity items such as high-fructose corn syrup, to speciality chemicals for the pharmaceutical industry. Some biotransformation processes, such as that producing vitamin C, produce thousands of tonnes of product per annum. The advantage of biotransformation over conventional chemistry is the specificity of the enzymes. Reactions can be:

- stereospecific; i.e. they produce only one optical isomer of a chiral compound, or

- regiospecific; i.e. they change only one part of a large and rather homogeneous molecule (analogous to only digging up one particular stretch of a motorway).

A key use for biotransformation is in 'resolution'. This is a biotransformation that takes a racemic mix of a chiral compound and converts one optical isomer into another compound. This means than conventional chemistry or separation techniques can now take what was a racemic mixture and produce an optically pure compound from it (*see* **Chirality**). The success of a biotransformation in making a chiral compound is measured by the enantiomeric excess of the product: the percentage amount by which one of the enantiomers (chiral versions) exceeds the other.

The most commonly used biotransformations involve:

- acylases (to resolve chemically synthesized amino acids)

- esterases and lipases (to make a range of esters, lipids, and to resolve fatty acids and alcohols)

- beta-lactamases and penicillin acylase (to make penicillins and cephalosporins)

- peptidases and proteases (to make peptides)

- steroid-transforming enzymes (to make steroid derivatives). These are always used in whole organisms, as many enzymes are involved in each biotransformation.

Biotransformations involving the use of **proteases**, **glycosidases**, and **lipases** are discussed separately.

Blood disorders

There are a range of diseases of the blood that biotechnology seeks to address. The main ones are:

Haemophilia. The blood will not clot because the gene for one of the proteins involved in clotting is defective. Several of the blood clotting 'factors' (Factors VII, VIII, and IX) have been cloned and are used as biopharmaceuticals to treat these inherited diseases.

Sickle cell disease, thalassemia (alpha and beta). These diseases are caused by a mutation in the genes for haemoglobin, the red protein in blood cells. Boosting blood production with erythropoietin, replacing the haemoglobin with haemoglobin made in yeast, and, ultimately, gene therapy to replace the gene, have all been suggested and tried on animal models.

Leukaemia and anaemias. There is a very wide range of disorders that mean that one of the many types of cells in the blood are not made in sufficient quantity. They can be treated by various transplant-type techniques, including transplanting genetically altered bone marrow cells to boost production of the missing cell type. Production can also be boosted by relevant growth factors and by haemopoietic factors (i.e. factors that boost haematopoiesis, the making of blood in the bone marrow): several of these factors have been made as potential biopharmaceuticals.

Blood products

Originally, these were biopharmaceutical products made from human blood, such as the blood clotting Factor VIII used to treat haemophiliacs. Such extracted products are usually made by a series of filtrations and solvent extractions. The major 'blood products' in this category are:

HSA (human serum albumin). This is the major human blood product by volume, and is used to produce blood substitutes and extenders for transfusion.

Human gamma globulins. These are antibody preparations, and are used medically to give people an extra high level of antibodies (immunoglobulins) when they might be exposed to specific, unusual diseases.

The term blood products is also used to refer to biopharmaceuticals that act on blood or the cells that make blood. They are also usually made by those cells, but in such tiny amounts that extracting them from blood itself is impractical. So, they are made by genetic engineering instead.

Among the 'blood products' category of biopharmaceuticals are:

Thrombolytics. Drugs such as tissue plasminogen activator (tPA) produced by Genentech, streptokinase, and eminase (made by SmithKline Beecham). These dissolve blood clots in the arteries and therefore are used as treatments for heart attacks. An enormous clinical trial spreading over several years in the 1990s (GUSTO) showed that tPA gave a slightly better clinical outcome in heart attack patients than did streptokinase: however, it costs ten times as much. The protein hirudin, originally extracted from leeches, is also being developed as an antithrombin compound, which prevents further clotting rather than breaking up clots that already exist.

Clotting agents. Factors VIII and IX to treat haemophilia, a disease where these proteins are missing. Baxter Healthcare and Mile Inc. are developing recombinant Factor VIII.

Erythropoietin (EPO). This stimulates the bone marrow to make more red blood cells, and was the subject of a fierce patent dispute (*see* **Patents**).

GCSF, GMCSF, etc. These are cytokines, materials made by the immune cells to regulate the immune system's function (*see* **Cytokines**).

Animal blood products, notably fetal and newborn calf serum, are also used in the biotechnology industry: serums are used as a supplement for the media used to culture a range of mammalian cells.

'Blots'

A range of molecular biological techniques are called 'blots'. They all share a common appearance. At the start, biological molecules are usually present in a jelly-like matrix, often the result of separation by **gel electrophoresis**. The contents of the gel are then transferred on to a porous membrane, often a chemical derivative of paper or a nylon mesh. This technique was traditionally done by allowing liquid to seep through the gel, through the membrane, and into a pile of paper towels, which acted like blotting paper—the biomolecules travelled with the fluid until they stuck to the membrane. Electroblotting (which uses an electric field to pull the molecules out of the gel) and vacuum blotting (which uses suction) are also used. Once on the membrane the molecules can be analysed by techniques that would not work in the original gel, such as antibody staining or DNA hybridization (*see* **DNA probes**).

Southern Blot

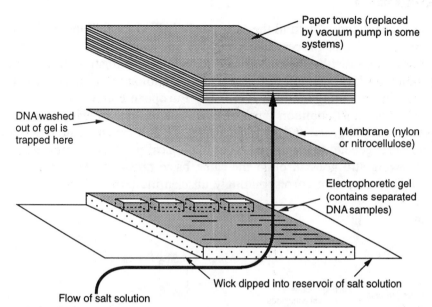

Paper towels (replaced by vacuum pump in some systems)

DNA washed out of gel is trapped here

Membrane (nylon or nitrocellulose)

Electrophoretic gel (contains separated DNA samples)

Wick dipped into reservoir of salt solution

Flow of salt solution

The variations on this theme depend on the molecules:

Southern blot. Named after Professor Ed Southern, the gel here is a DNA electrophoresis system and so the molecules transferred are DNA molecules.

Northern blot. Almost the same as the Southern blot, but the molecules are RNA.

Western blot. Here the molecules are protein, also separated by gel electrophoresis.

Southwestern blot. This is a variant of the Southern blot used to find protein molecules that stick to DNA molecules.

(Desperate attempts to get something (anything) named as the Eastern blot have not been generally successful.)

Dot blot. Here DNA, RNA, or protein is dotted directly on to the membrane support, so that they form discrete spots. Also slot blots (if the spots are not round).

Colony blot. Here the molecules (usually DNA) are from colonies of bacteria or yeast growing on a bacteriological plate. A variation (called the plaque lift) can also be used for viruses.

With the advent of PCR and the fall in interest in Southern and

Northern blots, the most commonly seen blot is now the Western blot, where proteins are separated according to size and then identified by reacting them with an antibody.

Brewing

The brewing industry (and the associated distilled spirit industry) is both a major source of process biotechnology and a user of new techniques. Brewing and distillation is carried out on a large industrial scale. Wine production has been affected less by biotechnology than beer brewing, remaining a 'cottage industry' in many European countries.

The two most common types of yeast used in beer brewing (*Saccharomyces carlsbergensis* and *Saccharomyces cerevisiae*) are mainstays of both molecular genetic research and biotechnological production processes.

Brewing still goes through the same basic processes defined in pre-Roman times, with correspondingly unscientific terms.

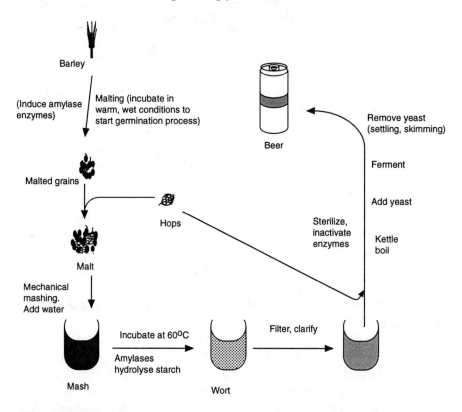

Malting. Barley is incubated in damp, warm conditions to make it sprout. The seeds make substantial amylase, which breaks the starch into simple sugars that yeasts can metabolize.

Mashing. After addition of hops (primarily for flavour), the plant matter is mechanically fragmented into a uniform suspension, a mash. The mash is then heated up to speed up amylase action. The result is called wort, which is filtered to remove most of the solids.

Kettle boil heats up the ingredients to inactivate enzymes in the wort, precipitates a number of high molecular weight impurities (including a lot of proteins), and concentrates the sugars. It is then ready to be fermented.

Yeast performs fermentation. The culture in traditional brewing is not aseptic. Boiling kills off the endogenous microorganisms, and the yeast outgrows any newcomers, suppressing their growth by ethanol production.

Clarification. This is the removal of proteins and other high molecular weight material that can cause 'haze' in the beer. This material could be removed by filtering or centrifugation, but the particles are extremely small, and so these processes would be expensive. So they are usually removed by precipitation (started by adding a polymer that binds to them, making them stick together) or cooling (which can alter their chemistry, again making them sticky).

Potable alcohol is also made in many other fermentations. Many cultures produce low-grade fermentations from grains such as wheat or rice, or from potatoes, swede, or any other cheap carbohydrate source, and then distil the result. The fermentation processes are similar to those in beer. After the fermentation step, the entire broth is distilled to separate an alcohol-containing liquid from the yeast and remaining solids from the grain. In fact, the liquid distilled off is usually almost pure ethanol and water, with little flavour and a kick like a mule. In whisky distillation, this is called grain spirit or grain neutral spirit. Some drinks, such as vodka and tequila, use the primary distillate as a drink. Others make it more potable by 'ageing' it to diffuse woody flavours into it (whisky), or using it to extract the flavour from juniper berries (gin).

The solid material left after distillation of the alcohol from grain-based fermentations is called 'distillers grains', and is often used as an animal feed component. 'Dark grains' is the same thing, but with some of the spent liquor still absorbed into it.

New biotechnologies are applied to all aspects of this process. The use of recombinant yeasts with enzymes that degrade more of the sugars (producing 'lite' beers with little sugar left), or which can be separated more easily, is particularly high profile.

BST (bovine somatotropin)

Bovine somatotropin is also called bovine growth hormone. This protein hormone is found naturally in cattle, and is the counterpart of human growth hormone, one of the earliest biopharmaceutical products. It has been cloned, expressed in large amounts, and is being marketed by Monsanto as an agricultural product to improve the growth rate and protein:fat ratios in farm cattle, and to improve milk yields.

There are animal welfare concerns about this, and health concerns about the possibility that BST will get into the milk or meat, and hence into people. In particular, the possibility that BST given to improve milk yield will end up in the milk given to children has proven a powerful weapon against Monsanto, one of the principal developers of BST for agricultural use. Monsanto has also been accused of reducing cows to unhappy milk-producing machines (*see* **Yuk factor**), as if 5000 years of breeding had not done that already. The debate has become highly polarized, with contenders on both sides seeing it as a trial case for the application of biotechnology to agriculture and food industries. The USSR, Czechoslovakia, Bulgaria, South Africa, Mexico, and Brazil have approved BST, but in many other countries the debate on its safety is holding up any approval. There is also a debate over whether BST will offer the consumer any advantage, especially in Europe where there is often a surplus of milk over the European Community's (EC) 'quota' for production. It would, however, allow the same amount of milk to be produced from fewer cows using less food.

Capillary zone electrophoresis

Also called simply capillary electrophoresis, this is an up-and-coming technique in many biochemical and biotechnological fields.

Gel electrophoresis is electrophoresis moving molecules around using electric fields, performed in a polymer material. The polymer does two things: it sieves the molecules by size, and it stabilizes the solution in which the electrophoresis is happening. Without it, any slight vibration or convection would stir all the molecules up, and the system's ability to separate very similar molecules (its resolution) would go down dramatically. However, the separation is a complex result of the molecule's shape, size, charge, and how it interacts with the polymer gel. This complexity can itself reduce the system's resolution.

Electrophoresis without gels has been used. It is called free zone electrophoresis, and uses a flow of water, or sometimes a density gradient column (such density gradients are discussed further under **centrifugation**). However, the stirring effects in these systems can be substantial.

Capillary electrophoresis performs free zone electrophoresis in a very fine capillary tube (a tube with an internal diameter of less than 1 mm). Here, stirring effects undoubtedly occur, but they only stir up volumes of solution less than the tube diameter (i.e. less than 1 mm), and so the effect on resolution is very small. The electrophoresis can be 'run' much faster than conventional electrophoresis, where making the molecules go faster means putting a higher voltage across the gel slab, which means that more current flows through the gel, more heat is produced in the gel, and ultimately the biological molecules denature (or the gel tank cracks or bursts into flames). The mass of liquid in a capillary tube is so small that even very high voltages produce tiny currents, and the heat produced can be radiated away from the tube rapidly. So, the electrophoresis can either be run very fast or be run on a very long capillary tube, so increasing resolution.

Capillary electrophoresis sometimes relies on electroendosmosis. This is the movement of a solution of ions in an electric field, when they are carried in a charged tube. It can be seen when salt solutions are exposed to an electric voltage in a charge gel (like agarose), or in a very small tube with charged walls (like silica). This latter is what happens in capillary electrophoresis, so that as well as the sample being pulled through the tube by the electric field, the whole liquid column in the tube is pulled through the tube by the same field. The direction of the flow depends on the charge on the tube walls, which itself depends on the chemical nature of the walls and the solution. Silica usually has a negative charge, but this can be 'converted' to a positive charge by coating the walls with positively charged detergent molecules.

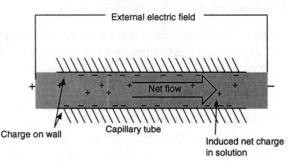

Electroendosmosis

There are several commercial systems for performing capillary electrophoresis on biological molecules for research.

Capillary electrophoresis systems can also perform capillary gel electrophoresis. This uses a gel inside the tube. It has the same advantages of high power and good resolution, and separates molecules in a way more similar to other, conventional electrophoresis techniques. Capillary gel electrophoresis is gaining acceptance as a tool for DNA sequencing.

Capital (money)

Most biotechnology companies are based on research or development, and do not make a profit as soon as they start up. Some do not expect to make a profit for years. So they have to get money from somewhere. This usually means raising capital in the investment market. Investors and biotechnologists have quite different ideas about the nature of the universe, so this can be a difficult process.

There are generally three types of investor. They put money into the company in return for shares (stock, equity) in its future.

Rich people. There are different names for these (private investors, business angels), but they are people with sufficient money of their own to make a substantial contribution to the running costs of a company for several years.

Venture capital companies. These are professional groups who invest in very early stage companies in the hopes that their value will rocket. Traditionally regarded as very fickle, mercenary groups (the standing joke is that biotechnology venture capitalists have no capital, show no liking for adventure, and know nothing about biotechnology), in reality they have been responsible for providing money to take many companies from initial start-up to flotation. They are looking for a good return on investment (ROI) and an exit route, i.e. some way that they can sell their share in the company and get hard cash back in 3–5 years. A key issue here is 'burn rate'—how fast the company is using up its supply of money vs. how long it will be before it either starts earning some by selling products or can get another 'round' of financing to top up the supply. Burn rates of $10–20 million a year are typical for medium-sized US biotechnology companies.

Public investors. When a company is stable enough, shares in it can be traded on a public stock exchange. Usually, the most important investors here are big institutions, like pension funds, who will have specialists who invest just in healthcare companies, environmental companies, etc.

The path to these riches is rocky. It usually involves:

Seed funding. A miserly few hundred thousand to get started.

Private financing. Any investment that gets substantial funds in, but which is not on a public stock market. There may be several 'rounds' of financing, i.e. the company may run out of money and need to ask more investors for more. This may or may not be what the company planned to do. They may include 'mezzanine financing', which is a sort of half-way top-up on the way to IPO.

IPO. Initial public offering, the Rubicon of funding, when the company is 'taken public' and its shares are traded on a stock exchange where anyone can buy or sell them. In the US there is a special stock market called NASDAQ for 'high risk' investments. Companies will sometimes say that they are 'quoted on NASDAQ' if their shares are being bought and sold in that market. The UK AIM market plays a similar role, but on a much smaller scale.

There can then be a quite bewildering range of methods of issuing more shares on a public stock exchange, which it is beyond the scope of this book to discuss.

There are several vexed issues here.

The value of technology. To know how much a company should be paid in return for a share, you have to know how much the whole company is worth, and that means valuing its most valuable asset, its science and technology. But how do you value a development that might be hugely valuable in 10 years, but might not work at all? Part of the calculation is to do with discounting, which is reducing the value according to the perceived risk. Part is calculating net present value (NPV), i.e. the value today of something that will only exist in 10 years time, when the same money could have been in the bank earning interest. There is no 'right answer'.

Technology push vs. market pull. Many biotechnology companies are set up around a new piece of science or technology, which the company then tries to exploit. This is known as 'technology push'—the product is pushed through a development 'pipeline' by the technology behind it. Investors prefer 'market pull', where there is a defined market need, and the company goes and develops the technology that will meet that need. A lot of the 'selling' of a biotechnology company to investors usually boils down to disguising technology push as market pull.

The downside of public ownership. Floating your company brings in lots of money and lots of kudos, but means that you are now compared with IBM or United Airways, or Amgen. Analysts scrutinize your financial figures every three months ('quarterly returns') and mark your shares down by 70% if your clinical trial fails. A private company can have a more stable, long-term, family-style relationship with its funders.

And once you are a public company, your value is defined by your share price, which is influenced by public news about you (any news, no matter how scientifically irrelevant), casual remarks made by brokers over a beer, the performance of unrelated companies that are seen to be similar to yours, the economy as a whole, even the phase of the moon. (The latter really is a well-documented effect.) Many scientists feel that this is a poor way to manage the long-term development of new technology, and in 1996 a survey of biopharmaceutical CEOs found that 50% of them thought that they could improve their clinical trials results 'by ignoring Wall Street'.

Catalytic antibodies

Catalytic antibodies, also called abzymes, are antibodies whose binding sites catalyse a reaction in their target 'antigen', rather than just passively binding to it. Antibodies do not normally possess catalytic activity.

In the 1940s, Linus Pauling suggested than an enzyme was simply a protein that bound to and stabilized the transition state of a reaction. By stabilizing the transition state, the enzyme made the reaction from substrate to product more probable, and hence the reaction faster. In the 1960s several workers suggested that an antibody that bound to the transition state of a reaction would catalyse that reaction.

However, it is not possible to isolate the transition state of a reaction. So, to raise an antibody against it is impossible. A near approximation is to raise an antibody against an analogue of the transition state. Since transition state analogues are often powerful inhibitors of enzymes (they mimic the transition state that the enzyme binds), quite a lot are known. Others can be synthesized from considerations of the reaction mechanism. By raising a monoclonal antibody against the transition state analogue, an antibody whose binding site catalyses the reaction concerned can be created. Reaction rate enhancements of 6×10^6 have been reported for some reactions.

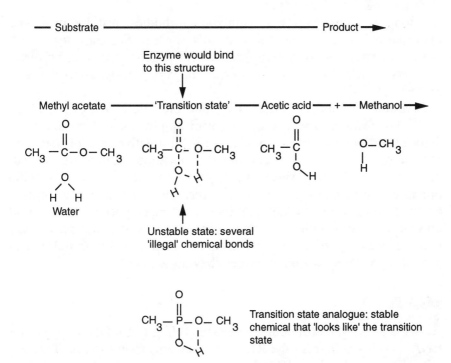

Catalytic antibodies can also work through reducing the entropy of reaction, i.e. bringing together two molecules in the right orientation to allow them to react. This can apply to two substrates for a reaction, or a substrate and a cofactor. Catalytic antibodies have been made that catalyse reactions through both these mechanisms. ('Entropy' in this case is chemical entropy, i.e. disorder. Two molecules aligned exactly right for a reaction represent a very ordered system—it is much more likely that they will collide in some unsuitable way, or, indeed, not collide at all. Thus the reaction has a high 'entropy barrier', which the catalytic antibody reduces by making the system more ordered—it brings the two reactants together in the correct way to react.)

As would be expected of a protein catalyst, abzymes are quite specific in the reactions that they catalyse, including selecting only one stereo-isomer from a racemic mix. Reactions catalysed to date include a variety of esterase and peptidase reactions. Abzymes have the advantage that, in principle, a specific abzyme can be created for any reaction. Although an enzyme could also be found for that reaction, finding it would be a major task. The technology of creating an antibody that recognizes a specific small molecule (hapten) is, by contrast, quite simple.

A lot of research work has generated catalytic antibodies that have some small catalytic effect. However, in 1996 it was found that albumin, a completely non-antibody-like protein from blood, can also catalyse

79

reactions at similar rates to catalytic antibodies, just by providing hydrophobic 'pockets' on the protein's surface. So, the jury is still out on whether the abzyme approach will produce genuine, powerful catalysts, or whether it is just detecting the very poor catalysis that any protein could do

Favoured targets for abzymes include biotransformations, and, in particular, resolution reactions, biosensor applications where the specificity of antibodies can be coupled with the relative ease of detecting enzyme reactions, and pharmaceutical applications. Drugs are particularly favoured targets, since an abzyme that acts as a very specific protease to cleave any protein in the body (such as a viral coat protein or an inflammation-causing cytokine) could be engineered, and would only act on that peptide. Drugs also promise the substantial market size that is needed to justify the very considerable amount of time and money needed to make even simple model abzymes work.

cDNA

cDNA is copy DNA (or complementary DNA). It is a DNA copy of an RNA, and is made from the RNA using reverse transcriptase. This is a gene cloning technology. There are two potential reasons for wanting to do this.

First, the DNA gene itself may be unknown. In this case, the cDNA is made from a messenger RNA that codes for a known protein (or for a protein whose activity can be measured, by antibody reaction or because it is an enzyme). Then the DNA gene can be found using the cDNA as a 'probe'. Secondly, the scientist might not want the 'original' gene. This is especially true if the purpose of cloning the gene is to express it in a bacterium. In this case, the scientist wants a section of DNA that codes for the protein concerned and nothing else. They do not want introns, other neighbouring genes, and so on in the gene clone. The cDNA is a much better approximation of this, consisting (in eukaryotes anyway) of a single mRNA with no introns that code for a single protein. Often, the cDNA can be spliced directly into an expression vector and some expression made to happen in bacteria.

cDNAs hit the headlines at the end of 1991 when Craig Venter at the US National Institute of Health (NIH) filed a patent claiming 337 new cDNA sequences that he had discovered using an automated DNA sequencer. The NIH's idea was to patent them as Venter churns the sequence out so that, if at any time in the future someone finds a use for those sequences, then the NIH can claim royalties on them. The Medical Research Council in the UK took the route of keeping its cDNA sequences generated by large-scale sequencing a secret, until the legality and acceptability of patenting cDNAs is established. It seems unlikely that the cDNA patent will hold up, as Venter admits that he has no idea what the cDNAs do or what use the knowledge of their sequence could possibly be put to.

cDNA is also made when you want to use PCR to amplify an RNA. PCR does not work on RNA, so you copy it to DNA using reverse transcriptase and then PCR amplify the cDNA. This is called RT-PCR (reverse transcriptase PCR).

Cell adhesion molecules (CAM)

Also called ICAMs (intracellular adhesion molecules). These molecules are present on a wide range of human cells, and are part of the mechanism used by cells to recognize each other. They are glycoproteins, and the sugar residues can be crucial to their function: the differences between some blood groups, for example, are the result of variation in the sugar residues on some ICAM molecules.

There are several families of mammalian cell adhesion molecules, with many variants of each family being seen in different cell and disease states. The major families are:

Integrins. These are cell-surface receptors for a range of other cell adhesion molecules, and for molecules found in the intracellular matrix (which is the material that mechanically supports cells, and is secreted by them). Integrins bind to proteins with the three amino acid sequence RGD in them.

Immunoglobulin superfamily. These are molecules with a molecular similarity to the antibody (immunoglobulin) molecules. They include ICAM-1, VCAM-1, LFA-1, and Mac-1 (found on leucocytes only).

Selectins. They bind to carbohydrate cell surface molecules, and are important in the initiation of blood clots and inflammatory responses, among other processes.

Fibronectin is another of the cell surface molecules, which, in this case, is a glycoprotein that links the cell to other proteins in the intercellular matrix, notably to collagen. It is also sometimes called large external transformation-sensitive protein (LETS), because it is often absent from cancer cells.

Cell adhesion molecules are important to drug companies, and hence to biotechnology companies, because they are the molecules through which cells control their interactions with each other. There has been particular focus on ELAMs (endothelial lymphocyte adhesion molecules), because of their role in immune disease. These are proteins on the surfaces of lymphocytes and the endothelial cells (flat cells) that line the surfaces of blood vessels. During inflammation, white blood cells leave the blood and invade damaged tissue to engulf any invading organisms. They also release a range of chemicals that cause inflammation of the tissue. The invasion is partly controlled by the ELAMs, which allow lymphocytes to stick to and recognize the endothelial cells. Modulating this interaction, then, is a potential route to controlling inflammatory diseases.

Cell adhesion molecules are important in many other processes in growth, development, and regeneration, and so are a hot research topic. They can also be the receptor that a virus binds to to gain entry to the cells it infects. Research into cures for viral diseases can also be concerned with CAMs. The best known example is the CD4 receptor, a member of the immunoglobulin superfamily of receptors, which is the target for the AIDS virus.

Cell culture

This is the cultivation of cells from a multicellular organism, usually mammalian cells. The term '**tissue culture**' is often used as a synonym, which strictly it is not: however, the intuitive nature of cell culture science does not lend itself to such pedantry.

Many biotechnological production processes and research programmes depend on the ability to grow animal cells outside their parent animal. Because animal cells are not free-living organisms, like yeast or bacteria, they do not function well when separated from their animal. They need to be treated with special care to enable them to proliferate. Among the conditions that have to be maintained are:

- pH—usually 7.5–8,

- osmotic potential—a total concentration of soluble compounds similar to that in blood or serum,

- temperature—between 35 and 38°C for mammalian cells,

- specific nutrients—most of the amino acids, sugars, salts, vitamins, and some specific lipids are the usual minimal requirements,

- growth factors—hormones, often proteins, to trigger the cells to grow and divide.

Often, the vitamins, growth factors, and some minor nutrients are all provided by feeding the cells with serum. Usually fetal calf serum (FCS; also fetal bovine serum, FBS) or newborn calf serum (NCS, NBS) is reckoned to be rich in growth factors. Pregnant mare serum (PMS) and horse serum is also sometimes used. There is an on-going debate about the quality and reliability of serum for cell culture, especially if it is to be used to grow cells that are making a biotherapeutic product such as a therapeutic antibody. This is a substantial economic issue for biotechnology, since about 500 000 litres of FBS are traded world-wide each year at \$200/litre for commercial cell culture.

The needs of some cells are so specialized that they will only grow on other, living cells. Such a layer of support cells is called a 'feeder layer'.

A key condition for cell culture is sterility. Yeasts and bacteria grow much faster than cultured cells, and so if just one bacterium gets into your cell culture it will soon outnumber the mammalian cells. The bacterium's metabolic wastes, and particularly the acid it produces, will then kill the cells. Thus, other organisms must be excluded rigorously. This is relatively simple to do on a laboratory scale, but much harder if the cells are to be produced in bulk. Any bacterial contamination is bad news, even if the bacteria are dead, because endotoxins produced by the bacteria can kill cells or contaminate their products.

Many animal cells do not all grow as bacteria do, in free suspension. Those that do not are called 'anchorage dependent', which means that the cells have to stick down to a surface to grow. Usually the surface has to be specially treated with polymers such as collagen or polylysine. Glass is a good surface for many cells. Anchorage independent cells can survive and grow drifting free in solution. Sometimes, anchorage independent cells will stick on to things anyway, but they do not need to in order to survive.

Mammalian cell culture is used widely in biotechnology. Monoclonal antibodies are manufactured in bulk using cell culture. A range of biopharmaceutical products are produced in genetically engineered mammalian cells, since these synthesize the correct glycoforms of the proteins.

Cells can be cultured in the laboratory in a variety of containers. The simplest is a Petri dish, which is called a tissue culture dish because it is made with a surface charge that cells can adhere to. Square-sectioned plastic bottles are also common. Larger cell culture is done in roller bottles, which are ordinary bottles that are rolled gently along their long axis to keep a gentle movement of culture medium over the cells.

Cell culture for production

This means the culture of mammalian cells (or occasionally the culture of isolated plant cells, as opposed to plant calluses), as opposed to bacterial or yeast cells, which is called fermentation. Cell culture is a difficult process, requiring sterility and careful control of the conditions (*see* **Cell culture**). Large-scale cell culture has to take several additional factors into account.

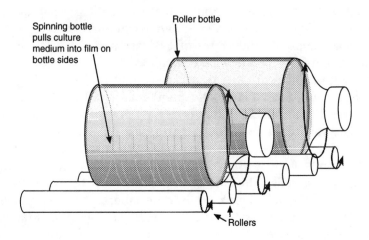

Fermenter design. Confusingly, large-scale cell culture can take place in large vessels called fermentors or bioreactors. As mammalian cells are quite fragile, the usual stirring methods will destroy them, so gentle methods are used. For very large scales, airlift fermentation is established. **Hollow fibre** methods are increasingly used, not only for pilot scale, but also for smaller production runs.

Small-scale systems. The favourite small-scale system is the roller bottle. This can be used in a research laboratory, and scaled up to a dozen litres simply by adding more bottles. Roller bottle culture is often used to 'grow up' cells for use in a larger fermentor. Mammalian cells can usually only grow in relatively crowded conditions, so you cannot seed a 100-litre fermentor with a few cells and expect them to grow. You have to seed it with the output of a 10-litre fermentor.

Cell cycle

Cells grow in a carefully regulated cycle of events. In mammalian cells, the stages of the cycle are quite distinct. They are: G_1, when the cell is making all the components for the new cell, except DNA; S phase, when the DNA is replicated; G_2, when the cell organizes itself to divide; and M phase (or mitosis), when the cell actually divides in two. M phase is the phase when the DNA is organized into macroscopic chromosomes, which can be seen under the microscope. During G_1, S, and G_2 the DNA is contained in the nucleus, which looks like a relatively undifferentiated blob of DNA, RNA, and protein under the light microscope. Normal cellular metabolism is temporarily suspended during M phase, and in many cases cells that are usually flat, extended, and attached to the support they are growing on become much rounder and much less firmly attached. Cells in mitosis can be stained and examined under a microscope to count the number of chromosomes they have, a rough indication that they are the cells you thought they were: such a chromosome count is called a karyotype.

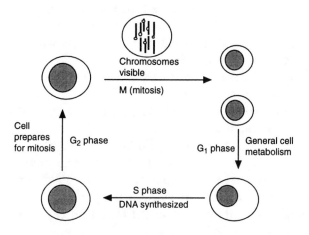

In cell culture, cells are at a mixture of stages in the cell cycle. There are a range of chemical tricks to arrest them at one point, so that all the cells accumulate at, for example, the start of S phase. If you then release the cells, they all march through the cell cycle together: this process is called synchronization, and the result is a synchronized cell population.

There is a lot known about the control of the cell cycle, which is important for the development of drugs that target abnormal proliferation of cells, such as cancer.

The position of cells in the cell cycle is often one of the measures used to characterize cells. The position of individual cells is measured in a flow cytometer measuring the amount of DNA in the cell. If the amount of DNA in the cell after mitosis is 1 n, then cells in G_1 have 1 n DNA, the cells in G_2 have 2 n, and those in S phase have an intermediate amount of DNA. (Mitosis is so short in comparison to the other phases—minutes as opposed to hours—that it can be ignored here.) The amount of DNA is measured by staining the cells with a dye that binds to DNA and is fluorescent. In general, cells spend most of their time in G_1. When the replication rate of mammalian cells speed up, it is usually because G_1 gets shorter.

FACS machines are also sometimes used to measure DNA content of individual cells. The FACS (fluorescent activated cell sorter, pronounced 'fax') separates cells according to their fluorescence. Cells are suspended in individual droplets, squirted past a laser that excites any fluorescent dye in them and past a detector that measures the fluorescence, and then into one of several receivers into which they are steered using electric fields. The detectors can detect one or two different colours of fluorescence (the latter being called two-colour FACS). Usually, one of the colours is provided by a fluorescently labelled antibody. Sometimes the other is a DNA-binding dye. Thus the experimenter can select for cells that are expressing a particular cell surface antigen (detected by an antibody) and are in S phase (detected by the dye).

FACS machines are widely used in immunology, where they are used to identify and select populations of cells bearing specific combinations of cell surface markers. In this case, the cell cycle position is irrelevant, and is not measured.

Some cells do not divide at all. These are usually said to be in G_0, effectively a dead-end sideline off the main cycle at G_1. After mitosis, cells that are not going to divide again go into this distinct G_0 state and stay there as long as their developmental programme dictates. If they are dividing, they must first switch back to G_1 before going on to S phase. This involves up-regulation of some aspects of cell metabolism, and switching on the signalling apparatus that tells a cell when to divide.

Cell disruption

Many fermentation processes produce products that are inside cells. Examples are many proteins produced by genetic engineering, enzymes, and large molecules such as the biodegradable plastic polyhydroxybutyrate. It is necessary to break the cells to get these products out. This process is called cell disruption. It is easy to break up mammalian cells, which spontaneously break open (lyse) if you put them in solutions of low osmotic strength, e.g. pure water. However yeast and bacterial cells are specifically 'designed' by evolution to be unbreakable. Thus, a lot of energy has to be put into breaking them, and there is the risk that that energy will also disrupt the product inside the cell. Plant cells in culture are not usually as tough as yeasts, but are much stronger than mammalian cells. Methods used are:

Autolysis. This simply alters the conditions so that the cell digests itself. This is the simplest possible method, but tends to be useless for protein products since the cell digests itself from the inside out, so breaking down the product before breaking down the cell wall.

Enzyme action. This is very effective; the cells are treated with an enzyme that dissolves some key component of their cell wall, which then simply falls to bits. Typical enzymes used are lysozyme(for bacteria), chitinase or glucanase (for yeasts), and cellulase (for plant cells).

Detergents, alkali, osmotic shock (i.e. pure water), plasmolysis (treatment with high concentrations of salt), organic solvents. Any of these treatments will knock holes in the plasma membrane, the thin layer of lipids inside a cell wall which actually holds the cell's contents in (the cell wall, by contrast, is meant to keep the outside world out). If the product is small enough (as is true of small proteins and of metabolites) or if there is no cell wall (as is true of animal cells) then the product leaks out.

Freeze–thaw. Freezing and thawing can break up any structure as ice crystals form inside the wet materials of which cells are made.

Mechanical methods. The most obvious method is to break the cells mechanically. There are many ways of doing this:

- French press, which forces the cells through a very small hole at high pressure. A large scale version of this is called a Manton Goulin homogenizer.

- mills, in which the cells are shaken vigorously with abrasives, or metal balls (a ball mill), or rods (a cylinder mill).

- blenders. Traditionally the laboratory uses a blender called the Waring blender (named after a 1930s New York dance band leader who invented or popularized it for making cocktails), but essentially this is a food processor with a powerful motor.

Many of the mechanical techniques are direct translations from larger industrial processes. Milling, for example, is used in ore processing in the metals industry.

A number of cell disruption techniques produce cells that are lysed, i.e. broken open, but not otherwise disrupted. These cell suspensions can be extremely viscous, mainly because the cells' DNA has not been broken up, and so expands out of the cell to form a dense interlocking network of molecules. Thus many cell lysis treatments include a nuclease treatment step. Nucleases are enzymes that break down nucleic acids, and the object here is to find a very non-specific nuclease that will break down any nucleic acid into very small pieces, ideally only a few bases long. The solution's viscosity then drops dramatically. This also breaks up the RNA in the solution, which is present in much greater mass than the DNA (although it does not contribute to the viscosity problem), and can be a problem in other purification steps if it is not broken down into small fragments.

Cell fusion

The fusion of two cells together results in a new cell that has all the genetic material of the two original cells, and hence is a new type of cell. The ability to fuse different types of cell (from the same species or from different species) has been used widely in biotechnological research. Common methods used include:

Electroporation. See separate entry.

Fusogen-mediated fusion. Some chemicals, such as PEG (polyethylene glycol), can make cells fuse together: they are called fusogens. PEG is a polymer that binds into the lipid membrane of cells and fuses it with any other lipid membranes around. Thus is can mediate the fusion of any cells that are bounded by a lipid membrane (i.e. all animal cells, and plant or bacterial protoplasts).

Virus-mediated fusion. Some viruses have lipid coats that fuse with the membrane of cells when the virus infects that cell. Thus they also act as fusogens. If the virus fuses with two cells at once, then it effectively joins the cell via a small bridge of membrane. Thus viruses have been used rather like PEG to fuse cells. Indeed, their cell-fusing capabilities were discovered before PEG's were, but PEG is preferred now because it is easier to get hold of and less potentially hazardous.

Cell fusion is used in a variety of techniques. Making monoclonal antibodies relies on making a fusion between lymphocytes and an immortalized cell line. Some plant genetic engineering has used cell fusion to generate hybrid plants, i.e. plants with all the genetic material of two different plant types brought together into one species, by fusing the protoplasts of the two 'parental' species and then regenerating a plant from the result. (This is a difficult trick to achieve.) Polyploid plants, plants with abnormally large numbers of chromosomes, can also be made by fusing cells from the same plant together.

Cell growth

The growth of isolated cells in culture follows a characteristic curve, shown in the figure. The phases of the curve are:

Lag phase. This occurs when the cells are introduced to their new growth medium, and is the time taken for them to adapt to it. If their new environment is identical to their old medium, then the lag phase can disappear: however, even the mechanical shock of moving some cells around can cause a lag phase.

Log phase. This is the main growth phase of the culture, when the cells are growing exponentially. When plotted on a logarithmic scale (on the right of the figure), the log phase shows as a straight line.

Transition. This is the period (which can be minutes to days) between log phase and:

Stationary phase. Here the cells have stopped growing; they have reached the capacity of their growth system to sustain growth. In most bacterial systems, this is a balance between some cells dying and others continuing to grow. Mammalian cells usually stop dividing in stationary phase.

Cell Growth curves:

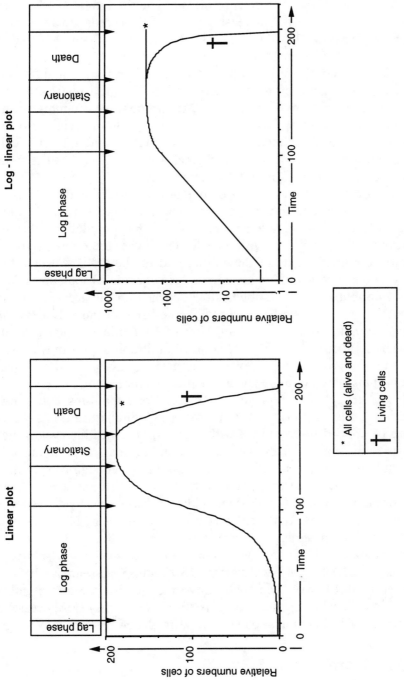

Linear plot

Log - linear plot

Death phase. If the cells are not then given fresh medium to start growing again, they will start to die off. The total mass of cells remains the same (top line) but increasingly few of those cells are alive (bottom line) in the sense that they could start a new growth curve if they were given fresh growth medium.

The length of the different phases varies enormously with different cell types. Thus, many common bacteria have a stationary phase lasting only a day or two before death phase starts. In contrast, mammalian nerve cells can last almost indefinitely in culture without dividing. A single mammalian cell isolated from skin or muscle and put into culture medium could take a week to divide, a single *E. coli* cell is unlikely to take more than 10 minutes to start growth.

The other key idea in cell growth studies is doubling time. This is the time that is needed for the cell population to double in number, and is equal to the time an 'average' member of that population takes to go through one complete division cycle. The higher the doubling time, the lower the growth rate of the culture and the longer it will take for an inoculum to reach stationary phase ('growth rate' is sometimes defined as 1/doubling time). Doubling time depends on the growth conditions and on the organism being grown; some bacteria (notably *Clostridium perfringens*) can have a doubling time of 10 minutes in the right culture medium. Strictly speaking, the concept of doubling time only applies to organisms growing in log phase, i.e. growing exponentially. A related concept is generation time, which is the time that a single organism takes to go through one reproductive cycle. For organisms that grow by division, this is pretty much the same as doubling time. However, it can be applied to organisms with more complicated life cycles, such as flies and people, where you can get more than a doubling of the population every generation if each parent has enough offspring.

Thus growth cycle is not the same as the overall life and senescence cycle of primary mammalian cells. Mammalian cells will stop dividing when they have exhausted one of the critical components in their culture medium, or when their surroundings are too crowded for their liking. However, if they are separated and put into new medium (a process known as 'splitting' the cells), then healthy cells will start to grow again. Senescence is what occurs when the cells have been split many times so that the total number of divisions they have gone through is 40–60. Then, they slowly cease to be able to divide any more, no matter how much new medium they are given.

Cell line

The term 'cell line' is usually applied to mammalian cells cultured *in vitro*, outside their original mammalian body. However, it can also be applied to plant cells. A mammalian cell line is a clone of cells, i.e. one that has been derived from one single cell. It is capable of being grown indefinitely, which mammalian cells taken straight from the body are not. Thus the cell has been 'immortalized', i.e. turned from a mortal cell (whose descendants are going to stop dividing after a few dozen divisions) into an immortal one. This can be by transformation of the cell with a virus, with DNA from an oncogene, or by mutagenesis of the cell and subsequent selection of anything that can continue growing (*see* **Immortalization**).

Cell lines should also be stable; i.e. they should not change their properties as they are grown. This can be difficult. Unlike normal cells, mammalian cells that have been immortalized often do not pass on their chromosomes very faithfully, and so can lose genes that are not essential to the survival of the cell. These may include genes vital to the biotechnologist, such as the genes making an antibody in a hybridoma cell line. Before a cell clone is described as a cell line, its inventor has to demonstrate that it is stable in this sense.

Favoured cell lines for general biological research and for production of recombinant proteins in cell culture include BHK cell lines (baby hamster kidney), CHO (Chinese hamster ovary). The names indicate where the cells to make these lines came from, and do not suggest that you have to cut up baby hamsters every time you need more cells.

See also **Strain**.

Cell line rights

While a protein can be patented and its ownership is fairly clear, the ownership of a living system is rather more vague. In general, the rulings seem to suggest that any organism that is patentable at all can be patented if it has been manipulated to do something useful, almost no matter how the manipulation is done or how large or small it is. Thus the 'oncomouse' transgenic mouse has one new gene out of about 100 000 but is still considered a 'new' entity. [For contrast, most mice (and people) probably contain at least half a dozen new, potentially physiologically significant mutations, never seen before, as a result of normal genetic change.]

Ownership of the new organism entity usually resides in who made it. It does not reside in the source of material for the new entity: the Moore case in the USA (when John Moore claimed that a cell line used in cloning interferon was derived from the hairy cell leukemia he was treated for in 1978, and hence was at least partly his) said that Moore had no rights to his own cell lines. In most countries, people do not have rights over organs removed in surgery: they only have the right to say what happens to their own, whole, body on death. (Perhaps the medics and lawyers have a point, as nearly everybody elects to have their body burnt or buried.)

Interestingly, if the Moore decision had gone against Sandoz and Genetic Institute (who now own the cell line), then many other people would have rights to a wide range of cells in research and industry. The descendants of Henrietta Lacks, originator of the HeLa cell line 40 years ago, would now have rights to a substantial part of all molecular biology and a mass of cells that probably greatly exceeds her own weight in life.

Centrifugation

This is one of the more common biochemical techniques, and it crops up very frequently in biotechnological purification schemes. Key terms are:

Zone vs. equilibrium centrifugation. Zone centrifugation places the sample at the top of a tube, puts the tube in the centrifuge and spins it for a limited time, and then takes it out. What you want has then 'sunk' a certain way down the tube, separating it from what you do not want. If you run the centrifuge too long, everything sinks to the bottom of the tube. Zonal centrifugation separates things essentially according to their size. Equilibrium centrifugation runs the centrifuge until the contents have come to equilibrium, floating, for example, at their buoyant density. Further running will not alter the separation. This is related to:

Density gradients. Here, the solution in the centrifuge tube is arranged so that it gets more dense as you go down the tube. This is done by dissolving something in it: colloidal silica ('Percoll'), to separate live mammalian cells, sucrose to separate bits of cells, caesium chloride to separate nucleic acids, etc. When centrifuged to equilibrium, the sample will be separated according to its density, more dense parts floating further down the tube in more dense solution. This is also called 'isopycnic' centrifugation.

Density gradient stabilization is also used in centrifugation, as well as in free zone electrophoresis and some other separation techniques. Here again the tube has an increasing density of solution in it, usually sugar solution. This, however, is not done to effect separation. Rather, it stabilizes the column of liquid against stirring. If a little bit of solution is stirred out of its 'proper' layer, then it will have a different density to the solution around it, and so will sink back to where it came from.

Rotors. Most centrifuges consist of a drive unit (which powers it, controls rotation speed, etc.) and a rotor in which the samples are placed and which goes round. The rotor is often removable, and fits in a bowl in the machine. In ultracentifuges (centrifuges capable of tens to hundreds of thousands of gravities) the bowl is armoured steel to protect the operator should a rotor 'fail' in a run. Legend has it that Svedberg, who developed ultracentrifugation for chemical and biochemical analysis, killed a couple of postdoctoral workers with bits of flying centrifuge.

Some rotors are 'zonal' or continuous rotors. Liquid is fed into the middle of them, and bacteria or other particulate matter is centrifuged out to the outside. These are of obvious use in separating microbial cells from their culture medium, but are an expensive way of doing this for large volumes.

Chaperones

Also known as chaperonins, these are a type of protein that help other proteins to fold up into their correct 3D structure. Some proteins will fold up correctly on their own as soon as they are made in the cell, forming a working protein molecule. Some, however, seem to do this very inefficiently, and to need other proteins to make them fold up properly.

This is important to the production of 'foreign' proteins in bacteria. If a protein folds inefficiently or very slowly, then it will have a greater chance of aggregating into an insoluble, inactive mass, from which it may be very difficult to recover active protein. If the folding can be speeded up with chaperone proteins, then the amount of usable protein that can be got out of the bacterium (as opposed to the total amount of protein, usable or not) will be greater. Whether the role of chaperones in protein folding can be used, as opposed to described, is still open to question. If it can, then it may be manipulated to produce protein complexes with functions we want that would not otherwise be produced.

Some chaperones are also known as heat shock proteins (HSPs), because they are made when a cell is 'shocked', i.e. stressed by sudden heat or cooling, some chemicals, or other insults. It is believed that the HSP chaperones are made to help to refold proteins or remove those that have been damaged during the stress. Several groups are looking at how to harness this response so as to boost the cell's ability to repair itself after damage. Targets include reducing the damage caused by stroke, iscaemic heart disease, and in Alzheimer's disease. HSPs can also be used as diagnostic markers for environmental stress. If an organism is making HSPs then it's chemical environment is poor.

Chemicals produced by biotechnology

A number of chemicals are produced commercially by biotechnology in large amounts (apart from drugs and other specialist materials). Chemicals produced in large amounts by fermentation include:

Chemical	Amount produced world-wide per annum (tonnes)	
Ethanol	75 million	
Acetone	5 million	
Butanol	1 million	
Citric acid	750 000	
Acetic acid	160 000	(much as vinegar)
Glutamate	400 000	
Lysine	80 000	
Other amino acids	20 000	
Nucleosides	5000	

Chimera

A chimera is an animal that is a mix of several other animals. The Chimera of mythology had a lion's head, a goat's body, and a serpent's tail, and breathed fire. More prosaic and realistic chimeras can be made by a range of methods that mix cells from two sources to make an early embryo, which then develops into an animal that has cells derived from two sets of parents.

Chimeras can and have been made by taking the cells from two very early embryos and mixing them together. This can be done at random, or the cells can be selected so that cells that are going to make specific areas of the body can come from one or other of the 'parent' embryos. The techniques of *in vitro* embryology are then used to put the embryo back into a pseudopregnant mother (i.e. a mother animal that has undergone all the hormonal changes necessary to prepare her for pregnancy but who is not carrying any embryos). A sheep/goat chimera was made in this way in the late 1980s (it was called a 'geep'), as was a cow/buffalo chimera. The former attracted such strong public disapproval that the latter was not publicized much (despite combining the resistance to tsetse fly of a buffalo with the milk production ability of good dairy cattle), and further research on this line has been minimal.

The animal that has been used in most research into chimeras is the mouse, where mice of different strains or carrying specific marker genes are used to make chimeras for research. The same method of joining cells from two distinct embryos into one embryo can be used on mice.

Another route is also available, to use cells call embryonic carcinoma cells (EC cells), derived from teratomas (a benign sort of cancer of the reproductive cells). These cells are totipotent, i.e. they can be encouraged to grow into a complete organism. This cannot be done in the test-tube (where the 'embryo' fails to develop for more than a few days) or by implanting the cells in a pseudopregnant mother (where they form a tumour). However, if a few EC cells are mixed with the normal cells of an embryo, then they can be incorporated into that embryo: the resulting mouse has cells derived from the EC cells in many tissues.

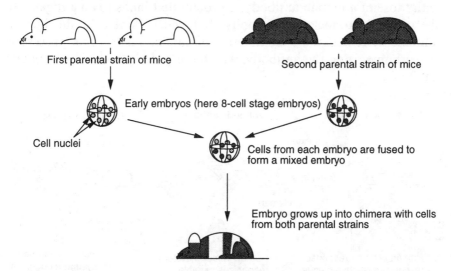

First parental strain of mice

Second parental strain of mice

Early embryos (here 8-cell stage embryos)

Cell nuclei

Cells from each embryo are fused to form a mixed embryo

Embryo grows up into chimera with cells from both parental strains

If some EC cells get into the reproductive organs, then the mouse can produce offspring derived solely from those EC cells. This is valuable for genetic engineering since the EC cells can be genetically engineered a lot more easily than mouse eggs can. Engineered cells can then be put into an embryo to form a chimeric animal, some of which is a transgenic animal. This has been proven as a route to generate transgenic mice, but EC cells are hard to generate and use, so the only other animal used to produce EC-based chimerics or transgenes is the pig (*see* **Transgenic animals**).

Chimeric/humanized antibodies

A problem with using antibodies in medical therapy is that monoclonal antibodies are foreign proteins, and hence, when they are injected, the patient will have an immune response to them. This does not matter for a one-shot therapy because the immune response is too slow to have an effect within hours of first encountering a foreign protein. But for longer term treatment it means that after a few days or weeks the patient will have their own antibodies that bind to and neutralize the immunotherapeutic one as soon as it is injected. This is known as the human anti-mouse antibody (HAMA) response (nearly all monoclonal antibodies are made in mice). It is extremely difficult to overcome this by making genuine human monoclonal antibodies as drugs: the technology of monoclonal antibody production works with mouse or rat cells, not human cells.

To overcome this, scientists engineer the antibody so that it 'looks' like a human antibody to the immune system. The species-specific parts of the antibody to which the immune system responds are in the constant regions. Thus, by replacing the constant regions of a mouse antibody with those of a human antibody, a protein that binds to an antigen like the original monoclonal antibody, but which 'looks' to the human immune system like a human protein, can be made. This process is called humanizing the antibody. The fused protein is called a chimeric antibody.

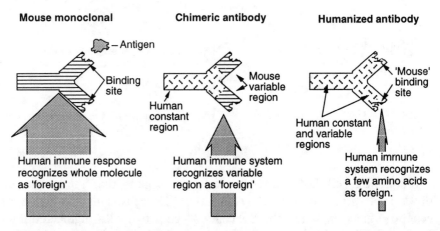

More sophisticated engineering can be done to produce a 'humanized' antibody. The ultimate engineering is to take only those parts of the antibody that determine the antibody's binding specificity (the complementarity determining regions—CDRs) and splice them into a totally human antibody. This is called CDR grafting. An alternative version is called 'antibody resurfacing' or 'antibody veneering'. This engineers all the surface residues on an antibody to 'look like' the human residues, so that the human immune system does not recognize them as 'foreign', but leaves the core residues and the binding site unaffected.

Engineering antibodies in this way has an added complication. Antibodies consist of two protein chains (heavy and light chains) and so two genes must be engineered into the producer cell to make a finished antibody. While this is possible, and various technical tricks to make it easier have been developed, it would be much easier to handle only one protein chain. This is one advantage of Dabs and SCAs: they are antibody-based proteins that contain only one chain.

The alternative approach is to make genuinely human antibodies by cloning them from humans and expressing them in bacteria or cultured cells.

Chiral synthesis

Chiral synthesis is the production of chiral compounds (*see* **Chirality**) in only one enantiomer, or 'handedness'. Since chiral compounds can be made in two (or more) physical arrangements that are virtually indistinguishable chemically, this is a difficult task for conventional chemistry. Biological systems make just that sort of discrimination all the time, however, and so have great potential for making chiral compounds.

To make chiral compounds in only one enantiomer, there are a range of chemical methods. These include:

Asymmetric catalysis. A catalyst that is itself chiral is used in a key step in the reaction. (Enzymes, of course, are one such catalyst; see below.)

Chiral chromatography. A racemic mixture of isomers is separated on a chromatographic column that is itself chiral, e.g. has a chiral compound linked to it or is made of a chiral material such as cellulose or protein.

There are several routes to chiral synthesis that use biotechnology. The success of all of them is measured by the 'enantiomeric excess', the percentage by which one of the enantiomers outweighs the other in the preparation. An enantiomeric excess of 100% means that you have one of the optical isomers absolutely pure.

Biotransformation. This is the synthesis of a compound using enzymes. Since most enzymes produce only one enantiomer as a product, they can be used to take symmetric (i.e. not chiral) starting products and produce pure enantiomers from them.

Bioconversion. This is the same idea, but uses whole organisms to convert one chemical to another. This may be better than using isolated enzymes if the enzymes concerned are not very stable, or if a number of enzymes are needed to make one conversion. The chiral drug ephedrine has been traditionally produced by bioconversion.

Fermentation methods. If you can obtain the chemical from a fermentation culture, either of a microorganism or of a plant or animal cell, then that chemical will almost certainly be produced as one enantiomer. Many amino acids produced for animal feed supplements have been produced traditionally as single optical isomers by fermentation, especially in Japan.

For all these processes, there are two approaches.

Stereospecific synthesis. This takes two, non-chiral starting materials and makes a chiral product from them. This has to be done using some third party to introduce the chirality into the system. This can be a third reagent, or a catalyst: often this chiral catalyst is an enzyme.

Resolution. This takes a racemic mixture (a 'racemate') of a chiral compound, i.e. a mixture in which all the various enantiomers are present as a mix, and removes one of them. A range of techniques can be used. One isomer may be bound to a material that is itself optically active (such as an optically active HPLC column, or an antibody), but because of their capacity to process only a few milligrams at a time these are usually used as analytical techniques rather than preparative ones. One isomer may be converted into another chemical (which can then be removed by conventional means) using another optically active chemical or, most effectively, an enzyme. The enzyme can either act on the compound you want (turning it into the product, or something more like the product) or on the one you do not want (turning it into something that is easy to remove).

Often, chiral synthesis is not used to make the final chemical itself. Rather, it is used to make a precursor to it that is easier to make using the available enzyme systems. This precursor can then be turned into the final chemical using conventional chemistry.

Chirality

Chirality is the chemical version of 'handedness'. Some molecules have distinct left- and right-hand forms, which, although containing the same atoms tied up in the same way, are not physically the same (just as your hands have the same number of fingers tied to the palm in the same way, but nevertheless are not the same). Such a chemical is called a chiral compound, and the two (or more) forms are called enantiomers (or optical isomers) of each other. Compounds that have two enantiomers are usually divided into L and D, or + and –, or S and R forms, so you have (L)-alanine or (+)-ephedrine. There are complicated rules about these nomenclatures for organic chemists.

Usually there is no chemical difference between the enantiomers of a compound, or indeed between the pure enantiomers and an equal mixture of all of them (called a racemic mixture). The only detectable difference is that they interact with polarized light in slightly different ways. However, nearly all the molecules that make up living systems are chiral. Thus, all the amino acids in proteins are L-amino acids, not their chemically identical D forms. Because of this, all the chemistry of life is chiral, and so how other chemicals affect life depends on which enantiomer you have, just as it is easier to shake hands left-to-left or right-to-right, not left-to-right, because both hands are chiral, whereas it is as easy to pick up a light briefcase with left or right hands (because, although your hand is chiral, the briefcase handle is not).

This has substantial implications for pharmaceuticals and agrochemicals. Different enantiomers of exactly the same drug can affect biological systems in quite different ways. Thalidomide is a case in point: an effective and safe antinausea agent, the teratogenic side-effects were not caused by the drug itself, but by its mirror opposite, the other enantiomer. However, the drug was given as a racemic mixture, so the patients got both therapeutic effects and side-effects.

Clearly, as legislative pressures grow for chemicals used in agriculture and medicine to be more specific, there is increasing pressure for any chiral product to be manufactured by these industries as one enantiomer and not as a racemic mixture for those applications. **Chiral synthesis** is a major aspect of biotransformation and bioconversion technology.

For biopharmaceuticals, of course, chirality is not a worry, since, as biologically derived proteins, they all have the correct 'handedness' anyway.

Chromatography

Many separation systems used in biochemistry, molecular biology, and biotechnological production are chromatography systems. Chromatography was originally developed as a way of separating pigments from plants by wicking them through paper, an experiment that many schoolchildren do today, and the same basic ideas apply to all chromatographic separations. A small sample is put on one end of a slab or wick of porous material. A solvent is then passed over the sample and up the wick or slab. Depending on whether the molecules in the sample stick to the solid wick or dissolve in the solvent, they either move up the wick or stay put. Most materials do a bit of both, and so move up the wick slowly: the exact speed varies with each component of the sample, and so they are spread out. The pattern of how materials are washed (eluted) off the end of the slab or wick is called the elution profile. This is, in fact, a two-phase separation, and so the two parts of the system are called the mobile phase (the solvent) and the stationary or solid phase (the solid material that the solvent moves over).

There are a lot of variations of chromatography. The more common are:

Gel chromatography/gel exclusion chromatography/size exclusion chromatography. These sift by molecular size. The chromatographic material has small pores in it, into which small molecules can enter but from which large molecules are excluded. (Different materials have different pore sizes, so that the size limits 'small' and 'large' can be adjusted to suit whatever the scientist is trying to separate.) As the mixture of molecules passes down the column, small molecules diffuse into the pores, where liquid is stationary, and so spend some of their time standing still. Large molecules cannot enter the pores, and so spend all their time in the moving phase. Thus large molecules move down the column faster than small ones.

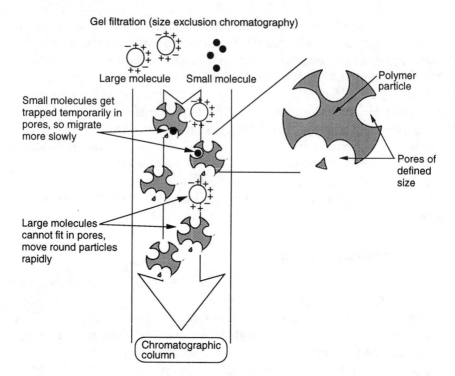

Gel filtration (size exclusion chromatography)

Large molecule   Small molecule

Polymer particle

Small molecules get trapped temporarily in pores, so migrate more slowly

Pores of defined size

Large molecules cannot fit in pores, move round particles rapidly

Chromatographic column

Affinity chromatography. Here a specific molecule is linked to the chromatographic material and molecules are separated by their ability to bind to it. If the bound molecule is large and the molecule to be separated is small, this is usually called **affinity chromatography**. If the bound molecule is small and the separated molecule large, the method can be called covalent chromatography, although this too is often called affinity chromatography.

Hydrophobic chromatography. This simply uses a hydrophobic material, such as untreated silica, as the stationary phase. Molecules stick to it depending on how hydrophobic they are, so it is an effective way of separating many metabolic products. Most HPLC systems are hydrophobic adsorption chromatography. So are adsorption fermentation systems, where a fermentation broth is passed continuously over some solid material that adsorbs the product of the liquid while leaving the microorganisms behind.

Gradient chromatography. Here all the molecules in a sample are bound to the solid support material, and then they are washed off one at a time with an increasing concentration of some solution, often increasing salt, acid, or alkali concentration.

Chromatography also varies according to the physical arrangement of the solid material (the stationary phase):

Column chromatography. This is by far the most common; the solid phase is packed as small particles into a tube, and the liquid passes over it. Column chromatography methods capable of purifying kilograms of materials at a time have been developed. A variant is high performance liquid chromatography (HPLC; sometimes also called high pressure liquid chromatography) in which the liquid is pumped slowly over quite a small column at very high pressure. This greatly increases the resolution of the method, i.e. how well it can separate similar materials.

Paper chromatography. This is essentially the same as the original method, and uses a paper wick as the solid phase. This is not as limited as it may sound since paper is a complex material and papers with widely varying properties are available.

Thin layer chromatography (TLC). Here the stationary phase is a thin layer of treated silica coated on to a glass plate.

Gas chromatography (GC). This is quite commonly used to analyse the volatile metabolites produced during fermentations, and also to analyse other biomolecules that have been made more volatile by chemical modification. GC passes a stream of hot gas through a very fine tube in a carefully thermostatically controlled oven. The volatiles partition between the gas phase and as liquid on the walls of the tube (hence this is sometimes called gas–liquid chromatography, GLC). The time that it takes them to get through the tube depends on this partition.

Gas chromatographs often have ionization detectors at the end to detect the molecules coming off the tube. The gas is fed into a flame, and the ions produced as the target molecules burn is detected electrically. This is a sensitive and widely applicable detection method, but obviously not directly suited to collecting the molecules afterwards for further work.

Lastly, there are the different materials that can make up the mobile and solid phase. In general the mobile phase is water, or some watery solution: this is because nearly all the materials that biotechnology is interested in are soluble to various degrees in water, and some, like proteins, are not stable in any other solvent. The solid phase gives greater flexibility.

Polysaccharides. Much loved by biochemists are polysaccharides such as cellulose (both as a granular material and as paper) or agarose. The popular Sepharose is a specific product of Pharmacia Biotech, but has almost achieved the status of a generic name like Hoover. They are used for gel chromatography and for affinity methods.

Synthetic polymers. Coming into increasing favour are synthetic polymers such as polystyrene, PMMA (perspex), and teflon, because they are easier to form into small, rigid, uniform spheres and are chemically more robust. Polyacrylamide is also used.

Silica. Chemically modified silica, especially silicas with chemically modified surfaces and silica materials with a porous structure (CPG, controlled pore glass) are used in many applications. For applications involving substantial pressures, such as HPLC (in which polysaccharide beads are likely to be squashed), silica is very useful.

Generally, chromatographic methods are used to separate several different chemicals from a mixture at once. The pattern in which they emerge from the chromatographic column is called the 'elution profile'. The efficiency of a chromatographic system is defined in terms of 'theoretical plates'. The column can be viewed as a series of plates, each of which act like a cell with sample in one side and fresh solvent in the other. (Strictly, a theoretical plate is that bed length that brings about the same relative change in concentration of a solute as would be achieved on equilibrating with fresh stationary phase.) The efficacy of the column depends on how many theoretical plates it is equivalent to, and this can be calculated if you know the properties of the material involved. In practice, much biochemical chromatography development is developed on a trial and error basis.

Cleaning-in-place

This is the cleaning (and sterilization) of a manufacturing system without dismantling it, so that the parts are cleaned as a whole: it also called '*in situ* sterilization' or sterilization-in-place (CIP and SIP). This is a much easier operation to perform than cleaning and sterilizing all the components separately and then trying to assemble them under sterile conditions, or having separate cleaning and sterilizing operations. However, it requires some specialist techniques and equipment.

In particular, bioreactor machinery much be designed so that there are no 'dead legs' (i.e. pipes that are blocked at one end), crevices, or 'shadowed' areas (i.e. areas where the bulk of some other piece of the apparatus prevents fluid from flowing) into which cleaning fluid could not flow. It is also useful for the equipment to be designed so that bits of it can be cleaned while the rest remains in operation .

CIP systems are a combination of the equipment used and the processes used in it. Thus a full CIP system will include protocols for washing, usually with industrial strength detergents, and rinsing out the wash agents.

Clean room

A clean room is a room that meets specific standards of cleanliness, especially with respect to what may go into and out of it and what the concentration of particles in the air might be. Clean rooms are central to pharmaceutical manufacturing, since it is by producing, formulating, and packaging drugs under rigorously sterile conditions that the sterility of the drug is assured. The same cleanliness conditions apply to a lesser extent to other biotechnological products, and also can apply to the research and development phase of recombinant DNA or plant or animal cloning work, where the objective is to stop contamination getting into the experiments.

Clean rooms are classified in the USA according to US Federal Standard 209D. Roughly, clean rooms are classed by a number, which is the number of particles of more than half a micrometre diameter that are allowed per cubic foot of air. Thus a class 100 clean room would have around 100 0.5 μm particles per cubic foot. (The exact number varies slightly from this.) At the moment, class 100 rooms are the highest category of cleanliness required by the pharmaceutical industry. Other countries have different rating systems (notably, most are based on SI units), but the level of air purity required is similar.

Clean rooms are kept clean by a variety of methods. Air going into the room passes through HEPA filters (high efficiency particulate air) to keep out even very small particles: ultra-clean rooms have several layers of filtering. Walls, floor, and ceiling are usually painted with materials designed not to stick any dust (and, obviously, not to peel or flake). People going into the clean room have to wear hats and overshoes, since hair and shoes are the most particle-laden parts of the worker, as well as the usual laboratory coat. For less rigorously clean areas there may be a 'sticky mat' just inside the door, which pulls loose dirt off the bottom of the shoes of everyone who walks in.

To provide greater cleanliness within a clean room, a laminar flow hood are provided. These are benches either made of or surrounded by open meshwork, and covered by a hood. Air flows up past the work-surface and into the hood, where it is filtered before being returned to the worksurface. Thus all the air passing over the workplace is in a separate flow from the air in the room, and has been extra cleaned.

Clean rooms use much the same air filtration technology as containment laboratories, but for a different purpose. Containment laboratories are meant to to keep possibly hazardous material in the laboratory, rather than external contamination out.

Clone

A clone is a collection of genetically identical individuals. They turn up in molecular biology and biotechnology in many contexts.

Clones of organisms. Clones of plants and some animals have been developed using many techniques. Members of a clone show less variability than collections of the same organisms bred by sexual methods, because their genes are identical. Cloning may provide a method for the rapid multiplication of some desirable individual without having to wait for breeding cycles. Cloning plants usually involves plant cell culture. A plant is broken up into very small pieces, even single cells. These are grown into larger masses in culture, and then these masses (calluses) are induced to differentiate into the tissues of a plant. This route is particularly useful for propagating plants with long life cycles, such as trees. **Animal cloning** is more difficult, and relies on some manipulation of their normal reproductive cycle.

Gene clone. This means a collection of organisms (usually bacteria) that all contain the same piece of recombinant DNA. By extension, it means the piece of DNA they contain (*see* **Recombinant DNA**).

Cell cloning. Some biotechnology methods produce a collection of single cells that are genetically different. Producing hybridomas is an example: the fusion step produces a large number of different fused cells. These variants are then 'cloned', i.e. separated out and individual cells grown up to produce a clone of cells.

Cloning cells and microorganisms is done by a combination of dilution and growth. The cells are spread out, so that they are spaced well apart. Then they are grown up into individual colonies, each of which is therefore well separated, and the relatively enormous scientist can physically pick out one clone. The separation may be done by spreading cells on to an agar surface in a Petri dish; this is the usual approach for bacteria. Another approach, popular with mammalian cells, is to dilute a suspension of cells into many small tubes (usually, the wells of a 96-well plate) at such an extreme dilution that most wells have nothing in them. Then, the few that do have something in them will, by statistical chance alone, have only one cell in, and so that cell will grow up into a clone. Because cells can stick together, this cycle has to be repeated several times to make sure that you have a true clone. This process is called dilution cloning or cloning by limiting dilution, and is the common way of cloning hybridomas to make monoclonal antibodies.

A novel approach is called gel microdrop cloning, where cells are encapsulated in a small droplet of a gel such as agarose. The cells cannot move between gels, so after some time for growth each drop contains a clone derived from one cell. This has been applied to fungal and mammalian cell cloning.

Coenzyme

The term cofactor is used almost interchangeably with coenzyme in most contexts. A coenzyme is a molecule that is needed by an enzyme to work, is part of the chemical mechanism of the enzyme, but which is not produced for its own sake. Rather, it acts as a shuttle molecule, taking groups between one enzyme and another or between one reaction and another.

The most biochemically common coenzyme set is the NAD set. These molecules shuttle hydrogen atoms around the cell. There are two flavours [NAD (nicotinamide–adenine dinucleotide) and NADP (NAD phosphate)], and they come as hydrogenated (reduced) or non-hydrogenated (oxidized) molecules: NAD or NADP = oxidized, NADH or NADPH = reduced.

Many cofactors and coenzymes are derived from vitamins. Thus NAD is derived from nicotinic acid.

Some coenzymes are tightly, even covalently, linked to their enzymes: it is these that are often called cofactors. An example is the FAD (flavine–adenine dinucleotide) moiety that is needed by the common diagnostic enzyme glucose oxidase. If the FAD is removed the enzyme does not work. Such a cofactorless enzyme is called an apoenzyme. It contains all the protein of the intact, functional enzyme (the holoenzyme), but does not catalyse its reaction.

Cofactors are important for the industrial application of enzymes. Many enzymes that perform industrially important reactions cannot be used in the appropriate industrial process because the cofactors they need are too expensive and are gradually lost from the reaction as the enzyme is used. Biotechnology therefore seeks either to replace them with other molecules that are cheaper or more stable (biomimetics) or to recycle the cofactors more efficiently. Both are difficult to achieve economically on an industrial scale.

Coenzymes are also of concern to biotechnology in two other fields. The first is that they are often inconveniently complicated and expensive molecules to make and store, and so research is looking for synthetic alternatives (*see* **Biomimetics**). Secondly, some abzymes have been made that use coenzymes to catalyse reactions (*see* **Catalytic antibodies**).

Combinatorial chemistry

Combinatorial chemistry is a growing area of synthetic chemistry that provides new chemicals for drug discovery. Traditionally, new chemicals have been made by chemists, more or less one at a time. Thus the chemicals that can be screened for their potential as new drugs is limited to what happens to have been made before and stored in a company's chemical library. Combinatorial chemistry seeks to make many chemicals at once in small amounts, both as a source of new chemicals and to test which of many variants of a compound is the best drug.

COMBINATORIAL CHEMISTRY

Combinatorial chemistry techniques usually rely on building variants of a limited number of types of chemicals. The chemist will start with a core or 'scaffold', and then add a selection of side groups or functional groups on to it. The result will be a series of new chemicals, which are different but related.

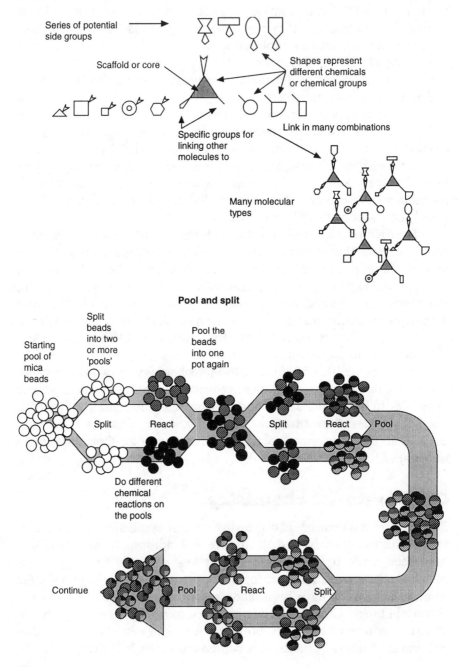

Series of potential side groups

Scaffold or core

Shapes represent different chemicals or chemical groups

Specific groups for linking other molecules to

Link in many combinations

Many molecular types

Pool and split

Starting pool of mica beads

Split beads into two or more 'pools'

Pool the beads into one pot again

Split   React   Split   React   Pool

Do different chemical reactions on the pools

Continue   Pool   React   Split

The simplest way of doing this is to build a robot that can do chemistry. Automated chemical synthesis machines can make up to a couple of hundred chemicals at once (when programmed properly), which eases the burden of repetitive labour. These machines often use solid phase synthesis. The growing chemical is linked to an inert solid (such as silica), so that the reagents used can be separated from the chemical easily. The gene machine DNA synthesizers use this approach.

The more radical way to generate lots of chemicals is to use small microparticles in a 'pool-and-split' scheme, illustrated here. At the end you have a huge number of small particles (often polystyrene), each with a unique chemical linked to them. These can then be tested separately for drug activity, or tested as a pool or set of pools. Once you find an 'active' particle, you then have to find out what chemical is attached to it. A neat way of doing this is to tag the particles physically, and tags like coloured stripes, DNA molecules, and even little radio transponders have been tested.

The other way to make huge numbers of chemicals is to miniaturize the robotic synthesis systems, so you can build a 'chemistry chip' that would make 10 000 chemicals at once. Such ideas are not new, but are fiendishly difficult to get to work. Companies such as Orchid Biocomputer are hoping to harness the microfabrication skills of the semiconductor industry to making thousands of specific chemicals at once.

The number of chemicals you could make using combinatorial chemistry is huge, but many of them are very similar. Some companies are therefore constructing 'virtual' chemical libraries in computers, i.e. simulating what a combinatorial chemistry synthesis would make, and then screening it for a set of really diverse molecules before actually making that, smaller, collection. This type of combinatorial chemistry arose from technology to make random peptides. Peptides are now usually made using **phage display** technologies.

Computational chemistry

This is a blanket term for using computers to predict or analyse the properties of molecules (as opposed to using computers simply to draw them, which is **molecular graphics**). Calculating the properties of molecules from first principles, which would be ideal, is impossible for practical purposes. Thus computational chemistry uses the known properties of chemicals to calculate the properties of similar molecules partly from empirical rules ('heuristics') and partly from 'rigorous' calculation. Typically, approaches use energy minimization, which calculates the total chemical energy of a molecular structure and seeks to adjust the structure to minimize it. Contributions to the energy include energy of solvation (binding water molecules), van de Waals forces between nearby atoms, and 'steric' constraints (which is a general term for how you can fit different parts of a molecule together without having two atoms in the same place at the same time).

One of the main areas of interest is predicting how proteins will fold up. In principle this should be predictable from their amino acid sequence, but this is not yet achievable. So there are a series of half-way houses. The most 'rigorous' is to model the peptide chain as a simpler molecule in which each amino acid is one 'unit', and then model that. Alternatively, you can use known protein structures as a model, build a rough structure by analogy with them, and then perform energy minimization on that structure to adjust the detailed position of atoms to the most stable structure. In recent years it has been found that many proteins use fairly standard structural bits, so if you can find a similar amino acid to the one you are trying to model, there is a good chance that its structure will be very similar too. So there is little need to do complex calculations in order to obtain a first approximation of a structure. The reason for doing this is to be able to predict the structural and functional properties of a protein. This is particularly important at a time when genome-scale DNA sequencing can generate gene sequence (and hence deduce protein sequence) very much faster than scientists can find out what the proteins do.

Although computational chemistry is distinct from molecular graphics, the two have a close link. The results of computational chemical calculations are often displayed as computer pictures of molecules. And one route round the enormous difficulty of computational chemistry is to use the human brain as a computer to analyse molecular patterns displayed on a screen.

Concentration

Biological products are usually produced in rather low concentrations, by fermentation methods or by extraction from animal or plant tissues. In order to keep the cost of purifying these materials down, it is useful to reduce the volume, i.e. to increase the concentration, as early as possible in the downstream processing stages of a biotechnological process. Many concentration methods also purify the product to some extent as well. The very best processes concentrate and purify in one step, but that is rarely possible.

Concentration is inherently expensive. Preparing any material from a diluted solution costs more. This is why extracting gold from sea water is not economic, whereas extracting salt is: salt is so much more concentrated than gold that it tips the balance (*see* **Purification methods: large scale**).

Methods used in concentration are based on:

The size of the molecules. In this category come various filter methods and reverse osmosis. In reverse osmosis, the sample is placed on one side of a semipermeable membrane, i.e. one that will let through water but not other materials. A very high pressure then pushes the water through the membrane, giving water on one side and much concentrated product on the other. This can be a way of purifying water too: it is sometimes used to make drinking water from sea water. It is the reverse of osmosis, the process by which water moves from one side of a semipermeable membrane to the other side if the concentration of soluble material is greater on the other side. Ultrafiltration is a similar technique. Here the molecules are filtered through a membrane with pores of molecular size. Large molecules stay on the sample side, while water and small molecules, including salt, pass through. Again, considerable pressure is usually needed to make this happen.

A closely related concept is dialysis, in which the protein solution is put on one side of the semipermeable membrane and salt solution or water on the other. Any small molecules in the protein solution diffuse across into the water. However, water also diffuses across into the protein, so that it usually ends up more dilute, not more concentrated.

The charge of the molecule. This usually means ion-exchange methods. Here a polymer is made with a charge on it: usually, this is a polymer with charged side groups. Molecules with the opposite charge to the one on the polymer will stick to the polymer. A large volume of dilute product can be washed over a small amount of ion-exchange polymer (or resin, as it is usually called), and the product concentrated on it. Product can be washed off again by washing with acid or alkali, or sometimes with concentrated salts.

The hydrophobicity or lipophilicity of the molecule. This is also related to charge. Hydrophobic molecules can be extracted from water by absorption on to a solid material or extraction into a solvent. As well as conventional extraction (essentially shaking two liquids and then letting them separate), this includes counter-current extraction methods, in which two immiscible liquids flow past each other, and the material you want is successively exchanged from one liquid to the other. Lipophilic molecules can also be adsorbed on to lipophilic plastics like teflon or nylon. The amount that sticks is not very great, however, forming at best a molecule-thick layer over the plastic's surface.

The volatility of the molecule. If the product is volatile, it can be separated easily by fractionation. The classic example of this is separating ethanol from yeast fermentations to make distilled drinks.

In the second and third methods, the crude extract can be mixed with or passed over an absorbent, which picks up the material you want. The product is then washed off the absorbent in a different solution, e.g. with a different solvent, at different pH or salt concentration. If the product is not a molecule but rather is whole cells, then methods based on the relatively large size of the cells can be used. These include:

Sedimentation. This simply collects the cells by allowing them to fall out of the culture medium. It works well for large fungal mycelia or animal or plant cells since these can settle out in a matter of hours. However, some bacteria could take days or weeks since they are very small, and ones that can swim by themselves would never settle out at all. Sedimentation can be speeded up in a centrifuge, but centrifuging large volumes of liquid can be expensive.

Flocculation. This means making the cells clump together and then letting them settle out as a visible precipitate. This is used quite extensively in the sewage industry.

Flotation. Cells can get stuck on to the walls of bubbles, and so be carried to the top of the liquid to be collected there as foam. This is a well-known technique from the mining industry.

Cows (and beef)

Agricultural biotechnology is concerned with cattle in three ways, other than the general programmes mentioned elsewhere.

Cows and sheep both suffer spongiform encephalopathies, and 'mad cow disease' has become a significant health concern in Europe. Both can also be used in producing pharmaceutical proteins through genetic engineering (*see* **Pharming**).

Cows convert cellulose to mammalian meat through rumination, the fermentation of their food in their four-chambered stomach. Since cellulose is the most common component of living things on earth (the most common protein is RUBISCO), we would like to either make cows more efficient or enable other organisms to ruminate. The cow's stomach consists of four chambers—the rumen (where anaerobic bacteria break down the cellulose and release short chain fatty acids, which the cow can metabolize), reticulum, omasum, and absomasum. Engineering the cow or, more plausibly, the bacteria to make the process more efficient is a medium-term goal. The bacteria can also be engineered to provide other abilities, such as the ability to break down toxins in forage plants: the rumen bacterium *Butyrivibrio fibrisolvens* has been engineered to break down fluoroacetate, a poison in many tropical shrubs, and may enable cattle to eat these plants. The abomasum of suckling cows is also a source of rennin, used to make cheese.

Both beef and milk quality are targets for cow breeding programmes. A cow genome project is under way, aiming to identify linked markers that will allow breeding cattle to be selected very early in life for genes that only show themselves in adults. Improved milk quality for cheese production, lower fat content, and other traits are being addressed.

Milk is itself a valuable product that is used to make a wide range of materials. Its main protein ingredient is whey protein or casein. The RFLP (restriction fragment length polymorphism) marker study was to improve the type of caseine produced.

Beef and milk are only two of the many products produced from cows. Others include mechanically recovered meat (MRM), which is obtained from the carcass after the high quality meat has been removed and is processed in a variety of ways to make 'burgers', pet food, and other lower value beef products. Part of that conversion now involves biotechnological processes such as partial proteolysis, use of collagenase to break up connective tissue, and manipulation of the content of chemicals such as amino acids and vitamins. Other products are gelatin (used, *inter alia*, in confectionery and in drug capsules) and bone meal for fertilizer. Manufacturing many of these products uses the tools and techniques of biotechnology.

Cross-flow filtration

This is a commonly used method of filtering the sorts of dense and thick fluids that have to be filtered in biotechnological separations in order to concentrate some material. If you try to filter (say) soup through a standard micropore filter in order to concentrate the particulate material, the pores will rapidly block up, and filtration will be brought to a halt. Cross-flow filtering does not seek to filter the liquid through the filter directly. Rather, it flows the liquid across the filter, allowing the carrier liquid to flow through. After it has passed over, the top (unfiltered) phase is more concentrated, and some of the fluid phase has passed through; meanwhile, the filter is not blocked up.

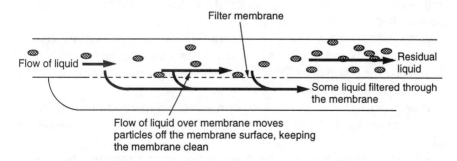

Filter membrane

Flow of liquid →

Residual liquid

Some liquid filtered through the membrane

Flow of liquid over membrane moves particles off the membrane surface, keeping the membrane clean

The build-up of material on a filter, blocking the pores or gaps, is called fouling, and is a major problem for any filtration process. Cross-flow filtration is one of the more effective ways of getting round fouling. Other approaches include scraping the filter clean or replacing it periodically, or back-flushing. In this last one, clean liquid is pumped back through the filter in the reverse direction, so that the accumulated material is pushed off the surface, and can then be cleared away. This is not as rough on the filter as scraping, and is simpler to engineer than cross-flow systems. However, it does not always work if the fouling material 'glues' itself to the filter.

Cryopreservation

This is preservation of things by keeping them cold. There are several variations of relevance to biotechnology.

Freezing. The most obvious route. Just putting something in the fridge or freezer is fine for many biological materials, but not all, since the process of freezing sometimes destroys what you are trying to preserve. This is especially true of cells because ice crystals forming in the cells break the cellular structures up, leaving you with dead meat.

Freezing in mixed solvents. To prevent damage to cells on freezing, they are often frozen in a mixture of a watery material (their usual growth medium) and another material called a cryoprotectant, which mixes with water and stops it forming ice crystals. Glycerol is a favourite for bacteria, dimethyl sulfoxide (DMSO) for animal cells. Sucrose can also be used: this is why ice lollies are not crunchy with ice crystals.

Bacterial cells preserved in this way can be kept in a conventional freezer, but animal cells need to be stored at liquid nitrogen temperatures if they are to survive more than a few weeks. This is often called storing them in liquid nitrogen vapour phase since the tubes of cells are kept in a flask of liquid nitrogen *above* the nitrogen itself, so they are not actually immersed in the liquid, only exposed to its vapour. Apart from anything else, this stops the tubes filling up with liquid nitrogen and then exploding when you warm them up. Even at these temperatures, mammalian cells eventually 'go off'.

Antifreeze proteins. There are proteins known that prevent ice crystals forming that are found in some arctic fish. In principle they could be used to replace the glycerol or DMSO (which are somewhat toxic), but this is rarely done in practice.

Freeze-drying. This is not a cryopreservation method really since the dried sample is not stored cold, but it is often classed as one (*see* **Freeze-drying**).

An odd application of cryopreservation was suggested by Robert Ettinger in his book *The Prospects of Immortality*, published in 1964. He suggested that dying people could be frozen until medicine had discovered a cure for whatever was killing them, whereupon they could be thawed out and cured. Both ice crystal formation and the inevitable decay of mammalian cells even at liquid nitrogen temperatures means that this surely cannot work, but if you are dead anyway, why not try it?

Culture collections

Many countries and institutions have set up places where samples of microorganisms and cell lines are stored. Other names are strain depositories and type culture collections, the latter because they are where the 'type specimens' (i.e. the definitive specimens, which describe that 'type' of organism) are kept. They have a triple function: they are a 'bank' for your valuable microorganisms (against the risk that your laboratory burns down); they are a centre from which other people can get samples of your organism (if you want them to) without bothering you; and they are somewhere where you can deposit an organism to prove you own it, a sort of biological patent office. Some patenting systems insist that you deposit a sample of any organism mentioned in a patent that cannot be created easily by someone else at a recognized depository so that, if there is a dispute later, there is some way of saying what your original organism looked like.

The best known depository is the American Type Culture Collection (ATCC), which collects all types or microorganism and cell line. ATCC is also the World Health Organization (WHO) international reference collection. There are a variety of other general depositories in other countries, some of which specialize in fungi, bacteria, or animal cells. There are also industry-specific depositories for the dairy industry, marine organisms, pathogens, etc. This can be confusing to someone trying to find a specific organism, so there are a number of centres and databases to help track down organisms. Europe has a culture collection of purely mammalian cells: the European Central Animal Cell Culture facility (ECACC) at Porton Down, UK.

Cyclodextrins

These are cyclic carbohydrates made of six, seven, or eight glucose molecules joined in a ring, to form alpha-, beta-, and gamma-cylodextrin, respectively. The cyclodextrins all form cylindrical molecules with their water-soluble groups on the outside of the molecule and a relatively non-polar hole down the middle. This hole can accommodate another molecule, known as the 'guest' molecule. This allows them to be used in a wide range of applications, including improving the solubility of drugs and biopharmaceuticals, and selectively binding materials that 'fit' into the central hole in affinity purification and chromatography methods.

Natural cyclodextrins are not used extensively in drug applications because they are not very soluble and are rather toxic to injection. However, they may be modified by adding alkyl or hydroxyalkyl groups on to the hydroxyls of the natural cyclodextrin, which reduces toxicity and can enhance solubility.

Cytokines

Cytokines are materials that stimulate cell migration, usually towards the source of the cytokines. Cytokines are studied in mammals because they are important to many processes that involve cells moving about, such as inflammation and development. Understanding them, and then isolating them and producing large amounts for therapeutic uses, has been a major research target of many 'genetic engineering' and pharmaceutical companies. Nowadays the enthusiasm for cytokines as drugs in themselves has waned.

The cytokines that act on the cells of the immune system are the best characterized. They attract the cells to sites of damage or infection where they can kill invading cells and, as a side-effect, produce inflammation, shock, and even death. So well understood are the immune system cytokines (compared with other cell mobility enhancers) that 'cytokine' usually refers exclusively to cytokines that act on lymphocytes and macrophages. Cytokines are also involved in the body's control of how many blood cells are made in the bone marrow, and so are of general interest as potential stimulators of blood production (haematopoeisis). A review of cytokines is beyond this book, but the ones known to date include:

Interleukins. There are over 15 known (IL-1 to IL-15). IL-2 has been used as a booster of the immune system in cancer and infectious disease therapy: it stimulates T cells to proliferate. IL-1 has several effects with the overall effect of stimulating the production of blood cells by the bone marrow, as well as stimulating non-immune cells to produce other cytokines. IL-4 is linked to the allergic response (IgE-mediated immunity), and so agents that affect IL-4 response have potential for modulating allergies.

Many of the CD antigens that allow scientists to distinguish different types of lymphocyte from each other are interleukin receptors: i.e. they are the proteins that interleukins bind to, and through which interleukins have their effect on a cell. CD (cluster differentiation) antigens turn up in a variety of contexts, most notoriously CD4 as the protein that the AIDS virus uses to bind to its target cells.

Colony stimulating factors (CSF). There are three varieties: G-CSF,

M-CSF, and GM-CSF, which stimulate granulocytes, macrophages, or both, respectively. They stimulate the differentiation of some types of white cell. Ten companies are trying CSFs as biopharmaceuticals.

Interferons (IFN). Well known as being one of the first proteins to be produced by the new biotechnology of the late 1970s, and touted as the wonder cure for everything, there are actually three classes of these cytokines. They are now consistently called interferons alpha, beta, and gamma. IFN-γ is a potent stimulator of the activity of macrophages, encouraging them to kill tumour cells and intracellular parasites. Interferon α (from Biogen) has been approved as a treatment of hepatitis C. Interferon β has been approved for treating several cancers and for multiple sclerosis: it is produced by several companies including Ares-Serono and Biogen. Bovine interferon has also been shown to help improve the pregnancy rate in sheep, because it increases the process of 'maternal recognition' by which the ewe's immune system learns that the developing fetus should not be rejected. This unusual use of a cytokine could be as widespread as medical uses.

Tissue necrosis factor (TNF). Slows cell growth and kills some cancer cells and cell lines. It is therefore a hot candidate for an anticancer drug, and as the 'toxin' part of an immunotoxin. It is also involved in the cell destruction that can occur in some inflammations, so finding ways to block TNF's action is also a hot pharmaceutical topic.

Several companies are developing genetically engineered cytokine preparations for drug use: Genentech (gamma interferon), Cetus (IL-2), Immunex (GM-CSF), Ares-Serono (beta interferon).

Dabs and other engineered antibodies

Dabs are antibodies in which there is only one protein chain derived from only one of the 'domains' of the antibody structure, and hence are called single domain antibodies or Dabs. Greg Winters at Cambridge, UK, has shown that, for some antibodies, half of the antibody molecule will bind to its target antigen almost as well as the whole molecule. Usually the binding site of an antibody consists of two protein chains (*see* **Antibodies**).

The potential advantage of Dabs is that they can be made easily by bacteria or yeasts. Whole antibodies have two protein chains, and therefore need engineering with two genes. Gene cloning vector systems exist to do this, but it is relatively tedious. Dabs offer a way to clone antibody-like molecules into bacteria, and hence enable the screening of millions of antibodies much more easily than it is possible to screen monoclonal antibodies.

Related ideas are single-chain antigen binding technology (SCA), patented by Genex, biosynthetic antibody binding sites (BABS), invented by Creative Biomolecules, and minimum recognition units (MRUs, or complementarity determining regions—CDRs), which is a more general description of the smallest part of an antibody that you need for it to bind to its target. SCAs and scFvs are terms for antibody-binding domains in which the two binding regions from light and heavy chains are linked by a short peptide, so they can be produced from one gene. This makes them much easier to produce in bacteria from recombinant DNA, since there is no need for the two chains of the normal antibody structure to be made separately and then assembled within the cell. Libraries of 10 trillion scFv clones have been made in bacteriophage, and antibodies that are claimed to be as good as good monoclonals have been obtained.

Bispecific antibodies are antibodies with two different binding sites, so that each molecule has one binding site for each of two different antigens. They are made by a combination of protein engineering and protein chemistry. Their potential uses include ADEPT-type therapeutics, where the antibody binds both to the cell surface and to the therapeutic molecule that you want to target to the cell surface. Equivalent terms are hybrid hybridomas and diabodies.

A more extreme version is to fuse the constant part of an antibody with another binding molecule entirely, for example a cytokine or integrin receptor. The antibody part still activates the immune system, and the receptor moiety binds highly specifically to a target cell. Such constructs are called immunoadhesins. This idea is related to that of an immunotoxin, where an antibody binding site is linked to a non-antibody 'effector' molecule.

Darwinian cloning

This means selecting a clone from a large number of essentially random starting points, rather than isolating a natural gene or making a carefully designed artificial one. From this mixture you select, by whatever means at hand, those molecules that look more like those you want than the rest. (How you select them depends, of course, on what sort of molecules you want.) You mutate these to generate a new set of variants, and reselect, make more variants, and so on until you have the molecule you require. This is a process akin to natural selection, and so is sometimes called 'directed evolution' or (incorrectly) 'molecular evolution'.

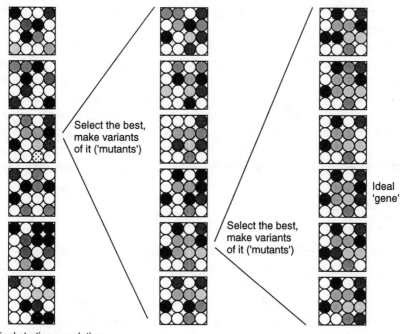

Select the best, make variants of it ('mutants')

Select the best, make variants of it ('mutants')

Ideal 'gene'

Original starting population of 'genes': generated at random

There are several classes of catalytic molecule suitable for this.

Catalytic antibodies. Indeed, all antibodies are evolved in this way: the body does the randomizing and selection procedures in the immune system (*see* **Catalytic antibodies**).

Random proteins. In principle, you could clone a totally random bit of DNA in an expression vector, measure enzyme activity, make alterations in the DNA of the clones that show the best activity by random mutagenesis, select again, and so on. The most practical version of this is done using **phage display** technology.

Antisense. The word 'aptamer' has been coined for antisense RNAs and DNAs that have been selected by Darwinian cloning to bind very tightly to a particular gene or RNA. The starting point here is a random chain of bases, which is bound to the target molecule. Those that do not bind, or only bind weakly, are simply washed away and discarded. The few molecules (out of billions) that are left are then detached and amplified using PCR. These can then be altered by mutation, selected, amplified again, and so on (*see* **Antisense**).

Catalytic RNA. RNA can also be selected in this way, but with the added advantage that RNA can be a catalyst in its own right. This sort of Darwinian selection has been done to make RNAs that will bind specific low molecular weight chemicals very tightly. The next step is to find one that binds a transition state analogue for a reaction (*see* **Catalytic antibodies**) which would, plausibly, make a new catalytic RNA.

The advantage of Darwinian systems is that they select a new catalyst from a vast number of possibilities. There are more possible 100-amino acid proteins than there are electrons in the universe, so screening them all is clearly impractical. However, this approach edges up to the catalyst you want one step at a time. If the catalyst you want has not been found in nature, then this could be an approach to getting it. A company, Affymax, has been founded specifically to take advantage of such technologies. Many other groups are using similar methods, all of which are still experimental.

DELFIA

This is a trade name for DELayed Fluorescence ImmunoAssay, marketed by Pharmacia. It is one application of a type of fluorescence detection called time-resolved fluorescence. The problem with fluorescence as a method of detection is that it is impossible to distinguish between the fluorescence of your 'marker' molecule (which you want to detect) and the fluorescence of everything else in your sample, including the sample holder (which you do not want to detect). One solution is to use a fluorescent material that has a long 'fluorescence half-life', i.e. one that goes on fluorescing for a long time after you have turned off the exciting light source. Then you look at the fluorescence after the exciting light has been turned off.

Deliberate release

This is putting something into the outside world ('the environment'), usually meaning putting a genetically engineered organism into field trials: they are often called GMOs (genetically engineered organism) or

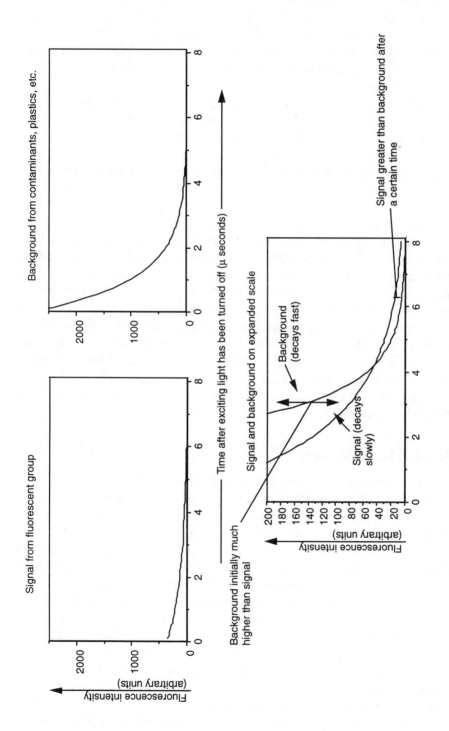

Signal from fluorescent group

Background from contaminants, plastics, etc.

Time after exciting light has been turned off (μ seconds)

Signal and background on expanded scale

Background initially much higher than signal

Background (decays fast)

Signal (decays slowly)

Signal greater than background after a certain time

Fluorescence intensity (arbitrary units)

sometimes GMMO (genetically engineered microorganism). A wide range of such trials have been suggested, and some have been done: the first was probably the trial of a genetically engineered frost-proofing bacterial strain in California in 1986. By the end of 1989 there had been 140 deliberate release experiments in the USA, with about half that number in Europe.

A wide range of social, scientific, and political pressure groups support and oppose such trials, on the basis that the organisms may be or are known to be dangerous. The biotechnology industry considers the fears to be widely exaggerated, and complains that every time they take a precaution to allay those fears, the opponents of the release experiments take it as proof that the organisms concerned really *were* dangerous after all.

Deliberate release experiments are the natural follow-on from laboratory trials, and then, for organisms involved in agricultural applications, greenhouse trials. Laboratories can include a range of barriers to stop genetically engineered organisms getting out: negative pressure rooms, sterilization procedures, and genetically engineering the organisms so that they cannot survive in the outside world. But, of necessity, none of these can apply to release into the outside world. Here, affected fields, animals, soil, etc. is kept isolated from surrounding farms, and affected material is destroyed after the trial (except for some Australian pigs, which accidentally got to market as human food in 1988).

Regulations about deliberate release vary widely (*see* **Regulation of organism release**).

Desulfurization

This is the removal of sulfur-containing materials, usually sulfides, from coal and oil. Sulfur residues in fuels end up as sulfur dioxide when the fuel is burned, causing acid rain. However, sulfur-containing fuels are often cheaper than 'clean' fuels. As a rough guide, a 'high sulfur' coal would contain 6% sulfur (mostly as pyrites), and cost $50–100/tonne less than a 'low sulfur' coal containing 1% sulfur or less. Thus there is an economic motive for removing the sulfur from oil and coal.

The same types of bacteria that are used in microbial mining can be used for desulfurization of coal. They oxidize sulfides (which are insoluble) into sulfates (which are soluble). The sulfates can then be washed away, with the bacteria. This does not work well on lump coal since the bacteria cannot get inside the lumps fast enough to be economic, but it can be effective for treating pulverized coal such as that used in electricity generating stations.

Crude oil also contains significant sulfur: between 0.1% (from the Far

East) to up to 3% (for some Middle East oils). Usually, oil has its sulfur removed by hydrodesulfurization, a physicochemical technique, but work on using bacteria to remove the sulfur shows that it has some potential.

Diagnostics

This usually means chemically based diagnostic tests for disease. Although it has its roots in traditional medicine (medical student lore has some ancient clinicians tasting urine to see if it was sweet, and therefore if the patient was diabetic), chemical diagnostics has come of age with the development of precise chemical methods. Diagnostics test for chemical changes in the body that are characteristic of specific diseases. Chemicals can be:

- simple molecules or ions, like sodium or potassium (as checks of kidney function), or glucose (an excess of which is definitive of diabetes),

- enzymes, such as creatine kinase (characteristic of heart attacks) or aspartate aminotranferase (released by damaged liver), or

- marker proteins, such as those made by viruses, bacteria, or cancers, or benign proteins such as the hCG hormone that marks pregnancy.

This third group now covers more than 40% of all diagnostic blood tests, and is performed by immunoassay. Other technologies used to measure these chemicals are:

- direct enzyme activity assay (measuring the catalytic activity of an enzyme), and

- enzyme assay of another molecule (using an enzyme to measure the presence of another molecule, e.g. cholesterol or glucose).

These assays can be colourimetric, can be linked to an enzyme electrode, or can be measured in a number of other, more complicated ways.

Biotechnology helps develop diagnostics in two ways. It finds out more about living systems in health and disease, so we can discover the 'markers' that a test subsequently detects. It can then develop reagents to do the tests, notably monoclonal antibodies. DNA probe-based tests are another rapidly growing area of diagnostics that are based on biotechnology.

Biotechnology can also take these techniques to other applications, for example veterinary or plant diagnostics. Usually these industries do not have the money to develop new diagnostic technology from scratch. However, the technology that has already been developed for the lucrative human healthcare market can be applied to less valuable testing of farm animals and plants.

The key measures of how useful a diagnostic is are its sensitivity and specificity (*see* **Assay**).

Differential display

This is a group of techniques for showing the differences between the way that genes are working in two samples, for example between two related bacteria or between normal and cancerous tissue. There are lots of technical approaches. The original differential display technology amplified fragments of cDNA from the mRNA of the target tissues, and then separated them on a polyacrylamide gel to detect the different sizes of fragments. The PCR primers have to be selected carefully so that they detect the majority of the mRNAs without giving so many bands that the gel that results is meaningless.

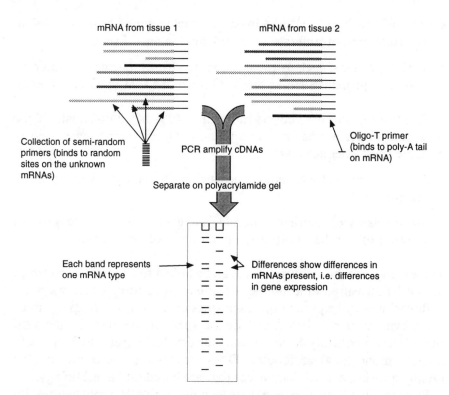

Virtual genomics. This is doing the comparison in a computer. You sequence every mRNA from two tissues. The mRNA from any one gene will be sequenced several times (depending on how much mRNA that gene is making). By comparing how many times you find the same mRNA in different tissues, you can estimate how often you find each mRNA in each tissue. Also called biology *in silico* or *in machina*, or the digital northern, this requires truly huge numbers of DNA sequencing reactions to work effectively.

Proteome. It is easier to separate a lot of proteins than a lot of mRNAs, so several groups are now looking at comparing the protein products of a cell's genes rather than the mRNAs. This general approach is called the **proteome** (contrasting with genome).

Digestor

This is a bioreactor that uses microbial growth to break down waste. The most common use is in disposal of sewage or agricultural waste, where the digestor is used to reduce the volume of biological matter to carbon dioxide or methane and compacted biomass (the cells).

Anaerobic digestors operate in the absence of oxygen, and generate methane gas, which can itself be used as a fuel. Anaerobic digestors for materials with high solids content, such as farm waste, can be very simple, and can generate useful amounts of methane. Typically 0.3–0.5 m^3 of gas, which is 50–70% methane, can be generated from 1 kg of compacted material. Most of the rest of the gas is CO_2: the amount of CO_2 that comes out of the digestor depends on the amount of liquid there is to dissolve this quite soluble gas. Digestors that handle material with more soluble organic matter in, such as sewage, need a more sophisticated design.

Anaerobic digestors can be batch fed or continuous. In the former, one batch of material is fed in, the unit sealed (to keep out air) and the digestion left to run to completion. In the latter, material is fed in continuously, and the flow must be balanced with the rate at which the bacteria in the digestor can cope with new material. A variant on this is the anaerobic contact digestor, where the living cells (active biomass) are separated from the material flowing out of the digestor and fed back into it to digest more substrate. The IRIS (Institute de Recherche de l'Industrie) digestor is of this design, but the separation system is integral to the digestor itself.

Commercial anaerobic digestors of both designs are used to dispose of sewage sludge, and to produce biogas from purpose-grown energy crops. The biogas has to be 'scrubbed' before use because sulfur compounds in the feedstock are turned to hydrogen sulfide, which is smelly and, at moderate concentrations, poisonous. Farm waste and some industrial effluents can also be reduced using digestion, but usually have to be pretreated to break down the cellulose in them. Generally these digestors operate at mesophilic temperatures (around 35°C), although ones using thermophilic organisms operating at over 60°C are also sometimes used.

Aerobic digestors are also used, notably the activated sludge process in sewage disposal. This destroys the organic content of waste more completely, and produces fewer noxious side products (sulfur is oxidized to relatively harmless sulfate, for example). However, the growing microorganisms need lots of oxygen, which has to be pumped into the digestor at considerable cost in energy. As well as the familiar trickle bed aerobic digestors, sewage waste and other organic materials are disposed in deep shaft systems. Here, oxygen or air is pumped into the bottom of a shaft at pressure, where it dissolves more rapidly than at atmospheric pressure. This allows high oxygen tension in the liquid, and so a high biomass content and fast metabolism of the waste.

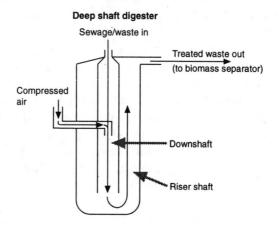

Deep shaft digester

Sewage/waste in

Treated waste out
(to biomass separator)

Compressed air

Downshaft

Riser shaft

Disulfide bond

This is a form of chemical bond in proteins much talked about by biotechnologists because of its role in stabilizing the 3D structure, and hence the normal function, of proteins. They form when two cysteine amino acids in the protein react to form one cystine residue. They link via their sulfur atoms, which therefore form a bridge of two sulfurs between distant parts of the peptide chain that fold close to each other in space. Once linked in that way, the chain is locked into that fold, since to unfold again would mean breaking the covalent bond.

Biotechnologists have used disulfides to stabilize genetically engineered proteins, by inserting new pairs of cysteines into the protein chains at points that will be next to each other when the chain folds up. They will then (the idea goes) link up to form a disulfide bridge and so hold the protein more tightly in its native conformation.

DNA amplification

This is the use of enzymes to take a piece of DNA and multiply it in a test-tube into many thousands of millions of copies. It is of enormous potential use in detecting when specific genes are there without using radioisotopes. The best known and by far the most commonly used in the **polymerase chain reaction** (PCR) system invented at Cetus. Other systems announced or being developed include the following. (I will not attempt to describe them all in detail here!)

Ligase chain reaction (LCR). Uses DNA ligase, the enzyme that joins two DNA molecules together, to link two oligonucleotides together if a target DNA is present. The linked molecules can then be templates for more molecules to link together, and so on. This technology is owned by Abbott Diagnostics, and was the first to be built into an automated test system reliable enough to be used in a routine medical diagnostic laboratory.

There is a non-cycling version of LCR, called oligonucleotide ligation assay (OLA), which links the two probes together but does not then separate the product. If one probe is linked to a solid support and the other to a fluorescent label, then the hybridization and ligation links the fluorophore to the solid support. This is a version of a 'traditional' **DNA probe** assay.

Nucleic acids sequence-based amplification (NASBA). This creates a new molecule of the DNA joined on to a promoter for RNA polymerase.

The amplification cycle occurs when the RNA polymerase copies this DNA on to RNA, which is then turned back into DNA by reverse transcriptase. The advantages are that this all occurs at one temperature, and that the RNA polymerase creates many RNA molecules from one DNA molecule, so it has the potential for being very efficient.

Branched multimer amplification. A system developed by Chiron Inc. builds a branched tree of separate nucleic acids on a single probe. At the end of each branch is a label enzyme. This is a probe amplification system: the amount of target is the same as before, but each single probe hybridized gives rise to many signals, and hence a good chance that the detection method will find it.

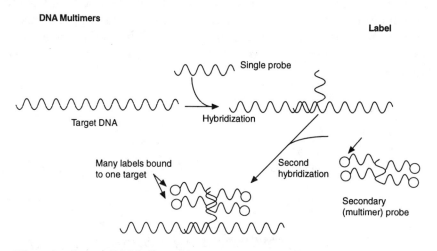

There is also an RNA-based system, the Q-beta system of Gene-Trak. The RNA of a small virus (Q-beta) is duplicated by the RNA polymerase enzyme, which the Q-beta virus carries. Add one molecule of Q-beta RNA to a tube of Q-beta replicase and the right chemicals, and the tube fills up with Q-beta RNA. The Q-beta replicase amplification system uses the enzyme to replicate RNAs that are related to the original RNA but have a probe sequence in them. Unlike the other systems mentioned above (which are target amplification systems), this is another probe amplification system.

All these systems are being developed to be used in medical diagnostics. All suffer by a greater or lesser degree from the problems of their extreme sensitivity to contamination (*see* **PCR**).

Most DNA amplification systems are qualitative: if there is target DNA there, they will amplify it. It is hard to turn them into quantitative assays, which could tell you how much DNA was there. The relationship between how much DNA there was to start with and how fast it is amplified is complicated by many factors. The branched multimer technology is the only commercially used one that is relatively straight-forward to quantify. Other amplification techniques need to use reference samples with known amounts of target in to get a comparative measure of the target levels.

DNA fingerprinting

DNA (or genetic) fingerprinting, or profiling, is a way of making a unique pattern from the DNA of an individual, which can then be used to distinguish that individual from another. Most DNA fingerprinting systems rely on DNA probes, short pieces of DNA that hybridize to the genes from an individual to identify specific pieces of DNA within the total collection of DNA. The original DNA fingerprinting probes, discovered by Prof. Alec Jeffreys, used 'minisatellite' DNA, DNA that hybridized to short runs of bases called minisatellites, which vary greatly between individuals. Because there are 50–100 of each type of mini-satellite in each person, the chances that any two individuals will have the same pattern of all the minisatellites is minuscule unless they are related. This type of repeat sequence is now usually called variable number tandem repeats (VNTRs).

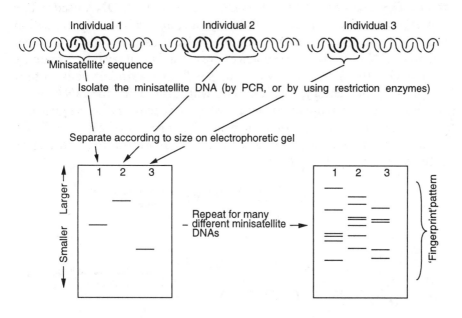

Different DNA fingerprinting systems use different probes. Some also use PCR to amplify the target DNA, so that smaller amounts can be analysed. It is also possible to create 'single locus probes'. Whereas the normal DNA fingerprinting probes create a pattern like an irregular ladder to be compared between individuals, single locus probes detect just one DNA sequence, i.e. one rung on the ladder. This makes comparison between two individuals easier.

DNA fingerprinting has been used extensively as evidence in paternity, rape, and murder cases to identify individuals. Until 1989 it was considered unassailable evidence, but since then several cases have questioned poorly collected or poorly analysed DNA fingerprinting data, starting with the case of State vs. Castro in New York, where supposedly water-tight DNA fingerprint evidence was refuted on technical grounds by the defence. This lead both to a better understanding of the strengths and weaknesses of DNA fingerprinting and to better quality control in DNA fingerprinting laboratories. DNA fingerprint evidence featured in the O. J. Simpson murder trial of 1996. The 'O. J. trial' showed that the value of DNA fingerprint evidence rests more on getting the samples carefully and agreeing what they mean than on the molecular genetic technology of actually making the fingerprint.

DNA probes

As well as being used as the genetic material to 'programme' cells to do things, DNA can be used as a reagent in its own right. DNA used in this way is almost always used as a DNA probe, also called a hybridization probe. One strand of the DNA double helix is used to bind to a target strand of DNA. If the base sequences are complementary (i.e. A matches with T, G with C), then the two strands will form a double helix. If they are not complementary, then no helix will form. Thus the DNA probe can be used as a reagent to detect when a specific DNA sequence is present among a mixture of sequences. This process of getting a DNA probe to bind to a target sequence is called hybridization, and can be used to detect DNA or RNA.

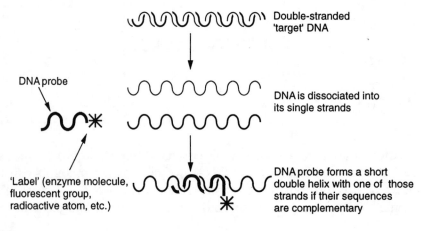

Double-stranded 'target' DNA

DNA probe

DNA is dissociated into its single strands

'Label' (enzyme molecule, fluorescent group, radioactive atom, etc.)

DNA probe forms a short double helix with one of those strands if their sequences are complementary

DNA probes have been used in genetic research for over 30 years, but only became common when DNA cloning meant that pure DNA probes derived from just one gene could be made. DNA probes are still the standard method for finding a DNA sequence among a mixture, often allied with the 'blot' technology of analysing complex mixtures of DNA molecules.

DNA probes are particularly used in medical genetics as a way of finding whether a person carries a particular gene or not (although in this application they are gradually being replaced by PCR-based techniques). They also have potential for use in detecting pathogenic bacteria, although this has not been realized as fast as was expected in the early 1980s. Probes are also the basis of **DNA fingerprinting**.

One common use of DNA probes is to find a gene similar to one you already possess. Thus if I have a clone of a gene that performs a useful function in one organism, I can use the DNA from that clone to identify the similar ('homologous') gene in a range of related organisms, which may be better suited to my practical application. (Actually, purists insist that 'homologous' has a different definition, but few technologists are purists.) This can be useful for cloning, say, heat-resistant enzymes from thermophiles if you have already cloned the gene from an organism such as *E. coli* which is easier to grow and manipulate but not so biotechnologically useful.

Traditionally, DNA probes have been made by cloning a gene and using its DNA as the probe. In the last few years oligonucleotides made on a DNA synthesizer have gained favour as probes. They react faster, so reducing assay time, they can be made more specific, so distinguishing between genes that differ by only one base, and they can be made in comparatively large amounts very cheaply. The primers necessary for such techniques as PCR could be considered as a form of probe (*see* **Oligonucleotides**).

DNA probes are starting to be integrated into 'biochips', large arrays of probes immobilized on to very small devices for use as diagnostic assays. Many DNA tests actually need you to conduct dozens or even hundreds of hybridization reactions to get all the information you need. The biochip idea allows you to do all those reactions on one small sample (because the chip is very small) and one chip (because it has all the probes on it). So far only Affymetrix has succeeded in reliably building DNA biochips, by chemically synthesizing the DNA on a glass 'chip' using a process akin to photolithography.

If you use enough DNA probes on a biochip, you can actually sequence the DNA with them, a technology called sequencing by hybridization (SBH). However, you need millions of probes to sequence a few thousand bases of DNA, which is scarcely practical. The biochip has potential, however, for resequencing genes to detect mutations in a sequence that you already roughly know.

DNA sequence patents

Biotechnology companies want to protect their intellectual property with patents, and since some of the leading edge of biology is concerned with gene sequencing, it is inevitable that researchers should try to patent DNA sequences that they have discovered. (Patenting invented DNA sequences, such as synthetic genes, has never been controversial.)

The debate was kicked off in 1991, when the NIH in the USA filed a patent claiming as their invention the sequence of 337 cDNA clones, copies of naturally occurring RNAs, sequenced by Craig Ventor. This broke new ground in claiming that a newly discovered sequence was actually an invention. They also claimed it before utility had been proven— no one knew what the sequences could be used for. Several other blanket claims on EST (Expressed Sequence Tag) sequences from NIH and the UK Medical Research Council (MRC) followed. The application was rejected in 1992, failing on the basis of novelty, utility, and obviousness. NIH and MRC dropped claims in 1994, but the debate over whether you can patent short DNA sequences of unknown utility still rages. Several companies, notably Human Genome Sciences and Incyte Pharmaceuticals, are basing their business in part on the patent-protectable value of the sequences they discover. HGS owns the patent rights to Crag Ventor's work (now at The Institute of Genomic Research, TIGR), and was still pursuing the patents in mid-1996. The US pharmaceutical company Merck is deliberately putting all the DNA sequence they discover into the public domain as a 'spoiling action' to stop anyone else from patenting it (because once it is described in public, it is no longer novel).

Meanwhile, several companies have applied for patents on genomic genes or gene variants. The first was Myriad Genetics, for the *BRCA1* gene (a variant of which predisposes about 5% of women to breast cancer). Again, there is fierce debate between those that think that this is essential and those that think it immoral.

In general, biotechnology companies believe that patenting DNA sequences is essential, and without the protection of a patent no one will put in the effort to develop products from the genes. This is especially relevant to DNA sequence, when you only need access to the Internet databases and a $10 000 DNA synthesizer to duplicate millions of dollars of science. However, nearly all religious groups and many lay ethical groups are deeply concerned that someone can patent a fundamental aspect of human nature. Scientists used to be adamantly opposed to patenting on the practical grounds that it blocked the free exchange of knowledge between researchers, but as commercial secrecy is proving far more of a block than patenting (which in fact is a method of publication), scientists as a group have backed away from this debate. Patent lawyers argue that the whole debate is based on a misunderstanding of what a patent is: it does not give someone 'ownership' of the patented invention, but rather allows them a monopoly to pursue practical applications of it for a couple of decades. In reality, practical solutions will be found to all this, and in 20 years' time the whole debate will look rather like angels and pins.

DNA sequencing

Determining the sequence of the bases in DNA (DNA sequencing) is one of the mainstays of gene cloning technology. The most common technology for this is the Sanger technique (dideoxy method, chain termination method). This uses enzymes to make a new DNA chain on the target you want to sequence, using the 'dideoxy' reagents to stop the chain randomly as it grows. A chemical technique invented by Maxam and Gilbert uses chemicals to break the DNA into fragments, but is not used much nowadays. In both cases the results of a series of reactions are analysed by polyacrylamide gel electrophoresis, to give information that can be read directly to give the sequence of the original DNA.

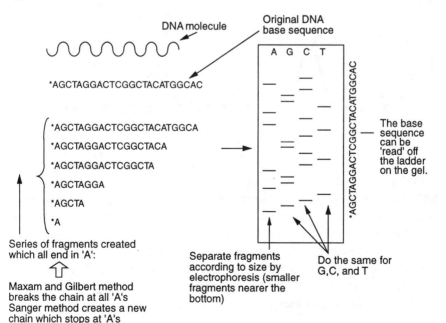

DNA molecule

Original DNA base sequence

*AGCTAGGACTCGGCTACATGGCAC

A G C T

The base sequence can be 'read' off the ladder on the gel.

*AGCTAGGACTCGGCTACATGGCA
*AGCTAGGACTCGGCTACA
*AGCTAGGACTCGGCTA
*AGCTAGGA
*AGCTA
*A

*AGCTAGGACTCGGCTACATGGCAC

Series of fragments created which all end in 'A':

⇧

Maxam and Gilbert method breaks the chain at all 'A's
Sanger method creates a new chain which stops at 'A's

Separate fragments according to size by electrophoresis (smaller fragments nearer the bottom)

Do the same for G,C, and T

DNA SEQUENCING

An associated technique is m13 cloning. m13 is a small virus that infects *E. coli*, and which is particularly convenient for making short sections of DNA to sequence. One favoured way of sequencing large pieces of DNA is to chop it up into random pieces, clone each piece by splicing it into the m13 virus, and then sequence viruses at random until you have covered all of the original DNA sequence. This is called 'shotgun' cloning or sequencing.

The human genome project, the project to sequence all three billion bases of DNA in humans, has lead to a lot of interest in building robots to sequence DNA. There are a number of machines that run DNA sequencing gels and detect the DNA in them, notably the ones made by Perkin–Elmer and Pharmacia Biotechnology. Other robots exist to perform the chemistry of DNA sequencing, but most researchers do this part of the process by hand. These machines all use fluorescence to detect the DNA, which consequently has to be labelled with a fluorescent 'tag' as part of the DNA sequencing reaction. The Perkin–Elmer/ Applied Biosystems (ABI) machine uses four different coloured dyes, so you mix the four different base reactions in one lane on the gel, resulting in greater efficiency. Sequence data are sometimes shown as traces of the fluorescence intensity detected by the machine, as shown below.

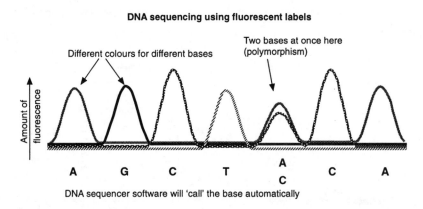

DNA sequencing using fluorescent labels

DNA sequencer software will 'call' the base automatically

DNA sequencing is starting to be used as a medical diagnostic technique as well as a research tool. Now, many genes are known, such as oncogenes, whose mutation is an important factor in disease. Because any mutation can cause disease, it is impractical to design a DNA probe to test for the disease-causing mutation. So clinicians would like to determine the whole gene sequence in their patients. At the moment this is very costly and complex, a situation the biochip might remedy (*see* **DNA probes**).

Downstream processing

This is a general term for all the things that happen in a biotechnological process after the biology, be it fermentation of a microorganism or growth of a plant. It is particularly relevant to fermentation processes, which produce a large amount of a dilute mixture of substrates, products, and microorganisms. These must be separated and the product concentrated and purified, and converted into a product that is useful.

There are three general steps to downstream processing:

- separation,

- **concentration**, and

- **purification**.

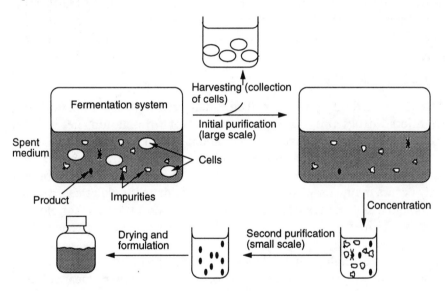

The first step separates the crude product from the microbial mass and other solid lumps, the second removes most of the water (and hence is often called dewatering), and the third takes the concentrated product and purifies it. The order of events can be different, but generally falls into this scheme.

Separating the microbial mass is necessary whether the product is inside the microorganism or outside it; the difference is that in the first case you keep the mass, in the second you throw it away. This can be done by centrifugation (expensive but guaranteed efficient), filtration methods, especially **cross-flow filtration**, or by flocculation (adding something to the microbes so that they clump together and settle out on their own). If the product is inside the organism then separation also concentrates the product: however, you then have to break open the organisms to get it (*see* **Cell Disruption**).

Some of the same methods can also be used for concentration. Simply drying large volumes of liquid is usually too expensive, so again ultrafiltration or reverse osmosis (both membrane methods that keep the product on one side of the membrane while most of the water goes through it to the other side) are popular.

Product concentration. The result of the above steps is usually a rather dilute solution of the product, which must be concentrated. This can be by reverse osmosis, adsorption methods, or extraction with another liquid.

Purification. Most biotechnology products are produced as mixtures by cells, but are required pure. Purification methods include chromatography, affinity methods, and various specific precipitation methods. If the product is produced by genetic engineering, then it may be engineered to have a molecular 'hook' which makes it easier to isolate.

Drug delivery

This is the method by which a drug is delivered to its site of action. For traditional drugs this is a different name for formulation, i.e. in what form the drug is to be given to the patient (tablet, capsule, syrup, etc.). Drugs can also be made as a prodrug, a compound that is not itself a drug but which the body metabolizes into the drug. If the metabolism only occurs in one tissue or cell type, then the drug is only released there. Although a skilled area of pharmacology, this has little biotechnological content. However, two aspects of biotechnology have needed drug delivery technology.

First, biotechnology has allowed the development of a range of new drug delivery systems, such as liposomes and other encapsulation technologies, and antibody-based drug targeting mechanisms (such as immunotoxins) that target a drug to a particular cell or tissue (*see* **Immunotoxin**). Secondly, biotechnology has created the need for new drug delivery systems to get biotechnologically derived drugs to their site of action. This is particularly acute in the case of biopharmaceuticals, protein drugs that cannot be taken by mouth since the stomach acid and gut enzymes would destroy them. Even if they could survive this digestive gauntlet, they would rarely get into the bloodstream, because the protein molecules are too big to diffuse through the intestinal walls.

The current solution is parenteral delivery (i.e. injection): this works well, and has been the way of giving patients insulin (a protein drug) for decades. However, it is invasive, expensive, and carries a continued risk of infection or tissue damage. Thus, several biotechnology companies have set out to find a better way of getting proteins into the blood. There are several approaches:

Transdermal delivery. Using a method of getting the protein across the skin without knocking visible holes in it. Methods used include iontophoresis (using electric fields to push the drug across the skin) and high pressure jets of fluid. However, the skin has evolved to resist exactly this sort of attack, so these methods are not very effective for proteins.

Oral delivery. Taking the drug by mouth, but with some other materials which help it to survive the gut. These can include protease inhibitors (to block the digestive enzymes) or carrier materials, which protect the proteins but which dissolve at the right time to make them available for absorption. Other tricks include linking the protein to something (like vitamin B_{12}) which is actively taken up from the gut, so the protein is absorbed along with it.

Nasal/pulmonary delivery. The cells lining the lungs and some of the nose (their epithelial cells) are very poor barriers to proteins compared with skin and guts, and so are potential 'weak points' for drug delivery. The nose is particularly attractive, because it has a large internal surface with lots of blood vessels, and is easily accessible.

Protein remodelling. This approach attempts to remodel the protein chemically to protect it against the difficulties of entering the body. This can be done by encapsulating it (as above), by embedding it in a variety of carrier materials such as dextrans, albumin, xanthan gums, or synthetic polymers such as polyethylene glycol (PEG), or chemically modifying it with these or other materials.

The blood–brain barrier. Many chemicals in the blood do not affect brain and spinal cord nerve cells. The nerve cells get their nutrients from surrounding cells and from the cerebrospinal fluid (CSF), which is not part of the circulatory system of the rest of the body. The cells form a barrier to penetration of drugs in the blood into the nerve cells of the brain. This can be a problem since taking drugs by mouth or even injecting them is much safer and easier than injecting them into the CSF. A substantial part of the effort on drug delivery is made directed at remodelling the drugs so that they can penetrate this blood–brain barrier.

So far, protein drug delivery systems have been much hyped, but not very effective. It is not clear whether they ever will be, or whether redesigning biopharmaceutical drugs to be chemically more robust and better suited to getting into the body will be successful before the drug delivery systems are ever launched into general use.

Drug development pathway

A substantial amount of biotechnology is concerned with developing new drugs, often biopharmaceuticals. As a consequence, some of the jargon of drug development and licensing creeps into biotechnology discussions. This entry summarizes the key points in the pathway a new drug candidate has to go through.

Preclinical research. This means all the research that goes on before you try the compound out on people, but is often taken to mean animal studies of the drug. Studies using biochemical methods, receptor screening, or cell culture assays are usually considered just 'research' since most of the drug candidates they throw up will not make it even as far as clinical trials. It also means animal testing for toxicology and ADME (adsorption, distribution, metabolism, and excretion, i.e. **pharmacodynamics**).

Phase I trials. These are the first trials in which a drug candidate is given to people. The only permission needed to do Phase I trials is the approval of the local hospital ethical board or committee (which, however, has to be convinced that there is some point in you doing the trial). The people are normal, healthy volunteers (often medical students), and the purpose of the trial is to establish the pharmacokinetics of the drug, and to find the minimum dose that will have some effect: thus the trial starts out with a very small dose and works up. Usually only a small number of people (between 10 and 20) are involved.

After Phase I the developer applies for an investigational new drug application (IND, in the USA), or the equivalent in other countries [e.g. the clinical trial exemption certificate (CTX) in Britain]. This is the regulatory hurdle necessary to go on to Phase II trials, and at this point the developer must show that they have complied with a wide range of GLP practices in their preclinical and Phase I trials (*see* **GLP/GMP**). For medical devices such as prostheses (whose development path is essentially the same), the IND is replaced by the 510(k) application in the USA.

Phase II trials. This is the first time the drug is tried on ill people. This trial is usually done at one hospital centre on a small number of patients, and looks for any evidence that the drug actually has a medical effect on the disease it is meant to treat. The drug is said to be being developed for one 'indication', i.e. one collection of symptoms or one disease type. The object of this and subsequent trials is to show that the drug has an effect for this indication. (Note that up to this point tests can be for any disease.) Again, the numbers of patients are quite small.

Phase III trials. This is where very big money is spent on drug development. The object of this phase is to see whether the drug is worth launching, because it is better than existing therapies, does not have severe side-effects, and so on. This requires hundreds or thousands of patients (all of which have to be followed in detail), usually at at least six hospital centres. The trial is done 'double blind', so that neither the people giving the drugs nor the people analysing the results know who has received the drug and who has received a placebo until the study is over. It is also usually a cross-over trial, i.e. one in which, half-way through, those who had been getting placebo get given drug and those who had been getting drug get placebo. (This helps to avoid problems arising from the differences that people show in their response to drugs.)

At the end of Phase III the drug is submitted for a new drug application (NDA, in the USA) or a product licence application (PLA in Europe). (For a medical device, the equivalent is the pre-marketing approval, PMA.) If this is approved, then the drug can be sold.

Phase IV trials. However, selling the drug does not mean that development is over. Phase IV trials, post-marketing surveillance or pharmacovigilance, then takes over to look for rare adverse reactions (adverse drug reactions, ADRs), to look for opportunities to decrease the dose (because the initial estimates, derived from Phase III studies, are often rather high), and to extend the range of indications for which the drug may be used. Extension of indications can come about because of off-label use, i.e. use of the drug by physicians for indications other than those for which the drug is licensed. There is nothing to stop people doing this, providing that they are very careful to emphasize to their patients that they are effectively doing an experiment on them. Successful experiments lead to new ideas for the use of the drug, and hence new clinical trials to see if the new indication is a suitable one for this drug.

Much of the legwork of clinical development can be contracted out to contract research organizations (CROs) (also sometimes called clinical research organizations).

Electrochemical sensors

These are types of biosensor in which a biological process is harnessed to an electrical sensor system, making a sensor. The most commonly discussed type of electrochemical sensor is the **enzyme electrode**. Other types couple a biological event to an electrical one via a range of mechanisms. Among the more common are:

Oxygen electrode-based sensors. These are sensors in which an 'oxygen electrode' (Clark electrode), a standard electrochemical cell which measures the amount of oxygen in a solution, is coated with a biological material that generates or (more usually) absorbs oxygen. When the biological coating is active, the amount of oxygen next to the electrode falls and the signal from the electrode changes. Typical coatings might be an oxidase enzyme (which consumes molecular oxygen to oxidize a particular substrate) or a whole cell (which consumes oxygen when presented with a range of substrates). This latter type of biosensor (a microbial- or cell-based biosensor) can be used to detect poisons since poisons damage the cells and hence reduce the rate at which they consume oxygen.

pH electrode-based sensors. Again, a standard electrochemical pH electrode is coated with a biological material. Many biological processes raise or lower pH, and so can be detected by a pH electrode. Examples could include hydrolysis of an ester to acid and alcohol, or, again, the metabolism of neutral pH substrates by a bacterium. Indeed, in one study, which was meant to measure the pH inside a volunteer's mouth by placing a very small pH electrode there, what the electrode ended up detecting was sugar. Bacteria grew over the electrode and, every time the subject ate sugary food, the bacteria metabolized some of it to lactic and acetic acids and the pH next to the electrode fell from 7 to 4.5.

Electroporation

This is manipulating cells by exposing them to a strong electric field. Initial studies (as one would expect) showed that while you could manipulate cells with strong electric forces, the cells never survived the treatment. However, by altering the conditions suitably, electroporation can be used to transform cells with DNA and to fuse cells. Electric field induced fusion of cells is called electrofusion.

Transformation of cells, getting DNA into them, can be achieved simply by exposing the cells to a suitable electric field while they are in a solution of the DNA. The electric field seems to modify the lipid membrane that surrounds the cells, and greatly increase the rate at which pinocytosis, a normal mechanism by which cells take up chemicals from solution, takes DNA into the cell. It is not widely used for animal or bacterial cells, where other methods have been developed which are quite reliable. However, electroporation is discussed fairly extensively when talking about getting DNA into plant protoplasts, and to a lesser extent into fungal cells. Some workers claim that electroporation or electrophoresis can even get DNA into intact plant cells (i.e. cells with their cell wall still there): the evidence for this is generally considered to be poor.

Fusing cells. This was the first application of electroporation. Protoplasts of plant cells or whole animal cells can be made to fuse by putting them next to each other and exposing them to a strong electric field. Typically, the cells are exposed to a low electric field, which induces a dipole in the cells and causes them to line up. They are then exposed to a quick pulse of very high voltage to make their membranes 'leaky' and so able to fuse. There seems to be no limit to the types of cells that may be fused together using this technology. Early studies uniformly resulted in dead cells: however, techniques have improved now so that cells may be fused to produce viable offspring using electrofusion. Applications in plant genetics include making hybrid plants and making polyploid plants.

Embryogenesis (in plant cell culture)

Embryogenesis is encouraging plant tissues to form new plants *in vitro*. The original experiments in the late 1950s showed that small pieces of carrot tissue could be grown back into whole carrot plants by culturing them in sterile conditions with the right chemicals. The new plants are usually very similar to the embryo plants that first emerge from the seeds, so this represents the cells returning to the 'genetic programme' at the start of the plant's life cycle. Although this usually only happens with the seed cells (germ cells), the embryogenesis referred to here is somatic cell embryogenesis, i.e. making embryos from cells outside the usual reproductive apparatus. Quite a number of plants occasionally generate embryos without generating seeds, so doing it in cell culture is exploiting a mechanism present in most, maybe all, plants.

The generation of embryos takes two stages: initiation and maturation. The former needs a high level of the group of plant hormones called auxins, the latter needs a lower level. Other chemicals have to be at suitable levels too. Thus the usual procedure is to take a piece of plant tissue and put it on a high-auxin medium, where the cells grow into a mass of 'callus'. This is then transferred to a maturation medium, where the callus starts to develop initial organs, ultimately growing a root and a shoot.

In plant culture circles, embryogenesis is used to describe the generation of new plants from bits of old plants. If you generate a plant from a single cell, it is organogenesis, although the techniques have many similarities. Embryogenesis is critical for plant cloning and micropropagation technologies.

Embryo technology

Embryo technology is a generic name for any manipulation of mammalian embryos, and is relevant to biotechnology in two areas. First, biotechnological methods and materials make some of the embryo technology possible. Secondly, biotechnological techniques such as transgenic technology rely on embryo technology to provide them with tools of the trade.

Embryo technology encompasses:

Cloning. This can be done in two ways, in principle: by embryo splitting or by nuclear transplant. Cloning mammals is very difficult (*see* **Clones**). Cold-blooded animals are easier to clone, and cloning as part of fish breeding is well established (*see* **Fish farming**).

In vitro *fertilization (IVF).* A widely used technique for animals and people, this means fertilizing the egg with sperm outside the body. Usually the fertilized egg is cultured outside the body for a few days before reimplantation into a female to make sure that fertilization has occurred. IVF has been the subject of intense and emotional public debate since it became possible to do it on humans in the early 1980s. A related technique is GIFT, which injects sperm directly into the fallopian tubes, and so is a half-way house to the fully external fertilization of IVF.

Artificial insemination (AI). This is simply fertilizing a female with sperm from a male without copulation. It is widely practised on people, farm animals, fish, oysters, and many plant species (although it is not usually called AI in the last case).

Gamete and embryo storage. This is the storage of eggs, sperm, or fertilized embryos outside their original source (an animal or person). Almost invariably this means freezing them at liquid nitrogen temperatures. There is also intense ethical debate about this.

Some of the other key points of debate about embryo technology are:

DNA-based genetic diagnostics. Because DNA probes can detect 'defective' genes whether they are actually doing anything or not, they can and have been used to detect whether a fertilized egg, an embryo, or a fetus is carrying an undesired gene. If it is, then it can be aborted before the gene has a chance to express itself. This is tied up with the abortion debate.

When is a fetus . . .? The ruling in the UK, following the influential and generally accepted Warnock Report, is that an embryo is not recognizably human until 14 days: before this it is classed as a 'pre-embryo'. After 14 days it becomes an embryo, and starts to acquire some rights as a human being. Sometime between then and around 15 weeks the embryo is renamed as fetus. The fetus is not usually considered capable of independent life until 24 weeks gestation (and even then only with heroic medical intervention, and a high risk of 'congenital' malformation). By 35 weeks gestation the fetus is generally capable of independent life if cared for in a specialist premature baby unit ('Special Care Baby Unit', SCBU, pronounced skiboo). Clearly, somewhere between fertilization and 35 weeks gestation the developing pre-embryo/embryo/fetus has become human. There is wide debate as to when this occurs, and whether it occurs at one time or as a continuing process.

See also **Yuk factor**.

Encapsulation

This is any method that gets something, usually an enzyme or bacterium, into a small package or capsule while it is still working (or alive). The capsule can be any size, but usually is no bigger than a few millimetres across. If it is too small to see fairly easily with the naked eye, the process is called microencapsulation.

Encapsulation is one method of immobilizing cells for use in a bioreactor. Encapsulating agents can be anything that will form a shell around something, but usually they are polysaccharides such as alginate or agar, because they are inert, allow nutrients and oxygen to diffuse into and out of the sphere readily, and are easy to convert from a gel (solid) to a sol (liquid) or solution form by altering the temperature or the concentration of ions such as calcium. Proteins such as collagen (gelatin) are also used. Cell encapsulation is also used to try to develop **artificial tissues**. Enzymes may also be encapsulated, although they are more usually immobilized on the surface of polymer particles (*see* **Enzyme immobilization**).

Drugs are often encapsulated to help their survival or release in the patient. A variety of sustained-release cold cure medicines, which come as little particles inside a capsule, are actually encapsulated drugs: each particle contains a shell of a slowly dissolving material around a core of solid drug powder. Only after the shell has dissolved in the intestines can the drug get out to the body. By having a mixture of shell thicknesses the pharmacologist arranges for the drug to be released over a period of time. Similar approaches have been tried for biotechnological drugs, although not always very successfully. Encapsulation of drugs is also a methods of protecting them from, say, acid in the stomach so that they can be taken by mouth instead of having to be injected. This has been something of a holy grail of drug delivery technology for biopharmaceuticals, but has not been successful to date.

Environmental biotechnology

Environmental biotechnology is a general term covering any biotechnological product or process that can be considered to be helpful to 'the environment'. Usually, this means control, reduction, or disposal of waste, removal of chemical pollutants, or reduction in power use, especially in industry. Because of the high political profile of 'the environment', a number of diverse biotechnology activities have been included under 'environmental biotechnology'.

Biotechnology is well placed to address some ecological and environmental concerns. As opposed to traditional heavy industry, biotechnology is likely to use potentially renewable resources, inherently low power processes, materials that are unlikely to be dangerous, and to produce products that can be justifiably labelled as 'natural'.

The most commonly discussed topics in environmental biotechnology are:

- **bioremediation**: cleaning up contaminated soil using biological processes.

- **soil amelioration**: improving soil quality through manipulation of its microflora.

- developing biodegradable replacements for plastics, and particularly developing biotechnological ways of making them (*see* **Biodegradable materials**).

- **waste disposal**: developing bacterial methods for disposing of waste, or at least of disposing of the biodegradable part of it more rapidly.

- creating alternative energy sources: specifically **biofuels**, **biogas**, and **solar energy** methods. These are much more speculative, and rarely are economic alternatives to oil, coal, or nuclear power.

Biotechnology is also providing tools for more conventional environmental agendas. The best-developed example is using immunoassay technology for environmental testing. Chemicals are usually detected in environmental samples by HPLC or GC, which require a central laboratory facility and expensive apparatus. The same chemicals could be detected by competitive immunoassay, so that the test could be done 'in the field'. Applications so far have been limited though, because the tests are expensive to develop, and the regulations usually demand that conventional testing methods are used.

Enzyme Commission (EC) number

All enzymes are given a systematic name and number which identifies them in the technical literature. (They may also have a 'common' name, like trypsin or rennin.) These names are given by the Enzyme Commission. The name and number are systematic descriptions of what the enzyme does. The number is a four-digit number. The first digit classifies the enzyme into one of six broad groups.

Number	Class
1	Oxidoreductases (transfer H atoms or electrons)
2	Transferases (transfer of small groups between molecules)
3	Hydrolases
4	Lyases (addition to double bonds)
5	Isomerases
6	Ligases (formation of bonds between C and another atom, using ATP as an energy source)

Each of the groups is subdivided into subgroups, each subgroup into sub-subgroups, and the last number is specific for the enzyme. The systematic name describes the reaction catalysed. Thus creatine kinase is EC 2.7.3.2 (2 because it transfers a group from ATP to creatine, 2.7 because the group is phosphate, 2.7.3 means the subgroup that transfers the phosphate to a nitrogen atom, and the last 2 is because this particular enzyme was the second classified in this group). Note that the dots are essential since some classes of enzyme have more than 10 members. The systematic name is ATP:creatine phosphotransferase: the enzyme that transfers a phosphate group from ATP to creatine.

Enzyme electrode

A type of biosensor, in which an enzyme is immobilized on to the surface of an electrode. When the enzyme catalyses its reaction electrons are transferred from the reactant to the electrode, and so a current is generated. (This is distinct from other types of electrochemical biosensors, where the enzyme generates a distinct chemical product, for example an acid, which is then detected by a separate electrode system.)

There are two types of enzyme electrode:

Ampometric. Here, the current generated by the reaction is measured, and the electrode is kept as near zero voltage as is practical. When the enzyme catalyses its reaction, electrons flow into the electrode, and so a current flows.

Potentiometric. Here, the voltage generated by the reaction is measured, and the electrode is held at a voltage that counteracts the voltage created by the enzyme's tendency to 'push' electrons into it. This may be done by actively adjusting the voltage or by not connecting the electrode to anything (as is the case in **ISFET** devices). The device's output is the voltage necessary to prevent any current flow through the electrode.

Usually, enzymes transfer their electrons inefficiently to the electrode, so a mediator compound is coated on to the electrode to help the transfer. The favoured mediators are ferrocenes, because they can easily carry a single electron at the electrode potential suitable for enzyme oxidations and reductions. A range of other organic chemicals have been considered, and the 'organic metals', i.e. organic compounds that conduct electricity, hold promise as electrode materials. Ionomers are also used. These are polymers that are not charged (and so stick to the electrode), but which have a charged group as a side chain.

The enzyme has to be immobilized on the electrode in some way. Common methods include:

Physical adsorption. The enzyme is encouraged to stick to the enzyme surface. Many proteins will stick quite avidly to some surfaces, held there by small patches of electrostatic charge or because they sit in a hydrophobic 'pocket'. This approach is simple, but the enzyme can simply leach off again unless it is held very tightly (which is usually not the case).

Chemical cross-linking. The enzyme is chemically linked on to the electrode surface. Rarely do the chemistries of the enzyme and the electrode match to allow this route.

Immobilization in a gel. The enzyme is mixed with a polymer such as agarose or polyacrylamide, and then chemically cross-linked to the gel to form a solid capsule around the electrode.

Capture behind a membrane. Here the electrode is inside a small sack which is permeable to the analyte but not to the enzyme. The enzyme is inside the sack.

A vast number of enzyme electrode systems have been developed in laboratories, and the early 1980s saw a boom in interest in their application. However, they nearly all proved to be hopelessly impractical for commercialization. The major exception was the glucose biosensor for diabetic monitoring: a few other medical sensors are now being commercialized.

Enzyme immobilization

This is linking an enzyme to a solid support material, typically polymeric beads in a column or bed. Typical immobilizing materials could be silica or carbohydrate polymers such as Sepharose, although Altus Biologics has cross-linked enzymes in pure crystals to stabilize them (to form cross-linked enzyme crystals; CLECs).

Immobilization has two general advantages:

(1) It enables you to recover the enzyme quickly and easily from the

reaction mixture, by separating the beads out. This is useful both for purifying the reaction mixture and for recycling valuable enzyme.

(2) It alters the chemical properties of the enzyme. Many enzymes are more stable when bound to a surface, which appears to 'lock' their conformation into an active form, so slowing down the inevitable reduction in enzyme activity that occurs with use. (Although in theory an enzyme is a perfect catalyst, emerging unchanged from the reaction it speeds up, in practice, a wide range of side reactions slowly break up the protein's structure, rendering it useless as a catalyst. Sensitive enzymes can be inactivated in a few minutes. Robust industrial ones usually take hours or days.)

Immobilization is useful in industrial biotransformation processes. The synthesis of the anticancer drug Camptosar, for example, uses a lipase to resolve a chiral intermediate (*see* **Chiral synthesis**): the process is only economic if the lipase is immobilized, to give it added stability and efficacy.

Enzymes are commonly immobilized in biosensor and bioassay applications. Here the main reason is to hold them in one place so that they do not wash off the assay strip.

Enzyme mechanisms

Since the use of enzymes is one of the most commercially important areas of biotechnology, understanding how they work is an important part of the research underpinning the technology. Indeed, one of the reasons that enzymes are so widely used is that their mechanism of action has been studied for most of this century, and the science of enzymology is correspondingly mature (as opposed to, say, the relatively new science of molecular genetics).

Specific aspects of how enzymes work and how this may be improved for a particular application are dealt with elsewhere. The fundamental research involved is beyond the subject (and scope) of this book. In general, modification aims to improve the enzyme's specificity (i.e. what reactions it catalyses), its kinetics (how well it catalyses them), or its stability (how long it lasts). However, there are several lines of research that are relatively new technologies to use in enzymology:

Chemical modification. Changing an amino acid in the protein into another by chemically reacting it. This usually results in a change in the enzyme's activity, and if it does the change is almost always 'for the worse', i.e. it reduces the enzyme's catalytic effect, specificity, or both. Sometimes the change can result in a commercially more useful enzyme, in which case the modified protein is used commercially. However the protein is changed, the result is usually interesting to the enzymologist.

Site-directed mutagenesis. Changing an amino acid into another amino acid by genetic modification. This is more flexible than chemical changes, because an amino acid, perhaps identified from protein sequencing work or X-ray crystallography, can be changed specifically to another, closely related (or totally unrelated) amino acid.

See also **Kinetics**.

Enzyme production by fermentation

Industrial enzymes can be made by extraction from a naturally occurring source, often part of a large animal or plant, or by production from a microorganism in fermentation. The former requires less equipment, but is more prone to interruption of supply by seasonal variation, the vagaries of climate, international trade, and (in extreme cases) war. Fermentation has the potential to provide a more uniform and reliable source of material.

The enzymes that account for the majority of production are essentially commodity products. Thus a substantial part of the cost of their production is the raw materials and power necessary to produce them. (This differs from enzymes used for research, such as restriction enzymes, which are produced in comparatively tiny amounts and whose production cost is dominated by the skilled labour necessary to make them.) Thus a successful fermentation process must use low-cost feed materials, an organism that does not require excessive heating or cooling, and one that makes a lot of the enzyme.

Typical feedstocks are hydrolysed starch, molasses, whey, and cereals for carbon, and soy flour, fish meal, blood, and cotton-seed meal for nitrogen. For high value enzymes (for example, for use as drugs), some of these are inappropriate since they contain insoluble dirt, which will have to be rigorously removed from the final product. The fermentation conditions that must be monitored to optimize enzyme production include pH, oxygen, CO_2, aeration, temperature, agitation, and, since some enzymes are denatured on surfaces or may concentrate at them, foaming. In addition, the production of many enzymes by bacteria is induced and repressed by specific chemicals (*see* **Induction**). Inducers must be present and repressors removed from the fermentation if the yield is to be satisfactory.

Many industrial enzymes are sold as fairly crude preparations, with a mixture of proteins in them. These have been prepared by separating the cells from the fermentation broth, and then partially purifying the protein from the liquor by precipitation, ultrafiltration, or similar technique.

Enzymes

The core of traditional biotechnology, and a key feature of the new biotechnology of gene cloning, is the use of enzymes. For practical purposes these can be considered to be catalytic proteins, although recent work has shown that RNA can act as an enzyme.

Enzymes are prepared from a huge variety of organisms, from viruses to whales. In general, they may be extracted from some organism that already produces the enzyme, extracted from a microorganism that is cultured under conditions where it produces the enzyme, or made from an organism that has been genetically engineered to produce the enzyme.

Enzymes are so widespread in biotechnology that they crop up in many entries in this book. Specific classes of enzymes covered are **glycosidases** and **glucose isomerase and invertase, proteases**, and **lipases**. Enzymes also crop up in **biotransformation, protein engineering, enzyme production by fermentation**, and **expression**, as well as, *en passant*, in many other entries.

The value of enzymes to the biotechnology industry can be estimated from the following table.

Industrial enzyme	Market value ($ millions)	Note
Pharmaceutical proteins	100	1
Detergents (proteases and lipases)	70	2
Dairy industry (mostly rennin)	50	
Research (a wide variety of enzymes)	42	
Starch processing	31	3
Diagnostic (a wide variety of enzymes)	16	
Textile processing	12	4
Drinks industry	11	
Baking (*see* **Glycosidase**)	4.5	5
Biotransformation	4.5	
Others	5	
Total (for 1990)	400	

Notes:
1. This includes enzymes like TPA (*see* **Blood products**).
2. Protease detergents are the traditional enzymes, although the fat-dissolving lipases are beginning to be used in retail as well as industrial detergents now.
3. *See* **Glucose isomerase and invertase, Polysaccharide processing, Glycosidases**.
4. Proteases and cellulases: cellulases and amylases used for whitening and softening cotton (for example, for producing 'stone-washed' jeans).
5. Variety of glycosidases for improving dough quality.

Enzyme stabilization using antibodies

This is a method of stabilizing proteins, usually enzymes, by binding antibodies to them. Some enzymes may be stabilized 200-fold by complexing them with an antibody, i.e. the 'half-life' of their enzymatic activity can be increased (from 5 minutes to 16 hours in the case of alpha-amylase). The antibodies have to be selected so that they do not block the active site of the enzyme, otherwise the protein is stabilized but becomes inactive as a catalyst: thus monoclonal antibodies, which bind to defined bits of the protein's surface, are usually used.

The process works because the antibodies bind to the active structure of the enzyme. If the enzyme now 'tries' to unfold into an inactive structure, it must not only overcome its own binding energy but also throw off all the bound antibodies. This needs more energy to do, and so is a correspondingly slower process.

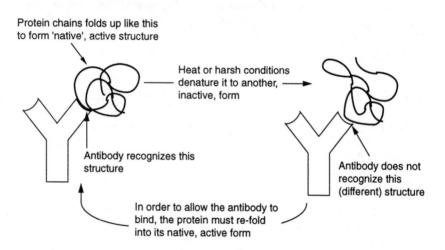

Protein chains folds up like this to form 'native', active structure

Heat or harsh conditions denature it to another, inactive, form

Antibody recognizes this structure

Antibody does not recognize this (different) structure

In order to allow the antibody to bind, the protein must re-fold into its native, active form

Antibody stabilization is used routinely to stabilize enzymes used in medical diagnostic assays. The antibodies are too expensive for it to be routine for enzymes used in large-scale processes.

Epigenetic

This is inheritance of biological traits outside the normal genetic mechanisms of DNA and genes. It is usually applied to biochemical aspects of a cell's metabolism. The first epigenetic traits known at a cellular level were the patterns of cillia (short 'tentacles') on the surface of pond-dwelling protozoa, which are passed down from mother to daughter organism even though they are not coded in any DNA: the pattern in one cell is copied on to its two daughter cells. Similar copying mechanisms also copy patterns from human mother cells to daughter cells, a form of cellular inheritance called epigenesis.

The best-characterized epigenetic phenomenon is DNA methylation. A small fraction of the cytosine bases in DNA are altered by adding a methyl group on to them, using an enzyme called DNA methylase. This affects whether the genes containing those methyl cytosines are active or not. When DNA is made during cell growth, the enzyme has to 'decide' whether a particular cytosine should be methylated or not. It does so by methylating all the cytosines in the double helix that are next to another methylated cytosine. Thus the pattern of methylation is copied from mother to daughter DNA. (This only works because the methylated cytosines are always in CG pairs, called CpG dinucleotides.) Thus, although the pattern of methylation is not 'coded' in the DNA, it is nevertheless inherited from one cell to the next.

Other aspects of the cell's structure may also be epigenetic rather than genetic. Known examples include the pattern of chromosome activity in women (who turn off one of their X chromosomes early in fetal life: which one their cells chose appears to be random, but all their descendent cells stick with the decision) and the pattern of some cell structures such as the centriole. The problem for research is that analysis of epigenetic processes is very hard, while analysis of true genes is only hard, so scientists naturally focus on the latter. However, it is becoming increasingly obvious that epigenetic phenomena are important in biology, maybe as important as genetic ones, despite the genome project.

Essential nutrients

There are a range of nutrients that mammals need in their diet: these are called essential nutrients. As well as minerals, mammals need varying combinations of essential amino acids, fats, and vitamins. The particular combination depends on the species.

Essential amino acids. These are essential components of our bodies, but we cannot make them, so we must get them from food. They are important to biotechnology because many microorganisms can make them, and so can be used to produce them as food supplements. Other mammals, such as farm animals and pets, share much the same list, and much of the food supplement amino acids made are for farm animals rather than people.

The essential amino acids for humans are leucine, isoleucine, valine, threonine, methionine, phenylalanine, tryptophan, and lysine. For children, histidine is also considered to be an essential amino acid. Food protein with a good supply of all these is called 'first class protein'. Many single crops are second class protein (they are notably deficient in some essential amino acids). One aim of crop genetic engineering is to boost the supply of specific amino acids in common crops like maize and cassava to make them into sources of first class protein.

People also need particular *fatty acids*, primarily to make fatty acid esters involved in cellular signalling. These are polyunsaturated fatty acids, i.e. fatty acids with several double carbon–carbon bonds in their backbone chains. These are increasingly produced from algae.

The third category of essential food additive is *vitamins*, which are a wide range of chemicals that we happen to need for our metabolism and are not able to make. Some vitamin requirements are common amongst mammals, some are rare; thus vitamin C is only an essential dietary requirement of humans, chimpanzees, gorillas, fruit bats, and guinea pigs, as far as is known. The vitamins sold as 'health supplements' are usually produced by fermentation.

Essential oils have nothing to do with diet. These are oils that have an 'essence', usually an aroma component. They are usually obtained from plants, and used as flavour and scent materials. Examples are mint oils (from mint leaves, containing a high concentration of menthol), lemon oil (from lemon rind, containing limonene), and so on.

Ethanol

More ethanol is produced by biotechnology than any other chemical. Yeast fermentation of natural carbohydrate sources such as grain and grape has deep historical roots, and is nowadays a highly technological industry. Aspects of its use are discussed in separate entries on **brewing** and **biofuels**.

Because ethanol is produced in such large amounts, a lot of effort has gone into optimizing fermentations to make it. Like many fermentation products, ethanol stops its own production once a certain concentration has been reached, because it kills the cells that are producing it. To get higher yields of ethanol, there is a great deal of effort going into developing alcohol-tolerant yeast, which can grow in the presence of high ethanol concentrations. However, such yeast often do not grow well in the presence of low ethanol, so they must be used in a two-stage fermentation process.

Another route to gaining high yields of ethanol, and other volatile metabolic products, is the experimental vacuum fermentation. The fermentor is held at reduced pressure, so that the volatile product 'boils' off into the head space and leaves a low concentration in the liquor.

Expression compartment (inclusion bodies)

Getting a protein made in a recombinant cell is relatively straightforward since a wide range of expression vectors exist to be used to clone the relevant gene. However, the protein is often produced in a form that is not useful to the genetic engineer. This is often a feature of where in the cell the protein is made.

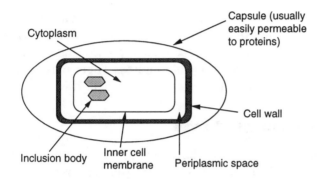

Inclusion bodies. These are condensed particles of protein that are formed inside bacteria and (to a lesser extent) eukaryotic cells when the cells are forced to make large amounts of protein. Inclusion bodies are the result of 'off pathway' aggregation or folding: proteins not folding according to the usual regular path, resulting in an ordered, soluble product, but following some other route resulting in an inactive product. The proteins are often cross-linked and/or denatured, making them useless for their intended purpose. Inclusion bodies were the bane of early recombinant DNA production methods, but the skills of manipulating bacterial physiology (i.e. how they grow) to avoid inclusion bodies are now much better developed.

Getting your protein as an inclusion body is not a catastrophe. Such proteins can be refolded by dissolving them in a detergent or 'chaotropic agent' solution, and then gradually removing the detergent by dialysis. Using the right buffers, this allows the protein to refold into its proper form. However, this is a bit of a black art, and does not always work.

Cytoplasmic expression. If you do not tell a protein where to go in a cell, it stays in the cytoplasm (the space inside the cell wall). Most proteins are expressed in the cytoplasm; however, this is where inclusion bodies form, and also where an efficient mechanism for breaking down aberrant proteins exists. As far as the cell is concerned your genetically engineered protein is aberrant, so it can be broken down very rapidly in the cytoplasm. (This is especially true for very small proteins or peptides: large ones tend to form inclusion bodies instead.)

Periplasmic space. This is the space between the cell membrane and the outer cell wall in bacteria. Many proteins that are secreted (see 'secretion' below) end up here. This has the advantage that it gets them out of the cytoplasm, but does not release them free into the medium (so that they can be harvested simply by collecting the cells). However, the periplasmic space has its own set of digestive enzymes, which can break down proteins, aimed at completely different types of protein molecule to the cytoplasmic ones.

Secretion. Some proteins can be exported from the cell entirely, to remain free in solution in the culture medium. This makes it much harder to collect them, but if they can be exported successfully they rarely precipitate as inclusion body-like lumps. There may, however, still be proteases in the solution that can break them down.

Expression systems

Expression systems are combinations of vector and host that provide a 'genetic context' which makes the gene function in the host cell: usually, this means make a protein at high levels. (Usually a cloned gene will be inert, i.e. it will not perform its usual function in the host cell since it is outside its usual genetic context.) The proteins made in such systems are called foreign or heterologous' proteins, because they are not native to the cell in which they are being made. Thus, this is 'heterologous gene expression'.

Because making lots of foreign proteins is often lethal to the host cell, there are several variations on the theme of the expression vector, which allow the level of the protein made from a cloned gene to be increased:

Inducible systems. Here the expression of the cloned gene is 'turned on' by an inducer, so that the cells can be grown up in bulk and then induced to make the protein (*see* **Induction**).

Amplification systems. Also called high copy-number vectors. Usually the plasmids and viruses from which vectors are made are present in only a few copies per cell. High copy-number vectors are present in hundreds of copies. More genes lead to more protein produced. The increase in the number of genes can be made conditional on, say, a rise in temperature, so that the host cells are grown at one temperature and then fill up with DNA and the target protein at another temperature.

Runaway replication plasmids. This is a logical extension of the amplification system. When the temperature is increased, the normal system that controls how much plasmid DNA is present is destroyed and the bacterium continues to make plasmid DNA until it runs out of precursors to make it from. This leads to a cell full of plasmid, and hence, in principle, the product of its genes.

Secretion vectors. These are vectors that allow the protein product of the cloned gene to be secreted from the cell. This can be very helpful for purification since all the other proteins in the host cell are removed with the cell itself, but it does not always work because the target protein is broken down in solution, is not stable, or is incapable of being secreted effectively (*see* **Secretion**).

Even with a host cell and vector that are consistent with the gene you want to express, actually obtaining high levels of expression can be difficult. Obtaining a fraction of a per cent of the cellular protein as the product you want is a research target and easy to achieve. However, getting 10% or more of the cell's protein as the product can be difficult. Bacterial cells usually do not handle large amounts of foreign proteins well, forming inclusion bodies (aggregated lumps of denatured protein; *see* **Expression compartment**) and breaking the protein down. Bacterial cells also do not add secondary modifications to eukaryotic proteins in the same way as eukaryotes would. Bacterial cells also produce endotoxins, which must be rigorously removed from medical products. For all these reasons, biotechnologists are switching to yeast, mammalian, or insect cells to make recombinant proteins commercially.

Obtaining the best performance from an expression system also requires considerable knowledge of how the host cell's internal machinery (its physiology) works. But there are no general rules about which expression system is best to make which protein.

A novel approach to expressing foreign proteins is to use transgenic animals. Here, rather than a bacterium or a yeast, a mammal is the carrier for the foreign gene, which is spliced on to the front of the gene for lactalbumin, a major component of milk. The animal expresses the gene construct in the mammary glands, and the recombinant protein is secreted, relatively pure, into the milk. GenPharm and Pharmaceutical Proteins Ltd are companies specializing in producing pharmaceutical proteins in this way (*see* **Pharming**).

Extremophiles

These are organisms that live in 'extreme' conditions, i.e. conditions substantially different from Western laboratories. They include:

Thermophiles. Organisms that live at high temperatures (*see* **Thermophiles**).

Psychrophiles. Organisms that live at very cold temperatures. They can be useful sources of enzymes that are active under cold conditions: most enzymes act only very slowly at temperatures much below the normal living temperature of the organism they came from. They can also be sources of heat-unstable enzymes, enzymes that are destroyed by moderate heating to 40°C. This can be useful if you want to 'turn off' an enzyme without affecting other components of a complicated process. Another interesting target for biotechnology is fish that can live in arctic seas at temperatures that should freeze them solid. They are sources of 'antifreeze proteins', proteins that prevent their blood freezing.

Halophiles. Organisms that can live in concentrated salt solutions. Halophilic bacteria are interesting because of their metabolic curiosities, including the purple membrane protein of *Halobacterium halobium*. Biotechnology is also interested in engineering plants to be halophiles, so that they can be watered with sea water.

Enzymes from extremophiles are potentially valuable in their own right because of their stability in industrial processes. They also have value for what they teach us about enzyme stability, and hence how we might make other enzymes more stable. They are sometimes called 'extremozymes'.

Fermentation

A central discipline of traditional biotechnology, and a key part of all biotechnology today, is fermentation. This is the growth of microorganisms, and encompasses a wide range of technologies, which are dealt with in different entries in this book. Fermentation includes:

- bioreactor design: the design of the container in which fermentation is to take place.

- substrates: what the microorganisms are to grow on.

- supports: whether the organism is to be on solid supports or in suspension.
  There is also a wide range of options in **downstream processing**.

Fermentation processes

Strictly speaking, fermentation is microorganism metabolism under anaerobic conditions on a carbon substrate. However, it has been extended to mean growing microbes in liquid under any conditions. Culturing bacterial cells in small amounts on a Petri dish or analogous small-scale mammalian cell culture is called incubation, and takes place (unsurprisingly) in an incubator.

There are three general ways in which fermentations are done, each with a variety of associated terms. In all cases there are come common terms in bacterial growth, such as the bacterial doubling time (the time needed to double the number of bacteria there): these are discussed further in the entry on **cell growth**.

Common terms. For all fermentation processes, the first thing that happens is that the fermenter (or bioreactor) is sterilized. This can be done with steam, chemicals, washing, or some combination of these. The fermentation is then started with an inoculum, a small, actively growing sample of the organism to be cultured. Fermentation then proceeds according to one of the schemes below.

Batch fermentation. Here the reactor is filled with a sterile nutrient substrate and inoculated with the microorganism. The culture is allowed to grow until no more of the product is being made, when the reactor is 'harvested' and cleaned out for another run. The culture goes through a lag phase (when the organisms adapt to their surroundings), exponential growth (when they grow in numbers), stationary phase (when they stop growing), and death phase. Depending on what the product is, the 'useful' part of the growth cycle can be any one of these four stages, although it is usually the growth or stationary phases.

Fed batch fermentation. Here the batch culture is fed a batch of nutrients before it gets to the stationary phase, so that it never runs out of nutrients. At the same time some of the fermentation is removed and taken off for processing.

Continuous culture. This is the logical extension of fed batch fermentation. The fermentor is fed continuously with nutrient and the culture medium removed continuously. This has some advantages over fed batch systems in that the culture conditions are always the same, but it is also harder to control. This is essentially a large-scale chemostat.

Cascade fermentation. Here the fermenting 'liquor' is passed through a series of ferments, so that more and more of the product builds up each time. Each step can then be optimized for a specific condition. A typical example would be in brewing, where the beer would be fermented in several stages to increase the alcohol content, each stage using a yeast adapted to working in alcohol of that concentration. At the end of each stage the yeast is separated from the beer and used again in that stage, while the beer goes on to the next stage.

Fermentations may also be classified according to when the product is made:

- Type I fermentation: product is made from primary metabolism.

- Type II fermentation: product is made from secondary metabolism at the same time as primary metabolism is going on (i.e. when the cells are growing).

- Type III: product is made by secondary metabolism at a different time from the primary metabolism (i.e. during stationary or death phases of the culture).

Lastly, fermentations may be classified according to how the culture is kept 'clean'.

Aseptic/sterile fermentation. All other organisms are excluded by the biotechnologist. This is by far the most common class.

Consortium fermentations. Here a group of organisms are growing together, rather than just one organism. For this to work, the organisms must be dependent on one another, otherwise one will outgrow the others and dominate the culture.

Protected fermentations. Here the culture is not aseptic, but is performed under conditions under which only one type of organism will grow. Thus fermentations at extremely high temperatures, extremes of pH, or with very hard to metabolize substrates will tend to be able to support only the organism that the biotechnologist is after, thus removing the problem of having to keep contaminants out.

Fermentation substrates

Many materials are used as food for growing microorganisms. These are referred to as the substrate. The substrate and the trace materials needed, together with chemicals added to make the fermentation easier (such as anti-foam agents to stop froth forming) make the culture medium.

The substrates can be divided according to the different essentials for life that they provide: a source of carbon, nitrogen, and (in the case of aerobic fermentation) oxygen. Usually carbon substrates cost the most, because you need the most of them. Among the common carbon substrates for large-scale fermentation are:

Molasses. A side product of sugar refining that contains most of the material from sugar beet or sugar cane that is not sugar; molasses is one of the cheapest substrates available. There are several varieties with different properties, components and costs:

- beet molasses: the molasses left from processing beet.

- black strap molasses: from an early stage in cane sugar processing.

- cane refinery molasses: from a later (more sugar rich) stage in processing.

- high test molasses: actually, concentrated cane juice treated with invertase.

Malt extract. Made from malted barley by soaking it in water.

Starch and dextrins. Polysaccharides often made from cheap crops like potatoes.

Cellulose. The world produces about 100 billion tonnes of cellulose a year, so it is a potential raw material for large-scale fermentation. But only a few organisms can degrade it.

Sulfite liquor, a by-product of paper pulp production, which contains much of the fermentable sugars from wood without substantial cellulose.

Whey. A side product of dairy processes, it is cheap but expensive to store or transport.

Methanol. A very cheap chemical from the oil industry, but it contains no nitrogen. Only a restricted range of organisms can grow on methanol. Similarly ethanol ('alcohol') can be used, but more usually ethanol is the product of a fermentation.

Oil, gas. Some organisms can use natural gas or some components of crude or refined oil as carbon substrates. However, their commercial use depends critically on the price of oil.

Nitrogen substrates include:

Ammonia. A very smelly gas produced as a bulk commodity for the chemical industry. Most organisms can use ammonia. Sometimes it is converted into ammonium salts or into urea for ease of handling.

Corn steep liquor. The liquid generated in the early stages of wet-milling maize to produce starch. Maize grains are submerged in water to soften them before milling. Low molecular weights sugars and peptides accumulate in the water.

Soy protein. The protein left over when you have taken the oil out of soybeans.

Yeast extracts. Made from waste yeast from industrial fermentations, they have everything necessary for microbial growth.

Peptones, casein hydrolysates. These are partially digested meat or milk proteins, respectively. The proteins used are usually waste material from the food industry: nevertheless, this can still be an expensive source of nitrogen.

Fingerprinting

In biotechnological terms, fingerprinting means making a characteristic chemical profile of something in order to identify it. There are many variants.

DNA fingerprinting. Generating a profile of DNA fragment (*see* **DNA fingerprinting**). Note that DNA fingerprinting can be applied to animals, plants, or bacteria in addition to its widely known application to potential rapists and murderers.

Protein fingerprinting. This generates a pattern of the proteins in a cell or organism, which provide a completely characteristic 'fingerprint' of that cell at that time. The technique usually used is 2D gel electrophoresis. This gives a complicated pattern of spots, which can be compared between cells. However, the comparison is difficult, and the pattern you get will depend on the cell's metabolic state as well as its origins, so this type of fingerprinting is now used only to compare the same cell at different times (e.g. with and without a drug added to its culture medium), and DNA fingerprinting is used to compare different cells.

Peptide fingerprinting. This can be one of two things. The first is use of peptides (i.e. the degradation products of proteins) to characterize a cell or organism, as above. The second is characterization of a single protein using its component peptides. Lots of peptide fragments can be generated from a protein, usually by partially digesting it with an enzyme. The peptides are easier to characterize and sequence than the whole protein. The pattern of peptides is used to identify the parent protein.

Chemical fingerprinting. Here, the low molecular weight chemicals in a cell are analysed, usually by GC or HPLC but increasingly by mass spectrometry, and the resulting pattern used to identify a cell or microorganism. Since some metabolites are very characteristic of groups of organisms (penicillins of some moulds, different chlorophyll types of different algae, for example), this can be a useful way of telling which organism is which.

Footprinting. This is a distantly related idea. This is primarily applied to DNA, although the same logic can be applied to other molecules. It is a method of finding how two molecules stick together. In the case of DNA, a protein is bound to a labelled piece of DNA, and then the DNA is broken down, by enzymes or (as in the diagram) by chemical attack. This produces a 'ladder' of fragments of all sizes. Where the DNA is protected by the bound protein it is degraded less, and so the 'ladder' appears fainter. Footprinting is a common technique for homing in on where the proteins that regulate gene activity actually bind to the DNA.

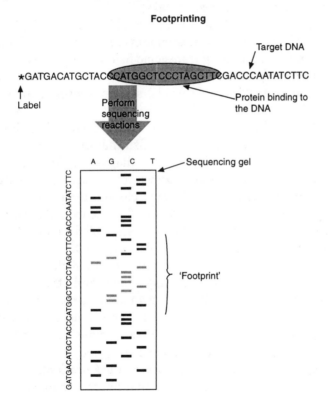

Fish farming

Fish farming is raising fish in tanks or holding nets, as opposed to hunting them at sea. Traditionally, fish farming has been a fairly undemanding technology: fish farming in some form was carried out by the Romans, and is the reason that many villages in England have a 'village pond'. However, biotechnology techniques are developing more effective technologies to improve yield, and particularly to improve the fish's genetics. The activity falls into three areas.

Sex manipulation. Fish's sex can be manipulated quite readily by exposing them to hormones: many fish will produce genetically male but physically female fish if they are exposed to oestrogen during development. This makes it much easier to develop inbred lines of fish. The genetically male females produce fertile eggs, some of which contain Y chromosomes instead of X ones; when fertilized by Y chromosome sperm, they produce 'supermale' YY fish. (This has been done for *Tilapia*, a popular fish in Asian fish farms, and the channel catfish in the USA.) When crossed with normal females the YY males produce only male fish, which grow faster and put more metabolic energy into meat (as opposed to eggs).

Transgenic fish. Fish cloning is well established, so a single 'desirable' fish can be cloned, some of its descendants sex-reversed, and a whole new strain developed. Generating transgenic fish is relatively easy (compared with transgenic mammals). However, very few fish genes have been identified that carry 'desirable' traits, so this is work is at the research stage.

Vaccines and diagnostics. In fish farms, controlling diseases is a substantial problem. (In the wild the fish that die of disease are simply not caught.) There is a growing need for diagnostics to tell which sorts of fish diseases are present, and vaccines to protect the fish from them. These are most easily derived through cloning the pathogens involved.

Food. Fish food is a complicated mix of nutrients that must not only be formulated for the dietary needs of the fish (which vary as the fish grow, sometimes quite dramatically), but also sink at the right rate and not dissolve in water. Several dietary ingredients are made using biotechnology, including essential amino acids, proteins, and other trace nutrients. Biotechnology also makes the pink pigment astaxanthin, which can be extracted from crab shells or from algae, and is included in salmon and trout feed to make the flesh pink. (The same material is fed to flamingos for the same reason.) Normally, the fish's diet of krill would supply enough astaxanthin and enough other trace nutrients: as with farm animals, fish farmers have to replace that natural source of food with a cheaper, more convenient, semi-synthetic food.

Flavour chemicals

An increasing use of biotechnology is to make flavour chemicals for the food industry. Flavour and fragrance chemicals are usually complex mixtures of very complicated molecules, and so are very difficult to reproduce with conventional chemistry. They arrive in food either as part of the ingredients or as chemicals created during cooking. Biotechnology can help both processes. (Most flavour chemicals are actually fragrances: we taste most of our food with our noses.)

Flavour chemicals in ingredients are usually secondary metabolites from plants (especially spice plants) or bulk ingredients like salt and sugar. Often, plants are unreliable industrial sources of chemicals, so biotechnology has sought to grow their tissues *in vitro* to make the chemicals (*see* **Plant cell culture, Hairy root culture**).

Enzymes are used extensively to help to mature the flavour ingredients of food during processing. Biotechnologists can assist the development of new food enzymes by finding or engineering enzymes that fit better with the other processes that the food must undergo, like cooking or canning. These improvements could include making them more stable to acid or heat, or making them easier to remove once they have done their job, by, for example, immobilizing them on beads or columns so they can be separated from liquid food or food ingredients easily. Examples of the use of enzymes in flavour development include:

- Using endopeptidases (proteases) to break down bitter-tasting peptides in low-fat foods (fat usually dissolves these hydrophobic peptides and so disguises their taste).

- Using a variety of oxidases to generate synthetic flavours to replace products such as cheese.

- Making synthetic peptides that mimic the flavour precursors in chocolate and cocoa, so allowing cheaper cocoa to be used in high-quality products.

Enzymes are also used to accelerate such processes as cheese ageing, which is, in part, a flavour-development process.

The same technology could also be used to make fragrance chemicals for the cosmetics industry. Biotechnological methods are often needed because fragrance chemicals are often complex, chiral molecules: the characteristic scents of oranges and lemons, for example, are caused by the different stereoisomers of the same compound, limonene. However, this application has not taken off. Cheaper fragrances can be made successfully from conventional chemicals. Top of the range perfumes are still made by traditional methods, such as extracting flower scents into candle wax or shaking seed pods with grain alcohol. The perfumes gain an appeal through their use of 'natural' and 'traditional' recipes of this sort, and so for them a step towards new technologies would, paradoxically, be a step backwards in marketing.

Flowers

Plant genetic engineering to alter flower characteristics is one of the two success stories for **ornamental plant** biotechnology (the other being pest resistance). Flowers are attractive targets for the genetic engineer, because the regulatory requirements for their release are less onerous than for something that is to be eaten, the value of a new flower is high compared to its mass, and new traits are easy to see, and hence both easy to select for and easy to 'sell' to investors.

As well as engineering pest resistance into horticultural plants, advanced genetics is seeking to target several other traits.

Enhancing flower structure. The number of petals and some aspects of their arrangement is controlled by only a few genes. These can be transferred from plant to plant to produce unusual and complex flower shapes. (However, the majority of such new flower shapes are still random mutations that are discovered by gardeners.)

Changing flower colour. Colour is partly determined by the enzymes that a flower makes to synthesize colour pigments. Most of the pathways are known, and the genetic engineer can place the appropriate genes into plants to change colours. Antisense genes can also block colour synthesis, producing white flowers or flowers that accumulate chemicals that have different colours from the normal end-products of colour synthesis. However, other aspects of the plant's physiology are also important, such as the pH of the vacuole (the blue colour produced by the accumulation of 5'-hydroxylated anthacyanin pigments requires a high vacuolar pH). One of the most saught after blue flowers is the blue rose; however, since no rose species produces blue pigments, this will need recombinant DNA technology to create.

Enhancing cut flower life. The lifetime of flowers can be enhanced by blocking ethylene production, which is the trigger for senescence. (The same trigger causes some fruit to ripen, and leaves to brown and fall off in the autumn.) Antisense can block the gene *ACCO* (aminocyclopropane carboxylic acid oxidase) which makes ethylene in plants.

Making ornamental versions of common plants. Examples are ornamental cabbages and potatoes, which are made by manipulating genes such as the *Agrobacterium rolC* gene which makes plant hormones resulting in dwarf, bushy plants that look like round clumps of leaves.

Fluidized bed

Many types of bioreactor use particles to support the enzyme molecules or cells. These particles form a 'bed'. There are a range of ways of making sure that the reactor contents flow smoothly and uniformly past such a bed.

Fluidized beds. Here the flow of material through a bed of particles is enough to shake them loose from their neighbours, but not enough to lift them. The bed of particles then behaves like a fluid, not a solid, and mixes very well with the material flowing through it. This technology is more widely used in the industrial processing of gases (such as burning coal dust in power stations), but can be used with liquids.

Entrained bed reactors. Here the flow is higher, and the particles are kept in suspension by the fluid flow. Some cell culture systems use variants on this approach, with the cells growing on the surface of polymer particles that are kept suspended in a flow of growth medium. Note that this is not the same as keeping the particles in suspension by stirring.

Packed bed reactors. Here the particles are packed into a bed, and remain essentially solid. This is the most common type of bed reactor, and is widely used in chromatography: a chromatographic column is in fact a columnar packed bed reactor. Fluid can flow unevenly through such a bed of solid particles, and so variants have been developed to encourage more even flow, such as the tapered bed reactor (with a conical base filled with particulate solids) and the disk bed reactor, where a disk of solid material has liquid fed in from the centre and flows out for collection at the outside.

Fluorescence

Fluorescence is a very widely used way of detecting things in biology and biotechnology. A fluorescent chemical absorbs light of one wavelength and re-emits it at once at another wavelength. (A rarer phenomenon, phosphorescence, is when it re-emits it again later.) Usually this means shining an ultraviolet (UV) light (which we cannot see) at the chemical and having it emit visible light (which we can see). The molecule is said to be excited by the UV, and to emit the visible light.

Because biological material is not generally very fluorescent, it is easy to see a fluorescent label molecule introduced to it. Fluorescent labels have been used to label a very wide range of biochemical reagents, including DNA probes, antibodies, and enzyme products. Common labels are fluorescein (which glows greenish yellow) and texas red. Increasingly used are red and near-infrared (NIR) fluorescent dyes, which are excited by red light (e.g. from a laser) and emit deeper red or infrared light. They are attractive because biological material has almost no background fluorescence in the red, and because solid-state lasers are cheap, compact power light sources for these applications.

DNA is very often stained with a UV dye, often ethidium bromide. This sits in between the bases of the double helix (it 'intercalates'), where it glows much more brightly than on its own. If an agarose gel is soaked in a dilute ethidium bromide solution, the DNA shows up under UV light as bright bands on a dark background.

An increasingly popular fluorescent label is green fluorescent protein (GFP), a protein that has a very strong green fluorescence. Since it is simply a protein, it can be made inside cells, and so one can engineer cells to make the GFP as a marker of gene activity. GFPs have been engineered to emit a range of colours between yellow and blue-green, so, in principle, one could have two or even three GFPs distinguishable within one cell.

As well as simply tying a fluorescent molecule on to an otherwise non-fluorescent reagent, there are other tricks that use fluorescence.

Delayed fluorescence effects. See **DELFIA**.

Fluorescence transfer. An excited fluorescent molecule can give away its energy as light, or by donating it to another molecule. If the other molecule fluoresces, this is called fluorescence transfer. The second molecule glows (with a different colour to the first), where it would not on its own. If the second molecule does not glow, this is called quenching: nothing glows. Both instances can be used to see when two molecules have been brought very close together: in the first case, the light output changes colour, in the second it stops. The molecules needs to be within a few nanometres of each other for this transfer to work.

Food

Biotechnology can be used to enhance existing foods or generate new ones. In the former category, biotechnology has been used to remove lactose from milk (for lactose intolerent individuals) and make rennin-free cheese (for vegetarians and people allergic to rennin). A very wide range of science and technology is also used more mundanely to monitor, control, and improve quality in all food products, and to understand how food can be kept fresh and palatable for the consumer. Amongst the enhancements to existing products are:

Genetically engineered plants. The most high profile is the Flavr Savr tomato (*see* **Antisense**), but other applications such as plants with altered oil compositions are in development or in the field.

Food chemicals. Biotechnology produces a wide range of chemical components of food, such as vitamins, some colouring agents, modified starches, fats and lipids, HFCS (High Fructose Corn Syrup) , etc. The pink colour in crab sticks does not come from crabs, for example (it comes from shrimp shells, plankton, or in some cases from red/pink algae). The most extreme example of this is single-cell protein (SCP), producing bulk protein for food use using fermentation.

There are few completely biotechnological foods. The nearest is probably quorn, a fungal protein product made from *Fusarium graminearum*, grown by fermentation and subsequently processed to look rather like meat. Flavour is added using synthetic flavour chemicals, and the result tastes better than some real meat products. Other organisms, including algae, are also grown as food products for processing in a similar way, but not as widely in the West. (In Japan, seaweed is a traditional staple food anyway.) In fact, many traditional foodstuffs such as tofu, cheese, soy sauce, and all alcoholic drinks are made by fermentation processes, but these have evolved over hundreds of years, rather than being the result of biological knowledge, and so are not really biotechnology.

Biotechnology also contributes methods and materials used in processing conventional food.

Food processing using enzymes

One of the major uses of enzymes is in the food industry. The food industry is traditionally conservative, preferring to retain existing processes and materials unless new ones provide overwhelming advantages. Nevertheless, biotechnology has provided a range of enzymes that are used in food processing. Among them are **proteases**, **lipases**, and a range of **glucosidases**.

Enzymes are used to control food texture, flavour, appearance, and, to a certain extent, nutritional value. Amylases are used to break down complex polysaccharides, which form viscous solutions or solid gels and do not have much flavour, to simpler sugars, which form more fluid solutions and taste sweet. Proteases are used to tenderize meat proteins, especially collagenase, which breaks down collagen, the major protein in connective tissue such as 'gristle' in meat. A widely used protease is rennin, which breaks down milk proteins and so causes them to coagulate, forming the basis of cheese: fungal rennins are now widely used in cheese making. Proteases are also used to clarify beers and condition dough for bread making.

These enzymes are often added to food during processing, so the amount of enzyme added and the stage of the process at which it acts can be controlled. These are 'exogenous' enzymes. Food also contains endogenous enzymes, enzymes that are naturally present in the food materials. These are also responsible for the changes in texture, flavour, and appearance of the food as it is processed, but are harder to control. Thus allinase helps to develop the characteristic odour of onions, but can also create a bitter flavour in the same food.

Genetically engineered rennin was the first enzyme produced by recombinant DNA to be approved for food use: it was cloned by Collaborative Research and marketed by Dow Chemicals. In the USA, as with pharmaceutical products, the FDA provides a rigorous regulatory gateway to using new enzymes in food, especially genetically engineered enzymes, and approval for a food material in the USA is generally taken as being a clear signal to European authorities that the new ingredient is safe. A much wider range of 'novel' food ingredients is approved for use in the Far East, including Japan, than is found in food in the 'West'.

Freeze-drying

This is a common technique, also called lyophilization, for preserving biomolecules and microorganisms. The sample is frozen, often in a solution containing another material such as lactose or trehalose, which acts to stabilize it (and is called the excipient). It is then put into a chamber attached to a vacuum pump and, while the sample is still frozen, the chamber is evacuated. The ice sublimes under vacuum (i.e. turns directly into vapour without melting), and the water vapour is removed and trapped in a 'cold trap'. After a while all the water in the sample has been removed, and what is left is a dry powder or pellet of material.

Commercial freeze-drying apparatus can control the temperature and pressure of the vacuum chamber very accurately, and can heat up the samples to be freeze-dried during later stages to drive off the last remaining water. However, simply connecting a frozen sample up to a vacuum pump often suffices for research freeze-drying applications.

Freeze-drying is the standard way of preserving microorganisms for long periods of time. It is also a favourite way of formulating biopharmaceuticals since these protein drugs are often not very stable in watery solution. A good freeze-dried preparation of a protein is a very light fluffy material which, when water or buffer is added, dissolves almost instantly.

Functional genomics

This is performing genome-scale gene mapping or sequencing together with biological experiments that add meaning to the mass of data generated in a 'genome project'. Increasingly it is clear that automated sequencers and high-throughput mapping programmes can generate genetic data far faster than scientists can understand it or turn it into practical results. So the 'functional genomics' movement is aiming to add high-throughput biology to the programme to identify what all these genes do.

The approach is to link gene identification and sequencing with a tool or set of tools that give clues about the gene's function. The tools used include:

Proteome technology. Looking at the proteins produced (*see* **Proteomes**).

Animal genetics. Cloning genes from animals which have known effects, and then seeing if related genes can be identified in the human genome. The mouse is the favoured organism here, although *Drosophila*, *Caenorhabditis*, and yeast can also provide insights.

Knock-out mutants. If you do not know what a gene is doing, knock it out in an animal (again, usually a mouse) and see what happens. Increasingly, the answers are very unexpected (*see* **Knock-outs and mutants**).

Cell biology. As a last resort, scientists are going back to cell biology and biochemistry and dissecting what the genes are doing one at a time in the laboratory. This really is a last resort: it can take a team of 20 scientists a decade to find what a new gene may be doing in even one cell type, and there are 100 cell types and 100 000 genes in humans.

Fusion biopharmaceuticals

Several biopharmaceutical proteins have been developed that are fusion proteins, i.e. they are the product of two genes, which have been fused together so that the proteins that they code for are joined end-to-end. The advantages of such proteins as drugs can be:

- They have two complementary or synergistic activities in one molecule. Thus, when the molecule binds to a cell, it does two things at once. To get the same effect with two molecules could need much more of both of them, to increase the chance that both would bind at once to one cell.

- The adverse effects or poor stability of one molecule are offset by the properties of the other.

- One molecule acts as a 'targeting' mechanism to bring the other to the site where it is meant to act (*see* **Immunotoxins**).

Examples of such fusion peptides are the CD4–IgG combined molecule that Genentech has developed as a potential AIDS treatment, and the Immunex GM-CSF–IL-3 fusion. The CD4–IgG blocks binding of the AIDS virus to cells, and is much more stable in the blood than the free CD4 molecule itself. GM-CSF and IL-3 have synergistic effects at stimulating bone marrow to produce white blood cells, so linking the two together produces a potentially more powerful compound than the two molecules separately. Immunotoxins can also be fusion pharmaceuticals.

Fusion protein

A fusion protein is a protein in which part of the amino acids come from one protein sequence and part from another. 'Biotechnology' is a fusion word, with the 'bio' of 'biology' and 'technology'. Fusion proteins are produced by splicing the gene for one protein next to, or into, the gene for another: the genetic apparatus reads the gene fusion as a single gene, and so produces a fusion protein.

Fusion proteins are used in a number of biotechnological applications.

● To add an **affinity tag** to a protein.

● To produce a peptide as part of a larger protein, which is then cut up after it has been made by cloning (*see* **Peptide synthesis**).

● To produce a protein with the combined characteristics of two natural proteins (for example, a chimeric antibody).

● To produce a protein where two different activities are physically linked (e.g. enzymes for substrate channelling or as a fusion biopharmaceutical).

In practice many proteins are expressed as fusion proteins during research. It is easier to splice the gene for a potentially interesting protein into the middle of another gene than to get it positioned exactly behind a promoter sequence in order to express it as a protein with no additional amino acids.

Gas transfer

One of the most important characteristics of a fermentation system is the rate at which gas can be transferred from the gas phase into solution. Often, the rate at which the organisms in the fermentor can metabolize is limited by how fast they can be provided with oxygen or have carbon dioxide, ammonia, or other 'waste' gases removed. Oxygen is poorly soluble in water, and so the liquid itself holds very little, which the organisms in a dense culture can use up in a few seconds. Therefore they must be constantly supplied with oxygen gas, either as pure oxygen (efficient but expensive) or as air. Many fermentor design features are aimed at optimizing this transfer rate.

There are several basic methods. Smaller bubbles of gas have a larger surface area per unit volume and greater internal pressure than large bubbles, and so gas diffuses out of them faster. Therefore the smaller the bubbles you can make, the faster oxygen diffusion will occur. However, creating small bubbles requires power, may cause disruption of the organism growing in the liquid, and can cause foaming, which fills the reactor vessel up with viscous foam. Anti-foaming agents can help this latter problem (which is also a problem when the organisms produce a lot of carbon dioxide gas). The sparger, the pipe system that delivers gas to the base of a tank fermentor, is responsible for breaking up the gas flow into the bioreactor into bubbles and making sure that they are evenly distributed in the reactor volume.

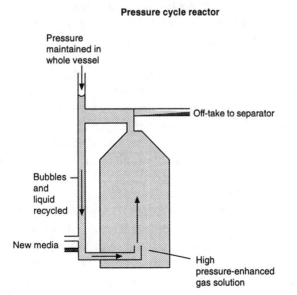

Pressure cycle reactor

Pressure maintained in whole vessel

Off-take to separator

Bubbles and liquid recycled

New media

High pressure-enhanced gas solution

There is a wide variety of methods for making small bubbles, keeping them suspended in the fermentation broth, and encouraging their content gases to dissolve. Among these are the deep jet fermentor, which injects the gas very fast into the bulk of the liquid, and the pressure cycle reactor. In the latter, both gas and liquor are recycled around the reactor loop at high hydrostatic pressure. At the base of the reactor the pressure is much higher, encouraging the gas to dissolve. The constant recycling ensures that bubbles stay in the liquid phase, and do not separate as froth.

Other methods ensuring good gas transfer rely on increasing the surface of the liquid in contact with gas. Bubbling the gas through the liquid spreads the gas out: other methods spread the liquid out, for example in a thin sheet (in an oxidation pond), or in a thin permeable tube (as in a hollow fibre bioreactor).

Gel electrophoresis

Gel electrophoresis is one of the most common analytical methods in biochemistry and molecular biology. Electrophoresis is the separation of molecules according to how fast they move in an electric field. In gel electrophoresis this is done in a gel, a polymer matrix. Samples are put at one end of a slab of polymer gel (any jelly-like material). An electric field across the gel pulls the molecules through it: smaller molecules can pass through the gel more easily and so move towards the other end faster. Thus molecules are separated mainly according to size.

A large number of materials are used to make the gel, but by far the most common are agarose (for DNA and RNA) and polyacrylamide (for DNA in DNA sequencing, and for proteins), which is often called PAGE (polyacrylamide gel electrophoresis). Various chemicals can be included in the gel to help the separation, such as the detergent sodium dodecyl sulfate (SDS) in protein gels, which unfolds all the proteins, and urea in DNA sequencing gels, which does the same to DNA.

A recent variation in DNA gels is pulsed field gel electrophoresis (PFGE) and its variants. These also use electric fields to separate molecules, but with several sets of electrodes: the electric field is switched between them, which encourages the DNA to wiggle its way through the gel matrix, now heading one way, now heading another. This helps the separation of very large DNA molecules, up to the size of whole yeast (but not human) chromosomes. Several terms such as orthogonal field gel electrophoresis and field inversion gel electrophoresis (FIGE) refer to how the different fields are arranged.

Variants on gel electrophoresis are isoelectric focusing (IEF) gels, which separate macromolecules on the basis of their isoelectric point

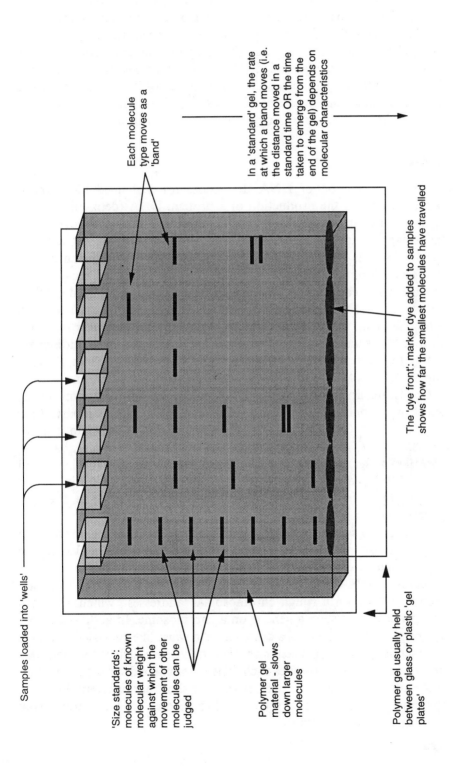

Samples loaded into 'wells'

Each molecule type moves as a 'band'

In a 'standard' gel, the rate at which a band moves (i.e. the distance moved in a standard time OR the time taken to emerge from the end of the gel) depends on molecular characteristics

The 'dye front': marker dye added to samples shows how far the smallest molecules have travelled

'Size standards': molecules of known molecular weight against which the movement of other molecules can be judged

Polymer gel material - slows down larger molecules

Polymer gel usually held between glass or plastic 'gel plates' (roughly the number of different charged groups they have) rather than on size. The gel has a pH gradient inside it, and the molecules migrate to the point where the pH of the gel is the same as their isoelectric point: at that pH they have no charge, so they stop moving and form a narrow band. O'Farrel gels (also called 2D gels) run an isoelectric focusing gel down one side of a gel slab, and then do a standard PAGE at right angles along the length: this produces a 2D pattern of protein spots which is as characteristic of a mixture of proteins as a fingerprint.

Gene

A gene is a section of DNA that codes for a defined biochemical function, usually the production of a protein. DNA (deoxyribonucleic acid) is made up of repeating units which vary in their chemical details (as a magnetic tape is basically the same all the way along, but varies in the details of its magnetization depending on what has been recorded on it). The parts of the DNA that differ are the bases, so called because they are the chemically basic (alkaline) part of the overall acidic DNA structure. In DNA there are two strands winding round each other in a double helix, so the bases come in pairs. In RNA there is only one strand. Molecular biologists use 'base' and 'base pair' fairly indiscriminately to mean the length of a piece of DNA or RNA since RNA is copied off DNA base for base in the process of transcription.

Genes are arranged on long DNA molecules called chromosomes, which may contain a few dozen genes in a few tens of kilobases (one kilobase = 1000 bases) in the chromosome of a virus, to tens of thousands of genes in hundreds of megabases (1 megabase = 1 000 000 bases) of DNA in the chromosomes of higher plants and animals. In eukaryotes (living things with cell nuclei, as distinct from prokaryotes, like bacteria) the chromosomes can be seen under the light microscope when the cell divides.

Chromosomes can be chemically treated so that some regions take up dye more easily than others and so stain as darker 'bands', reflecting the genetic structure of the DNA. Each band can have hundred of millions of bases of DNA in it, so bands are not precise landmarks to genes. However, they form rough chromosomal 'addresses', which are often used to describe where a gene is on a chromosome. In addition, somewhere on a chromosome is a centromere, a waist-like constriction that is an essential part of the mechanism that makes sure each of two new cells gets one set of chromosomes each. The centromere divides the chromosome into two, usually unequal, parts. The short part is referred to as p, the long part q, so 16p2.3 refers to chromosome 16, long 'arm', band number 2, sub-band number 3.

Preparing cells so that the chromosomes can be seen and hence counted is called karyology, and the resulting count of chromosomes is called the cell's karyotype. Preparing the more complex chromosomal band analyses is covered in the broader term cytogenetics.

All the genes (and hence, inevitably, all the chromosomes) in an organism are called its genome. The human genome is about three billion bases long. The record size is that of the lungfish, which has a 140 billion base genome, nearly all of which appears to be 'junk' DNA.

Animals and all larger plants have two sets of chromosomes in each cell, so that there are two copies of each gene there. Exceptions are the germ cells (sperm and ova in animals), which only have one chromosome set each. Thus, when a sperm and an ovum get together, the result is two chromosome sets again, a new genetic individual. Sometimes the two copies of a gene can be slightly different, in which case the copies are called alleles. If one of your genes is present as two different alleles in your cells, then you got one from your mother and one from your father. Organisms with two chromosome sets per cell are called diploid, and include animals, higher plants, and many fungi. Organisms with only one chromosome set per cell are called haploid, and include yeasts and bacteria. Some species have many chromosome sets: these are called triploid (for three), tetraploid (for four), and so on, or more generally 'polyploid'.

In bacteria, genes that are regulated together (i.e. that are turned on at the same time and by the same stimulus) can be arranged in a tight cluster called an operon. This has one control region at one end, and then a series of coding regions, i.e. regions of DNA that code for single proteins. This whole cluster is transcribed as one RNA, which is then decoded into multiple proteins by the enzymes of the cell. This operon structure is virtually unknown in higher organisms.

So that all genes are not active all the time, genes need control regions attached to them to regulate their activity. In a bacterial operon, these regions are located at one end of the gene. In eukaryotes the control regions (or control elements, since they are usually very short sections of DNA) are mostly at the start of the gene, but can also be spread quite a way away from that start, both within the gene itself and away from it (*see* **Gene control**).

Gene control

This is the genetic control of how a gene acts. Genes are pieces of information coded in DNA—on their own they do nothing. There must be a control system to make sure that their information is used at the correct time.

Prokaryoyes (bacteria) and eukaryotes (yeasts, plants, animals) do this differently. In both types of organism there is a promoter region, which essentially says 'gene starts here' on the DNA. The promoter region can contain DNA sequences that control when it is active, for example making sure that the gene is active under particular chemical conditions or (in eukaryotes) only in particular cells or at particular times during the organism's development . In prokaryotes such elements include operators and attenuators. In eukaryotes they include the hormone responsive elements (HREs) and a host of others.

Eukaryotes also have enhancers, pieces of DNA that alter the activity of a gene but which are not next to the promoter. In general, in prokaryotes the promoter region is no more than 50 bases from the start of a gene. In eukaryotes it can be over 200 bases or more away. The enhancers can be up to several thousand bases away from the start of the gene in either direction, i.e. it is possible for a gene to contain its own enhancer.

All these 'elements' are short pieces of DNA sequence, 5–30 bases long, each of which is recognized by a protein. The proteins are called regulatory proteins or transcription factors. They need genes to make them: in prokaryotes such genes are called regulatory genes, as opposed to the structural genes that code for the proteins that make the cell work. In eukaryotes this terminology is not used, because so many eukaryotic genes are to do with control at so many different levels.

Gene family

Many genes are not unique, but are closely related (anything from 30% the same to over 90% the same) to some other gene or genes. The whole collection is called a gene family. Gene families can be clustered into superfamilies, with even more members that are even less similar to each other. Gene families and superfamilies are often related in function as well as DNA sequence, which is why people want to find them. Showing that a new gene is a member of the immunoglobulin-like superfamily, for example, would strongly suggest that it has something to do with the immune system.

There can be any number of members of a gene family, from three or four to thousands. Other DNA sequences that are not genes (i.e. they do not have a recognizable function) can also come in families of similar sequences.

Repetitive sequences. Also 'SINES' in mammals, transposable elements in other organisms. These can be regarded as 'junk' DNA in any other than a genetic context.

Pseudogenes. These are sequences that have nearly the same DNA sequence as a real gene, but do not work as genes because some critical bases are mutated. A variant is a processed pseudogene, which is a pseudogene that has lost all its introns.

The 'average sequence' of a gene family of any sort is called a consensus sequence. It is a sequence which, at each base, resembles the family members the most. Like that fiction 'the average man', a consensus sequence is an abstraction which represents the family as a whole: it may not represent any one member of it. You can also have a consensus sequence of any other bits of DNA, like promoters (the sections of DNA at the start of genes).

The relationship of members of a gene family to each other is often shown as a phylogenetic tree, in which the length of the branches between each 'leaf' shows how different the sequences are.

Phylogenetic tree

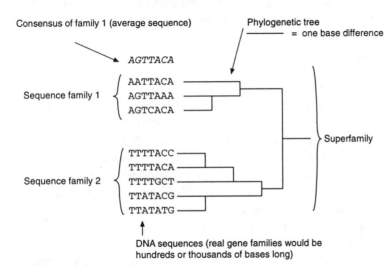

DNA sequences (real gene families would be hundreds or thousands of bases long)

Gene library

A gene library is a collection of gene clones that contains all the DNA present in some source, but split up and joined on to suitable vector DNAs. It is also sometimes called a gene bank. If the original source of the DNA was the original DNA from a living organism, then the library seeks to include clones of all that DNA: it is called a genomic gene library, because it contains all the DNA from that organism's genome ('genome' simply means all the genes, or DNA, in an individual). If the DNA is from some other source such as copy DNA (cDNA) made by enzymatic copying of RNA, then the maker of the library seeks to include representative clones from all that source, and, in this case, it would be called a cDNA library.

Gene libraries are not organized like book libraries, and can only claim to be complete because the number of clones in them is sufficiently large for us to be very confident that all the clones we want to be there are there, i.e. that there is a very small chance that anything has been left out. Usually genomic gene libraries are meant to be 95 or 99% complete, so that there is a 95 or 99% chance that the gene you are looking for is in there somewhere.

The number of clones needed to get a complete gene library depends on how big the cloned pieces of DNA are. Thus, if you are using a lambda phage vector to make a genomic gene library from human DNA, you need about 500 000 clones. However, cosmid cloning vectors can hold substantially more DNA: you would only need 200 000 of them. YAC (yeast artificial chromosome) vectors hold 10 times as much DNA again, so you only need 10 000 of them. This is why people use YAC vectors to make genomic gene libraries since screening 10 000 clones for the one you want is almost invariably easier than screening 500 000.

Gene synthesis

This is the complete synthesis of a gene using a DNA synthesizer ('gene machine'), rather than cloning it or assembling it from cloned fragments of DNA. Because most genes are longer than the maximum length of DNA that can be made conveniently on a DNA synthesizer, genes are usually assembled from a number of oligonucleotides. Each section of the gene is hybridized to the next section, and when the whole assembly has been hybridized together the DNA sections are joined together enzymatically to make one double helix. This requires the oligonucleotides to be designed carefully, so that they only hybridize to their correct partner and not to other oligonucleotides in the mix.

Other concerns involve making sure that the same sequence is not repeated too often within the gene (repeated sequences can be targets for rearrangements of the DNA within bacteria) and making sure that the codons used are suitable. Different codons that code for the same amino acid are not used with equal frequency, and generally the more frequently used codons are translated faster than rare codons. However, which codon is used most frequently depends on the organism in which the gene is to be expressed.

Other features of the gene are the presence or absence of restriction sites and suitable 'sticky ends' so that the final gene can be cloned into an expression vector easily.

Gene therapy

Gene therapy is changing the genetic make-up of a human. There are two approaches, germ-line gene therapy and somatic cell gene therapy. The former changes the 'germ cells', the cells that make sperm or ova. This has a permanent effect on all the individuals that are the descendants of whoever had the therapy. Somatic cells are the other cells in the body, i.e. the muscles, bone, nerves, etc. Changing them does not affect the germ cells, but does affect the engineered person.

Gene therapy of the germ cells of animals is called **transgenic** technology. Germ-line gene therapy of humans is considered unethical, and has not (legally) been tried.

Somatic cell gene therapy can be aimed at correcting a genetic or a non-genetic defect. Current therapeutic targets include both categories.

A relatively easy route to somatic gene therapy is bone marrow therapy since bone marrow is relatively easy to take out and replace, and generates itself inside the body. An engineered stem cell can reproduce itself inside the bone marrow, creating engineered blood cells as it does so. Targets for bone marrow therapy include SCID [severe combined immunodeficiency disease, a very rare genetic disease caused by a deficiency in the enzyme adenosine deaminase (ADA)]. W. French Anderson and Michael Blaese used a gene therapy treatment for SCID on a four-year-old girl in late 1991, with reasonable success.

Other targets include several cancer treatments, including introducing cells engineered to produce more tumour necrosis factor (TNF) or interleukin (IL-4) into cancer patients, where it is hoped they will be able to assist in destroying the cancer.

There have been several proposals for ways of getting DNA into cells while they are still in the patient's body. Routes suggested include:

Using viruses. This is the most common approach (*see* **Vectors for gene therapy**).

Biolistics. As well as delivering DNA into isolated cells, biolistics can be used to put DNA into cells that are still part of an animal (*see* **Biolistics**).

Injection. Simply injecting DNA complexed with calcium phosphate into liver or muscle causes some cells to take the DNA up and express the genes in it. This has attracted a lot of attention, because it offers the hope of a gene therapy treatment for muscular dystrophy, one of the most common genetic diseases.

Using lipsomes. DNA that has been encapsulated into liposomes and injected is taken up by the liver and, to a lesser extent, by the spleen, and any genes it carries are expressed briefly. Liposomes can also be targeted to other tissues (*see* **Liposomes**). DNA can also be complexed with negatively charged lipids the molecules form nanometer-scale clumps, which are taken up by cells.

Using cationic lipids. Positively charged lipids will form many-molecule aggregates with nucleic acids that are fairly lipid soluble and electrically neutral. These are taken up by cells much more readily than 'naked' DNA. The same approach has been used to get antisense oligonucleotides into cells, using a lipid formulation called lipofectin.

Use of peptides. Rather than use whole viruses, one could complex DNA to proteins that bind to specific cells. For example, Genzyme are trying to link peptides with the RGD sequence in them to DNA. RGD binds to integrins (*see* **Cell adhesion molecules**).

After a wave of hype in the early 1990s, it is now accepted that gene therapy is extremely difficult to get to work, for a variety of technical reasons. Thus it has fallen from favour as a method for curing disease. A few companies are pursuing it for applications where it may be unusually appropriate: Genzyme, for example, are developing a gene therapy system for replacing the defective gene in the lungs of cystic fibrosis patients. Others are pursuing it as a long-term development programme: Rhone Poulenc Rorer have set up a consortium of companies called GeneCell to develop all the components needed to make gene therapy work. Much of the basic technology (how to get engineered cells to function inside patients) will also be needed for xenograft and cell therapies.

See also **Transfection**, **Genoceuticals**, **Gene therapy: regulation**.

Gene therapy: regulation

Applying gene transfer techniques to humans, usually called gene therapy, has been a major problem for legislators and regulators as well as for scientists. From the Martin Cline experience in 1980, there has been substantial reluctance to let anyone put genes into anyone else, no matter what for. Cline, a UCLA researcher, wanted to put the genes for beta-globin into patients suffering from thalassaemia, a genetic disease caused by defects in the beta-globin genes. Refused permission to do so in the USA, he performed the medical parts of his experiments in Israel and Sardinia (which have much higher incidences of thalassaemias). This brought universal disapproval and a determination to make sure that any future gene therapy experiments were tightly regulated. (The experiment was a complete failure.)

Every agency or pressure group with an interest in biomedicine wants to have a say in whether gene therapy can take place. In late 1990 the first gene therapy experiment, giving an SCID patient the gene for ADA, took place. Before it could do so, the experimenters sought approval from the following agencies, any of which could have stopped the experiment.

- National Institute of Health (NIH), Biosafety Committee. Concerned with the technical safety aspects of the experiment.

- National Cancer Institute (NCI) review board.

- National Heart, Lung and Blood Institute review board. This body and the NCI were funding the experiment.

- NIH's Recombinant DNA Advisory Committee (RAC). This advises on whether any experiment involving recombinant DNA could be allowed. There is an RAC subcommittee on gene therapy, which also had to give its permission.

- The acting director of the NIH.

- Food and Drug Administration (FDA) external advisory committee (since this was an experimental therapeutic procedure).

Although the little girl patient receiving this experimental treatment was released after it was over, making this a formal case of the release of a genetically manipulated organism (GMO) into the environment, the Environmental Protection Agency was not consulted.

Regulations have relaxed since then, but are still stringent.

Genetic code and protein synthesis

The genetic code is the code used by living cells to turn the information in DNA into the information needed to make protein. How this works is not essential to an understanding of much of biotechnology—the genetic machinery can be treated as a 'black box' for even quite advanced discussions.

The information in DNA is held in the sequence of the four bases of DNA (adenine, guanine, cytosine, and thymidine). This information is transcribed into base sequence in RNA, and then translated into amino acid sequence in protein, the latter occurring on the ribosomes. The RNA is made starting at the 5'-end, and is translated starting from that end too: protein is made starting from the amino end (N-terminus). The sequence that codes for a protein starts with AUG or (less commonly) GUG, followed by a sequence of bases that is read as triplets, called codons. Of the 64 possible triplets, 61 code for an amino acid, the remaining three being termination codons (i.e. they code for 'stop'). This first AUG (or GUG) is called the initiation codon.

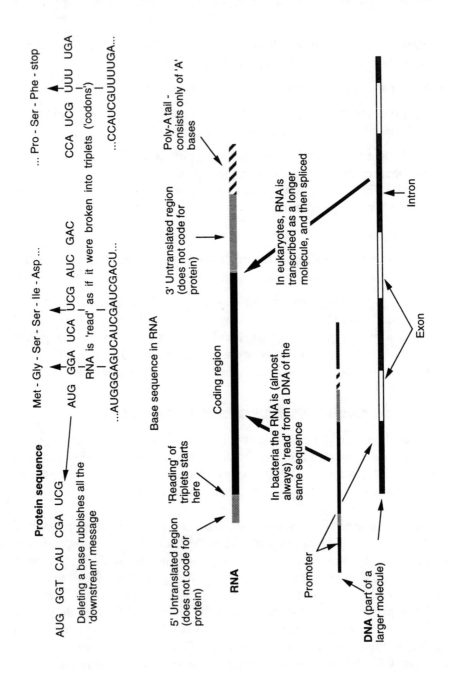

Protein sequence

AUG GGT CAU CGA UCG

Deleting a base rubbishes all the 'downstream' message

... Pro - Ser - Phe - stop

CCA UCG UUU UGA

...CCAUCGUUUUGA....

Met - Gly - Ser - Ser - Ile - Asp ...

AUG GGA UCA UCG AUC GAC

RNA is 'read' as if it were broken into triplets ('codons')

...AUGGGAGUCAUCGAUCGACU...

Base sequence in RNA

Poly-A tail - consists only of 'A' bases

3' Untranslated region (does not code for protein)

Coding region

In eukaryotes, RNA is transcribed as a longer molecule, and then spliced

'Reading' of triplets starts here

In bacteria the RNA is (almost always) 'read' from a DNA of the same sequence

5' Untranslated region (does not code for protein)

RNA

Promoter

DNA (part of a larger molecule)

Intron

Exon

Because there are 20 amino acids and 64 triplets, some amino acids are coded by more than one codon. Once the start code has been found, the cell reads off the other triplets starting from the AUG (or GUG). The way in which the cell reads the message off is called the 'reading frame', as if the cell overlaid a frame of squares three bases long over the RNA and read what was in each box. Clearly, the loss of a single base would then throw out all the cell's reading of the subsequent triplets. Such a mutation is called a nonsense mutation, because it makes a nonsense of the rest of the protein.

Although most of the code is shared between all living things, there are some differences: mitochondria, which have some of their own DNA, do not have quite the same genetic code as the cells in which they reside, for example.

The DNA sequence in the chromosomes need not be the same as the RNA sequence that is eventually translated into protein. There is a substantial amount of editing of RNA. The most common version is when segments called introns (which occur in most eukaryotic genes, also sometimes called intervening sequences) are removed, a process called splicing, rather like cutting the ads out of a taped TV movie. Introns were classed as 'junk' DNA, i.e. DNA that was just there and had no function, but the amount of it and the fact that removing it can affect gene activity argues against this. Some genes are over 95% intron: the gene that is mutated in muscular dystrophy is 2.4 million bases long, all of which is copied on to RNA (the 'primary transcript'), of which more than 2.35 million bases are then spliced out and thrown away.

Other, rarer versions of RNA editing include adding uracils into specific sites inside the RNA. There are even a couple of known cases of joining bits of different RNA molecules together, known as trans-splicing.

These complications have two implications for biotechnology. First, it is often not possible to express a eukaryotic gene in a prokaryote. Even if prokaryotic promoter sequences are spliced on, the prokaryote will be incapable of carrying out eukaryotic post-transcriptional modification to the RNA to make it readable. For this reason, many protein expression projects prefer to start with a cDNA clone (a cloned DNA made by enzymatic copying of the final RNA) rather than the original gene. Secondly, although sequencing DNA is easier than sequencing protein, it is not always safe to extrapolate from the DNA sequence to the protein it may code, because of variations in post-transcriptional modification of the RNA and variations in the genetic code.

Genetic disease diagnosis

A genetic disease is one that is caused by a gene, so we 'inherit' the disease from our parents. For a true genetic disease anyone with the correct genotype (collection of genes) will display the phenotype (the physical manifestations of the genes). In practice, quite a lot of genetic diseases have incomplete penetrance: i.e. the genes do not always cause the effect they are meant to. This makes detecting them rather complex.

Molecular genetics has made huge advances in medical genetics, particularly through making available DNA probes to detect the genes that 'cause' genetic diseases even when they are not causing them; for example, when a gene is present in a carrier, or when a gene that causes a disease late in life is present in an infant. These probes have been used both to identify the gene and to diagnose carrier status in people who carry the gene but do not have the disease.

Identifying the gene can be done in two ways. The 'traditional' way is to know how the disease is caused, and hence what protein defect causes it, and so clone the gene from knowledge of the protein. The 'reverse genetics' approach is to use the gene probes to localize the gene whose defective allele causes the disease to a particular chromosome, an approach also called positional gene cloning. This is usually done by linkage analysis. The gene itself may be cloned by one of a variety of methods such as chromosome walking or chromosome jumping. Essentially these are ways of using a piece of DNA that you have cloned to identify bits of DNA from nearby spots in the chromosome. A number of companies (such as Millennium and Myriad) have been founded to use such approaches to clone medically important genes.

The genetic diseases for which cloned probes (i.e. probes that identify the gene itself) have been isolated include the haemophilias, thalassaemias, sickle cell disease, Duchenne and Becker's muscular dystrophy, retinoblastoma, and cystic fibrosis. A large number of cloned probes that detect loci closely linked to other disease genes, and which can therefore be used for medical genetic diagnosis, have also been cloned. These are all diseases caused by single genes, also called monogenic diseases. Many diseases are influenced by many genes: they are called polygenic diseases.

See also **Predisposition analysis**.

Genetic engineering

A general term for the directed manipulation of genes, and usually used synonymously with genetic manipulation or genetic modification. A wide range of technologies are involved in this, but most involve the **recombinant DNA** techniques.

Genetic engineering falls into several different categories depending on what is being engineered.

Bacteria, yeast. This is 'traditional' genetic engineering (i.e. over 10 years old). Using recombinant DNA techniques, genes are put into microorganisms to make them produce something we want, be it insulin, or better beer, or protein for food.

Animals. Genetically engineered animals are usually called 'transgenic animals'. They are produced by a combination of *in vitro* fertilization techniques (IVF) and recombinant DNA technology, and produce animals that pass on their genetic modification to their offspring: they have a germ-line modification.

Plants. Genetically engineered plants are also sometimes called transgenic plants. They are created through the use of plant cloning technologies, which involve growing plants from isolated plant cells.

Humans. Although the genetic engineering methods applicable to cows or mice are, in theory, applicable to humans, they have not been applied for obvious ethical reasons. Some experiments treating disease have been performed: these do not modify the germ cells, only the 'somatic cells'. This is generally called gene therapy or somatic cell therapy, rather than the more spectacular (and potentially publicly alarming) term 'genetic engineering'.

Genetic information

Projects such as the Human Genome Project and the development of tests for genetic predisposition to disease have lead to a lot of talk about how human genetic information should or could be used. This contrasts to genetic information about animals, plants, or microorganisms, which is not thought to have the same ethical status: debates about who 'owns' the human genome focus on high ethical philosophy, those about pig genomes centre around the patent courts.

Several countries have suggested legislation on the use of human genetic information, with DNA-based methods particularly in mind. Denmark resolved to introduce legislation banning the use of genetic information for insurance, pension, and employment purposes in 1991. In the USA, California, Texas, and Oregon have introduced similar measures, and New York has a scheme for regulating genetic testing laboratories. The USA also has a Genetic Information Act, which prevents the use of genetic information in hiring federal employees.

So far, no one has resolved the problems of copyright and the ownership of proprietary 'DNA probes' in human genetics. Indeed, this is probably one of the most confused areas of the regulation of applied recombinant DNA work. This is in part because of a confusion with the debate over abortion, in part because of the history of the 'eugenics' movement in Europe (and not only in Germany, by any means, although it is primarily Germany that is sensitive on this issue). Also, as has been the case for almost every biotechnological development since 1970, there is a general belief that 'it will never really happen'. Since human genetic testing is already widespread, this is not very far-sighted.

Genetic map

A genetic map is a description of how genes lie along a chromosome. It looks rather like a ladder with very erratic rungs. There are several types.

The map can show the distance between any two points in terms of how much DNA lies between them (usually measured in megabases, Mb). A genetic map of this sort is called a physical map, because it represents what the genes are physically like.

A map can also represent the recombination distance between two genes. This is the chance that two genes at those two points will be separated by recombination during meiosis. (This is basic genetics. In summary, genes on the same chromosome are usually inherited together as a package between parent and child. When the sperm and ova are being made, a shuffling process called meiosis can shuffle these packages. The further the genes are apart on the chromosome, the more likely they are to be shuffled during meiosis. The shuffling process is called recombination. The chances of recombination between two genes can be described as a distance, and the distances used to build a map.) The resulting map is called a genetic map or a recombination map, with distances measured in centimorgans (cM), which is the distance apart genes have to be to have a 1% chance of being recombined during meiosis (named after American geneticist and Nobel Prize winner Thomas Hunt Morgan). The distances on the physical map and the genetic map are rarely the same, but the order of the genes is the same. (Just like a subway map vs. a town map: the distances on the subway map do not represent real-world distances, but the order of the stations is the same as the order you visit them in the real world.)

Genetic maps are important for breeding experiments, where they allow us to predict how much effort it will take to get a particular combination of genes into one organism, and determine the best strategy for doing so. They are also important for gene cloning experiments, where they tell us how near one gene is to another, and hence how our cloning strategy should approach that gene.

The landmarks on genetic maps are not always genes. They are more generally known as 'markers', and can be any piece of DNA, identified as a gene or by DNA probes or PCR amplification (*see* **RFLP**). The critical feature of a genetic map is its density: how many markers there are on it. The more markers there are, the more use it is in guiding our experiments, and hence the interest in the Human Genome Project in producing a high density genetic map with lots of markers on it.

True genetic maps can only be made by genetic experiments, i.e. breeding. Physical 'genetic' maps can be made by a variety of DNA techniques, such as looking at RFLPs, making physical maps of a whole lot of cloned DNAs and then stitching them together, or *in situ* hybridization to chromosomes to see where a piece of DNA lies physically on these cellular structures. Ideally, the maps made by these different techniques should match. Usually, they do not, quite.

The ultimate physical map is the DNA sequence of a chromosome. This tells where every single possible piece of DNA is. However, it may still not tell you what they do, and so is not the ultimate genetic map.

Genoceuticals

A vogue term for a version of gene therapy where a gene is placed into a cell and there produces a pharmaceutically active protein. So far several studies have shown that DNA can be placed into the cells of adult mice and rabbits, and that the DNA can work there to produce proteins. This has two potential applications: however, both are speculative at the moment, not having been tried properly even in animals.

'Genetic antibiotics' are genes that have some antibacterial or (more usually) antiviral activity. The genes are placed into the cells that are the potential targets of the parasite. For example, a gene for a toxin could be linked to a controller gene that is activated by a virus: when the virus infects the cell the toxin gene is turned on, toxin is produced, and the cell dies.

The other application is to insert genes that will themselves make biopharmaceuticals. For example, calcitonin has been suggested as a treatment for osteoporosis, a bone-wasting disease suffered most frequently by older women. However, calcitonin is a protein, and difficult to get into the body: consequently it has to be injected frequently. A genoceutical approach to osteoporosis would be to 'transfect' the gene for calcitonin into some suitable cells in the individual: these would then produce the hormone steadily for weeks or months.

This is a version of gene therapy, and shares the technical problems of that field. These include delivery barriers (it is very hard to get genes into adults reliably and reproducibly), potential problems with side-effects (the genes you put in might disrupt an oncogene, triggering a cancer, and this only need happen in one cell), and the considerable social concern over the use of gene therapy for any application.

Genome project

A genome project is a project to determine the exact genetic structure of an organism's genome, i.e. the DNA sequence of all its genes. There are quite a few.

The **Human Genome Project** is a project to determine the base sequence of all the DNA in humans.

Bacterial genomes. Complete sequencing of the genome of *Escherichia coli* is well under way by a consortium of laboratories, and will probably be complete by the time this book is published. In 1995 two complete bacterial genome sequences were completed, for *Haemophilus influenzae* and *Mycoplasma genitalium*, both disease-causing bacteria. *Mycoplasma genitalium* has a simplified genome, because it relies on its host for much of its metabolic apparatus. Even so, it is 580 067 bases round. *Haemophilus influenzae* is a more typical bacterium, and has a genome of 1 830 121 bases and 1749 genes. Scientists could assign a potential function to most of those genes by comparing them with genes from other species whose function was known.

The complete genome sequence of the yeast *Saccharomyces cerevisiae* was completed in 1996, by a consortium of academic laboratories. The project took longer to complete than expected because the research groups held on to the sequence until the last bit was checked. Since the whole genome is 12 146 000 bases long, this took a while.

Other genome projects include mice, cows, pigs, the weed *Arabidopsis thaliana*, the worm *Caenorhabditis elegans*, the fly *Drosophila melanogaster* and a number of crop plants: rice, wheat, maize, and apple among others.

GLP/GMP

These stand for Good Laboratory Practice and Good Manufacturing Practice. They are a code of practice that is designed to reduce to a minimum the chance of accidents that could affect a research project or a manufactured product. Sometimes referred to as cGMP (current GMP) or cGLP (clinical good laboratory practice).

The GLP and GMP prescriptions are quite voluminous, but boil down to a few key points. The essential point of both GLP and GMP is that everything is recorded, and only established procedures are used by people who have been trained to use them. This may seem obvious, but extends to everything: in a true GLP laboratory, for example, only staff who have been trained to use a balance may use it, every weighing has to be checked by another person (who has also been trained for that specific balance) who has to sign to say that they have checked that the weight of material is correct, weighing has to be done according to a written standard operating procedure (SOP) for using that balance, the protocol used has to be noted in the record of the experiment, and so on. All records are kept, and have to be archived on microfiche, magnetic tape, or (more recently) write-once CD-ROM. Samples of every batch of material used in the experiment or the manufacturing process also have to be archived, so they can be referred to in future should this become necessary.

Using these types of procedures, it is possible to trace exactly who did what at every stage of an experiment or a manufacturing process. Thus, if there is a problem afterwards, the GLP or GMP user can either point to a specific material or standard operating procedure that caused the problem, or can point to exhaustive documentation that shows that the problem is not their fault. This can be very important in pharmaceutical development and manufacture. (GLP was set up after some severe side-effects of a drug were overlooked during preclinical research because the protocols for an experiment were faulty.) Many biotechnology companies claim to work to GLP or GMP (depending on whether they are concerned with R&D or manufacturing). In practice, many who claim to do so do not actually work to GLP. It is enormously difficult to do innovative research to GLP, where you have to define a set of standard operating procedures, train the staff formally, etc. just to do one experiment that might take half a day. GLP is more relevant to pharmaceutical development (where a large number of very similar experiments are performed). GMP is a standard requirement for pharmaceutical production, and for a number of other industries.

GMP also stands for Good Microbiological Practice, a code of laboratory practice for performing basic microbiology safely. In this sense, GMP is simply a way of reducing the chance of contamination problems (contamination of the sample or of the laboratory) during a microbiological experiment to a minimum.

The International Standards Organization (ISO) also has a 'good practice' standard, ISO9000. There are a number of substandards within ISO9000, dealing with 'good practice' in manufacturing, product development, provision of services, etc. Organizations can be certified to ISO9000 as a mark that they provide a consistent quality of result to their customers. Certification requires that the company be regularly inspected by external, qualified inspectors: it is not a 'once and for all' event. ISO9000 certification and GMP status are not formally linked, but have many of the same requirements for documentation and traceability.

Glucose isomerase and invertase

Glucose isomerase is probably produced in larger amounts for industrial use than any other single enzyme (although the largest category of enzymes by far is the general class of alkaline proteases, used in detergents). It catalyses the interconversion of the two sugars glucose and fructose. Since fructose is slightly more chemically stable than glucose, a mixture of glucose and fructose with the enzyme will end up almost entirely as fructose. This is valuable for the food industry since fructose is substantially sweeter than glucose, and so you get more sweetness per gram than from glucose.

The usual use for glucose isomerase is to take glucose made by hydrolysis of corn starch and turn it into a mixture of mostly fructose with some glucose. The corn starch is broken down using amylases. The result is called 'high fructose corn syrup' (HFCS).

Invertase takes sucrose ('sugar') and turns it into glucose and fructose, a mixture called invert sugar or invert syrup. Invertase was the first enzyme to be used in industrial production, in the late 1940s. Thus, in conjunction with glucose isomerase it can convert sucrose into HFCS. Invertase can also be used in its own right to convert the easily crystallized sucrose into the less easily crystallized glucose–fructose mixture. 'After Eight Mints', for example, have invertase in their centres: it turns the hard sucrose core (which the chocolate coat was poured on to) into the soft centre we finally eat.

Glue

Biological glue is one of the many areas where biotechnology and medicine can overlap. Doctors are always interested in new medical techniques for repairing wounds. An obvious way is glue: however, the glue must have unusual properties. It must be able to set (cure) in a wet environment, not be broken down by watery liquids, not irritate or poison the body, not induce an immune or allergic response, and the body must be able to break it down after a time if its function is only temporary, like stitches.

The most widely used and discussed 'glue' is the protein fibrin. The body itself produces fibrin, a component of the clotting proteins in blood: however, it is not a very strong glue, and, unless derived from human blood (with its concomitant risk of viral contamination), causes a strong immune response. However, it is a natural human product, and is used in several commercial medical glue applications.

Several marine organisms produce glues that could fit these criteria. Mussels and barnacles produce protein-based glues which could, in principle, be produced by more convenient organisms using biotechnology. Genex has produced a yeast that makes the protein (which has a very unusual amino acid composition, making it difficult for the yeast cell to make it efficiently). The protein also needs extensive and rather specific post-translational modification, which the yeast cannot perform. Thus, these proteins are some way from commercialization yet.

Many other organisms make materials that glue them on to things or things (like eggs or nest material) on to other things. However, these have not been investigated nearly well enough to make them attractive targets for medical glues.

Biotechnology is also investigating semi-synthetic glues made from natural materials. Proteins can be cross-linked into a glue-link material, which can be used in wound healing and surgery, being fairly rapidly re-adsorbed by the body.

Glycation

Glycation is the non-enzymatic reaction of sugars with proteins. Many proteins have sugars added deliberately by the body (*see* **Glycosylation**), and there are enzyme mechanisms to make this happen. However, sugars can also react with the amino groups in proteins in an uncontrolled, chemical reaction. Since every part of mammalian bodies has sugar in it, this means that all proteins get glycated after a while. This is much accelerated by heating or if the sugar levels are very high. Hence chemical glycation is important to protein processing, and hence flavour formation, in food. Chemical glycation is also very important in the damage done to diabetics when their sugar levels rise above normal, and to all of us as we age. Indeed, one school of thought holds that much of the damage that causes ageing is a result of the effects of glycation. In particular, glycated proteins can continue to react and form complex, stable cross-links with sugars and, through them, with other proteins. These complexes are called advanced glycosylation end-products (AGEs). The body seems to be unable to remove them specifically, and so they accumulate, cross-linking collagen into a rigid, inflexible net, damaging critical proteins in long-lived nerve cells, and maybe even directly mutating DNA.

Glycobiology

Glycobiology is the study of sugars and their role in biology. Usually this is taken to mean the study of complex sugars and what their functions are, and not the metabolism of how sugars are put together and taken apart.

The twin thrusts of glycobiology are the study of glycoproteins, which are proteins with sugar residues attached, and the study of molecules that interact with sugars and affect sugar metabolism, especially the synthesis of those glycoproteins (i.e. glycosylation). Some glycoproteins have as much sugar in them as protein by weight, and the effects of this sugar on the protein can be substantial. Current theory suggests that the sugars on glycoproteins help in protein–protein binding (important for the mechanism by which cells recognize each other and by which viruses bind to and gain entrance to cells).

From this, glycobiology is interested in how complex sugars on their own interact with glycoproteins, glycolipids (lipids with sugars attached), and each other. It is also interested in lectins (proteins that bind specific sugars) as tools.

In living systems sugars, both as simple sugars and as blocks of sugar residues, are joined on to proteins at specific amino acid sites by glycosyltransferase enzymes (a process called glycosylation). Glycolipids can also be joined on to proteins by specific enzymes, a process called glypiation, producing glycolipoproteins. These complex entities are an important part of the surface membrane of cells, and so may be the 'docking molecules' that viruses use to attack cells: consequently, biotechnology researchers are interested in them because their study may lead to better antiviral drugs, and to markers for aberrant cells such as cancer cells.

The application of glycobiology is sometimes called glycotechnology, to distinguish it from biotechnology, a discipline that concentrates more on nucleic acids and proteins. Companies such as Oxford Glycosystems and Glycomed have been set up to exploit the potential of glycobiology. Carbohydrate-based drugs are a popular target. Thus Oxford Glycosystems are developing a carbohydrate-based anti-AIDS drug (which acts by blocking HIV's mechanism for latching on to cells as it infects them), and Glycomed has a drug aimed at blocking the effect of the glycosylated ELAMS (*see* **Cell adhesion molecules**). Other uses of glycobiological expertise are in manipulating glycosylation in expression systems, and in analysis of carbohydrates and glycoproteins.

Glycosidases

A group of enzymes that break up complex sugars (such as starch or sucrose) into simple ones (such as glucose or fructose). About 12 000 tonnes of glycosidases are made per annum, almost exclusively for use in the food industry.

The major glycosidase enzymes are amylases (which break down starch) and glucose isomerase (which is used to turn glucose into the sweeter fructose). Amylases break the long chains of starch molecules and similar polymers into shorter segments, and ultimately into glucose. Amylases are commonly extracted from barley, beans, potatoes, and a variety of fungi.

Other enzymes produced from bacteria and fungi for polysaccharide breakdown are isoamylases and pullulanases. These break off the side branches of starch, and are sometimes called 'debranching enzymes' for this reason. Since molecules that are single, unbranched strings of units have a quite different rheology from molecules that are branched like a tree, the debranching enzymes can be valuable to the food industry to alter the flow properties, or 'mouth feel', of food.

A third group of these enzymes are the cellulases, which break down cellulose. Since cellulose is probably the most common biological material in the world, using it as a raw material makes economic sense. However, it is very difficult to break down into its glucose monomer units. A range of cellulases and hemicellulases are used to process food ingredients, for example, enhancing the macceration of fruit pulp to make fruit juices. Gist brocades, a major provider of these enzymes, claim that enzymes can increase the amount of juice you can squeeze out of apples by 25%, through opening up the cell structure.

Glycosylation (glycoprotein)

Glycosylation is putting sugar molecules on things, almost always other molecules and usually proteins: glycosylated proteins are called glyco-proteins. Most of the proteins present on the surface of cells, viruses, and in the blood of animals are glycosylated, and so it is considered likely that some biopharmaceuticals will also have to be glycosylated to have the same function as their natural counterparts. (It is not clear if many of the peptides produced for biopharmaceuticals actually are more effective or more stable in the body if they are glycosylated.) Bacteria do not glycosylate their proteins (or rather have quite different types of peptide–sugar linkages compared with animals), and so genetic engineering techniques have been developed for yeast and eukaryotic cells that *do* glycosylate. Of course, they do not always glycosylate in exactly the same way as human cells do.

Sugars can be linked on to the proteins through the amide groups of asparagine in the short peptide sequence Asn–X–Ser/Thr, or, more rarely, through the hydroxyl of serine and threonine. This means that how much a protein may be glycosylated can be predicted to a limited extent from its amino acid sequence, and hence from the sequence of its gene. Whether this has practical implications, as opposed to being a *post hoc* rationaliza-tion of the sugars you find on the real protein, is still debated.

Such glycosylation is a form of post-translational modification, i.e. modification of the protein's chemistry after the protein has been 'translated' from RNA. Other protein glycosylation is chemical, and occurs whenever a protein sits in sugar solutions for a long time. This is also called **glycation**.

Other molecules can be glycosylated, especially cell surface lipids. The resulting glycolipids are important as tags to allow the body to recognize its cells, especially cells in the blood. Thus they may be important functional components of liposomes, enabling the maker of liposomes to fool the body into thinking that they are cells. Proteins can also have lipids linked on (forming lipoproteins) or even glycolipids. The results cause very different responses from the immune system compared with the unmodified protein; however, making such complex derivatives is much more difficult than making relatively simple glycoproteins.

Although proteins have well-defined places where sugars can be coupled on to them, whether sugars are coupled on, and what sugars are coupled, depends on many things. Among them are the cells the proteins are made in, and the metabolic state of the cells. Thus, proteins come in variants with different sugars linked on to the same polypeptide chain; these variants are called glycoforms. One cell can make a mixture of different glycoforms. The different glycoforms have detectably different functional properties in many cases, and are 'seen' as different by the immune system. Viruses, in particular, come as a population of different glycoforms and not as a single chemical entity: thus HIV (the 'AIDS virus') has different sugar moieties on its surface depending on the cells it is grown in, and exactly what strain of virus is grown in them. These variations certainly bind anti-HIV antibodies differently, and may affect the immune system of an HIV-positive person differently. Cancer cells often produce different glycoforms from normal cells, usually glycosylating their cell surface proteins less. Many tumour markers are in fact glycoprotein differences that are specific to the cancer cells, and hence are potential ways of diagnosing the cancer or targeting drugs to it.

Gold and uranium extraction

Gold and uranium are mined in commercial quantities using microbial leaching methods. Unlike other metal extractions using bacteria (*see* **Leaching**), gold and uranium are extracted using bacteria because of the high added value of the metals and some specific features of the elements.

Gold is usually found as metallic gold mixed with other materials. Crushing the minerals releases the gold metal, which can be separated physically, often by washing ('panning'). However, substantial sources of gold are ores in which the gold is extremely finely divided and so cannot be released by conventional crushing or milling, called refractory ores. Many such ore types, with widely differing chemistry, can contain gold, but it is often associated with sulfides, especially pyrites and arsenopyrites, both of which can be oxidized by bacteria. To release the metal, the sulfide must be removed chemically. Bioleaching methods digest the refractory gold ore in a tank fermentation system with a bacterium, usually *Thiobaccilus ferrooxidans*, which oxidizes the sulfide to sulfate. This is usually soluble, so the gold particles are released for mechanical collection. Gold extraction using biological processing is gaining support because the alternatives (oxidation of the sulfur to sulfur dioxide or dissolving the gold out of the mineral using cyanide) are increasingly considered to be environmentally unacceptable.

Uranium mining follows more conventional bioleaching lines, with ores that are low in available uranium being incubated with an oxidizing bacterium to release the metal. The tetravalent insoluble uranium is oxidized by ferric ions (generated by the bacteria) or directly by the bacteria themselves to soluble uranium(VI) ions. These can then be recovered from the nutrient mix running off the ore heap.

GRAS

This stands for 'generally regarded as safe', and is an important category for acceptance of biotechnological products in Western countries, and especially the USA. It means that the product has a long history of use for a specific purpose.

For microbial or genetically engineered products, regulatory approval for general release is much easier if the product is made in an organism that is GRAS, since the only unknown is then the new product, not the organism as well. For isolated materials, to be accepted as GRAS in one application (e.g. as a foodstuff) can greatly help getting approval for another application (e.g. cosmetics). The exceptions are usually pharmaceutical applications, where any new product, even if it is believed to be chemically identical to a previous product but made in a new way, must go though a full set of toxicological and clinical trials before it is released.

GRAS status varies from country to country. Quite a few of the biotechnological products used traditionally as foodstuffs in Japan, such as seaweed-derived materials and natural colourings like astaxanthin, were not seen as GRAS in the West because there was no history of their use here.

Growth factors

Growth factors are materials (apparently invariably proteins in mammals) that stimulate growth. They are of great interest as potential drugs (biopharmaceuticals) because they could be used to assist wound healing or even encourage tissue regeneration. Growth factors not only stimulate cell division, but also influence differentiation of cells and, in some cases, select which cells in a mixed population divide or differentiate.

The ones most often discussed are:

Epidermal growth factor (EGF). Stimulates a variety of cells in the upper skin to divide and differentiate. Could have a use in helping wounds to heal.

Erythropoietin (EPO). A factor that stimulates the cells that give rise to red blood cells. Thus it is used to boost the number of red blood cells in the blood, which is useful for leukaemia or kidney dialysis patients. It is also rumoured to be used by marathon runners to increase their blood's capacity to carry oxygen, a use that almost certainly dangerous. Erythropoietin is being made by Amgen and Genetics Institute, who are involved in a patent dispute about the protein.

Fibroblast growth factor (FGF). Stimulates growth of the cells common in connective tissue and the 'basement membrane' that many cells are attached to. It has been suggested as a stimulant to the healing of burns, ulcers, and bones.

Haemopoietic cell growth factor (HCGF). Stimulates the production of many of the haematoietic cells, i.e. cells made in the bone marrow and ending up in the blood.

Neurotropins. See separate entry.

Platelet-derived growth factor (PDGF). This stimulates connective tissue to grow, and is associated with wound healing.

Stem cell factors. Proteins that stimulates the 'stem cells' from which all the cells in the blood are made. The stem cells reside in the bone marrow. (Actually, many tissues have their own 'stem cells': these ones are specific for the blood, they are haemopoietic stem cells.)

Hairy root culture

This type of plant culture consists of highly branched roots of a plant. A piece of plant tissue (an explant, usually a leaf or leaf section) is sterilized to remove the bacteria on the surface, and then treated with a culture of the bacteria *Agrobacterium rhizogenes*. Like its cousin *Agrobacterium tumefaciens*, *A. rhizogenes* transfers part of its own plasmid DNA to the cells of an infected plant. This causes alterations in the plant's metabolism, including alterations in hormone levels. This in turn causes the explant to grow highly branched roots from the sites of infection. The roots branch much more frequently than the usual root system of that plant, and are also covered with a mass of tiny root hairs, hence the name of the culture system.

The hairy root cultures do not require hormones or vitamins to grow, unlike explant cultures or cell cultures of plant cells, so they can grow on simple media of salts and sugars. They are also genetically stable, again unlike explant or cell cultures, so they can be cultured in bulk without the culture changing. Their most significant feature, however, is that they produce secondary metabolites at levels similar to those made in the original plant. Thus they can be used as a replacement for plants for making such compounds as food flavours or fragrances. As such they are the target of considerable interest and research, although there are no products yet.

Hairy root cultures have been grown in several large laboratory fermentor systems as well as small pilot plants. They look like a mass of fibres when growing as an unstirred mass: they can be grown in a stirred tank reactor, but they are rather sensitive to being broken up by the stirring machinery. However, because they grow and metabolize more slowly than bacteria and do not have nearly such high oxygen demands, stirring is not usually necessary to obtain a successful culture.

Harvesting

Harvesting in biotechnological terms usually means collecting cells or organisms from a growth system. If the organisms are very large (like, say, trout) this is not difficult. However, most biotechnology uses single-celled organisms like bacteria or yeast, which have to be actively collected. Among the ways of doing this are:

Centrifugation. Expensive, but guaranteed to collect any particle. It can be used in small volumes to purify viruses, and anything as large as a bacterium can be processed quite easily. There are two variants. The mixture can be centrifuged in a 'pot', where the biomass sinks to the bottom and forms a pellet. Alternatively, the mixture can be spun in a pot with holes in it, and the filtrate is spun out while the solid remains behind. This has the advantage that more liquid can be added, so the centrifuge can be run continuously. This is called basket centrifugation.

Filtration. There are a range of filtration systems. This is again cheap and effective, but usually they have limited capacity. The reason for this is that the filter is, of necessity, full of holes that are just a bit smaller than the cells you want to collect. So, after a while, the cells fill up all the holes, the filter is fouled, and filtration stops. **Cross-flow filtration** can help to solve this.

Flocculation. This is popular. By adding an additional reagent to the reaction mix, or by altering conditions, you get the cells to stick together and settle out like snow. (Any loose association of particles into an irregular clump is called a 'floc'.) This is often the only practical way of removing cells from really big fermentations, and is essentially how yeast is removed from brewing vats after fermentation is complete. Many types of cells form flocs spontaneously under the fermentation conditions being used: this can be valuable (to separate the cells from the liquor) or disastrous (because the cells will not stay in suspension) depending on the needs of the fermentation.

Hydrocyclone. This is also called a whirlpool separator. Here the liquid flows relatively slowly round a bowl and the cells collect in the middle, like grounds in a stirred coffee cup. Cells are collected from the centre, usually to be recycled to the fermentation. Spent liquor, largely free of cells, is collected from the periphery. This is a highly energy efficient method of separation.

Hydrocyclone

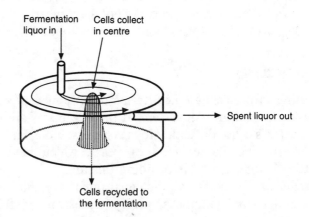

Fermentation liquor in

Cells collect in centre

Spent liquor out

Cells recycled to the fermentation

Healthcare reform

Reform of healthcare systems has been high on the political agenda of most Western countries in the last decade. This affects many aspects of biotechnology directly. The main concern is to reduce the rate of increase of the cost of healthcare. This has been a particular issue in the USA, where the Clinton administration promised a complete overhaul of the healthcare system, but was unable to deliver it. A particular target is Medicare, the government-supported healthcare in the USA for those unable to pay for the more widely used private healthcare. Biotechnology is affected in several conflicting ways.

One of the aims of the reform programmes is to introduce price controls, limiting what companies can charge for therapies. Since biotechnology product development is very expensive, this may hit biotechnology companies hard. However, most of the cost of healthcare is the cost of hospitals and doctors. Therefore, if treatment were more effective then people would spend less time in hospital and the costs would go down, even if the cost of the treatment itself were high. The real target of cost reduction should be the overall cost of treatment (the argument goes), not just the cost of the pills. (This is the complex area of pharmcoeconomics.) This is an argument for expensive, highly targeted biotechnology drugs.

Most of the cost of healthcare is directed at keeping old people alive for a bit longer, rather than at preventative healthcare. Biotechnological drug development is usually focused at the high-tech end of the therapeutic spectrum. Preventative medicine (basic nutrition, sanitation, vaccination programmes, and contraception advice) is not what the biotechnology industry usually provides.

In an environment where everyone is talking about cutting healthcare costs, investors are less willing to provide money to biotechnology companies. In the early 1990s in particular, this and a series of very public failures of clinical trials of biotechnological products caused investors to flee biotechnology, with severe consequences for new companies struggling to raise capital.

Other aspects of healthcare reform are a greater concern with respect to quality of life and ethical issues. The effect of quality of life can be estimated by discussing quality-adjusted years of life saved (QUALYs) rather than simply years of life saved (YLS), but, ultimately, what constitutes quality of life is a matter of judgment. Biotechnology-based therapies are usually claimed to provide better quality of life than less well-targeted, conventional drugs. This is part of a range of social issues around the public acceptance of biotechnology, usually misnamed 'ethics', which have been a significant brake on the industry.

Herbicides and resistance

One of the earliest targets of genetic engineering on plants was to make them resistant to common herbicides. If a broad spectrum herbicide was sprayed on to a field planted with such resistant crops, then all the plants except the crop would be killed, thus providing an effective method of weed control without having to develop herbicides specific to each weed type.

The tolerance mechanism has to be designed to fit the herbicide: consequently, different companies have been working on engineering resistance to their particular herbicide. There are two general approaches: alter the enzyme that the herbicide normally attacks so that it is no longer a target for that chemical, or add a system for detoxifying the herbicide in the plant.

There were concerns that such engineering would lead to increased use of the herbicides, at a time when it is generally accepted that chemical use should be kept as low as possible. However, field trials have vindicated the original view that being able to use pesticides effectively actually reduces the total amount used. The farmer does not have to use indiscriminately large amounts of herbicide before sowing to kill *all* weeds, only enough to suppress their growth relative to the crop.

There is also a concern that the genes used could spread to weed plants, resulting in herbicide-resistant weeds. It is not clear whether this could happen, or whether it is so genetically implausible as to be unrealistic.

The groups of pesticides that biotechnologists have examined so far are:

Glyphosate. Marketed by Monsanto as Roundup, this is a very commonly used herbicide which stops amino acid synthesis. Plants resistant to glyphosate have been created by giving them new, resistant enzymes and by selecting resistant cells and cloning them into whole plants. Monsanto is developing glyphosate-resistant cotton plants, expecting them to be ready for general farm use in the late-1990s.

Phosphinothricin (PPT). Produced by Hoecsht, this herbicide blocks amino acid synthesis. Resistant alfalfa has been created by isolating alfalfa cells resistant to the herbicide and cloning whole plants from them. Plant Genetic Systems have also engineered tobacco and potato to resist phosphinothricin.

Sulfonylureas. These block amino acid synthesis. Mutant genes from *E. coli* have been put into plants to confer resistance.

2,4-Dichlorophenoxyacetic acid. A compound that mimics plant hormones, and so disrupts their growth. Bacterial genes that break it down have been put into plant cells.

Triazines (Atrazine, Bromoxynil). These disrupt photosynthesis by binding to a protein (the Qb protein) in the chloroplast. Natural mutants that are resistant to triazines have an altered Qb: a resistant crop plant could therefore be made by putting that altered Qb into the plant. Getting this altered gene product into the chloroplast is, however, a major problem. Ciba–Geigy is working on an alternative route, putting enzymes that detoxify atrazine into several crop plants: because the detoxification enzymes work in the cytoplasm, this may be a simpler route for the genetic engineer.

High throughput screening (HTS)

This is an approach to drug discovery. It searches for a chemical that will act on a defined target, such as a receptor or an enzyme. Scientific research will have identified that target as a part of the mechanism of a disease: for example, it may be a cell surface molecule through which cells of the immune system become activated to start an inflammatory reaction, and so be a target for anti-inflammatory drugs.

There are two approaches to finding a chemical that will block that target. The first is a rational drug design approach, designing a molecule that you believe will interact with the target. This could be its natural ligand (insulin was 'designed' in this way) or it could be a synthetic molecule. Alternatively, you can search a huge range of essentially random molecules to find one that works fairly well as a blocking agent, and then optimize it. This is the HTS approach.

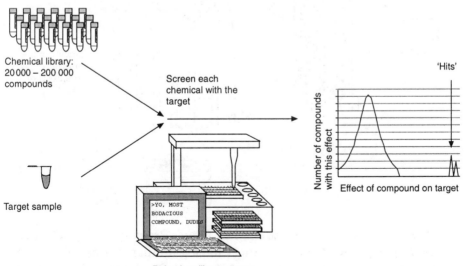

Chemical library: 20 000 – 200 000 compounds

Screen each chemical with the target

'Hits'

Number of compounds with this effect

Effect of compound on target

Target sample

>YO, MOST BODACIOUS COMPOUND, DUDE

Robot handling systems

HTS requires a large chemical library to screen, and a highly auto-mated method of performing biological assays on each of the com-pounds. The chemical library can be a 'compound bank' (a collection of chemicals built up over years of chemical synthetic work), a natural products library (a similar bank of chemicals of natural origin, often not purified), or a combinatorial chemical library. Researchers here are seeking 'molecular diversity', a collection of molecules as different from each other as they can be. The assay can be anything a biologist would do in the laboratory, but must be amenable to being done by a robot, and hence usually it must also be done in small volumes of liquid. Performing HTS on half a million mice is not practical. Automation nearly always means using assays in the 96-well 'microtitre plate' format.

Hollow fibre

Hollow fibres are tubes of a porous material. The tubes are very small, typically having an internal diameter of a fraction of a millimetre, and so their ratio of surface area to internal volume is very large. This has had two types of application.

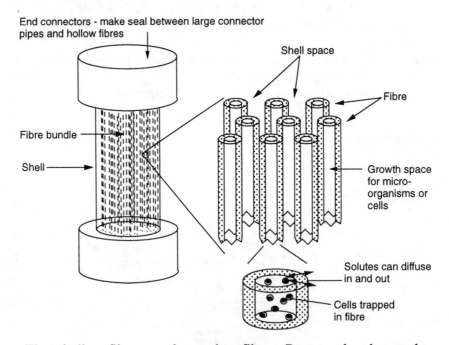

End connectors - make seal between large connector pipes and hollow fibres

Shell space

Fibre

Fibre bundle

Shell

Growth space for micro-organisms or cells

Solutes can diffuse in and out

Cells trapped in fibre

First, hollow fibres can be used as filters. Because they have a huge surface area, they take much longer to clog up than normal filters. The filters used in artificial kidney machines are often hollow fibre bundles. The second use is the hollow fibre bioreactor. This is a widely used type of bioreactor in which cells are kept inside the hollow, porous fibres, and the culture medium is circulated outside the reactor. The fibres have pores large enough to let the nutrients in and the product out, but not to let the cells out. The fibres are kept inside a shell: the space outside the fibres but inside the shell is called the shell space.

Hollow fibre bioreactors enjoy very general use in some applications. They are very effective for maintaining mammalian cells in culture because they have a very large surface area for the cells to grow on, without needing a large reactor to hold them, and because the nutrient reaching the cells can be kept fresh: mammalian cells can be very susceptible to changes in the medium in which they grow. The reactor also provides an easy way of removing the product the cells are making: this has meant that hollow fibre reactors have been particularly useful for making large amounts of monoclonal antibodies.

Hollow fibre reactors are less use when the cells themselves have to grow, because it is hard to get at the inside of the fibre to remove surplus cells, and it is hard to monitor exactly how many cells you have inside the fibres. This has meant that hollow fibre reactors have limited use for bacterial cultures.

Hollow fibre bioreactors are versions of dialysis bioreactors or dialysis fermentors. This is any system where the cells in a fermentation are retained behind dialysis membrane (a membrane that lets small molecules pass through it, but will not let large molecules or organisms pass). The reason for using fibres as opposed to a single sheet across the bioreactor is that fibres have a very much larger surface area for fermentation substrates to diffuse across. A variant on the idea is the adsorption fermentor.

Homologous recombination

Homologous recombination is a biological process by which a living cell joins two pieces of similar DNA together. It is a version of the general genetic process of recombination, by which any two pieces of DNA are spliced together in a living cell. Recombination occurs in all living things: recombinant DNA technology was called that because of the similarity of its gene splicing technology to natural recombination processes.

Homologous recombination is recombination between two pieces of DNA that are almost exactly the same, i.e. they are 'homologous'. This occurs much more readily than recombination between DNA that is completely different. It is used as a mechanism for ensuring that a cloned gene that the experimenter wishes to put into the chromosomes of a cell is inserted into those chromosomes at a specific point (i.e. at the point where the cell's DNA is the same at the cloned DNA). Because of this use, homologous recombination is sometimes called gene targeting.

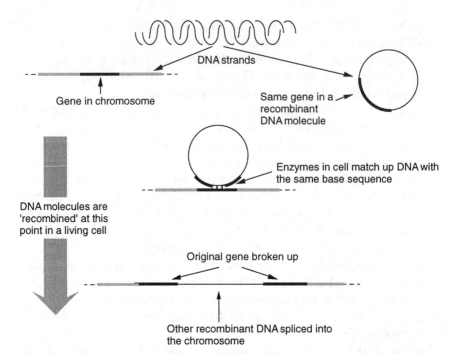

Homologous recombination is used in biotechnology in three areas.

In generating new mutants of many organisms, but particularly yeast, homologous recombination is a method for targeting a specific bit of DNA. A piece of yeast DNA is joined into your plasmid: when the whole is introduced into a yeast cell, the piece of yeast DNA can undergo homologous recombination with the corresponding bit of DNA in the original yeast's DNA. The two are spliced together and, because the plasmid is all one piece, this means that all the other bits of DNA in the plasmid are also joined into the yeast DNA. This can be used to join a plasmid into the yeast chromosomes, or, if the yeast DNA is from a known gene, to disrupt that gene by putting a large piece of plasmid in the middle of it. The resulting mutant is called a '**knock-out**' mutant.

The second role is in manipulating large plasmids such as the Ti plasmid of *Agrobacterium tumefaciens*, which are too big to alter using recombinant DNA techniques. Genes can be spliced into them in exactly the same way as they are spliced into the yeast chromosome.

The third application is in making transgenic animals (and potentially in gene therapy). Here, again, homologous recombination is used to carry a foreign gene into a cell's chromosome. However, the reason for doing it in this case is to avoid disrupting any genes in the target cell, and to make sure that the foreign gene arrives in a suitable chromosomal environment. The DNA that surrounds genes in mammalian cells (and many other types of cells) affects how that gene will be expressed. Thus it is important to target a foreign gene to a suitable place in the host cell's chromosomes, so that the gene functions correctly, and it is essential that the gene is not targeted to a site where it will damage the functioning of other genes. Homologous recombination offers a route to doing this, and hence makes the production of transgenic animals more reliable. It also offers the possibility of useful gene therapy in humans since one of the main problems with gene therapy ideas at the moment is the threat that the 'therapeutic' gene introduced into a patient's cells will do as much damage as the original disease.

Human Genome Project

The Human Genome Project is the descriptive name for the programme to map and then sequence all the DNA in the human genome. It is under the international umbrella of the Human Genome Organization (HUGO), and is funded primarily by the Department of Energy (DOE) and National Institutes of Health (NIH) in the USA, the European Commission (EC) in Europe, and a range of charities and national research councils, including the Wellcome Trust in the UK.

The project started largely because molecular biologists realized that they could sequence the whole human genome, given the money. It is supported strongly by the biotechnology and pharmaceutical industries because it will provide a database of information from which companies can obtain the DNA sequence, and hence protein sequence, of all the proteins in humans, including those that are the potential targets of new drugs, and because it could be of substantial assistance in medical genetics, including the diagnosis of inherited predisposition to disease.

In order to make the sequencing of the three billion bases of DNA in a human genome feasible, the genome projects have set up a number of less ambitious milestones along the way. The first is a complete genetic map of humans, defined by RFLPs. The second (which looks like being the one that will be completed first) is a complete sequence of all the cDNA in humans. As an intermediate step to sequencing all cDNAs, many groups are completing banks of ESTs (Expressed Sequence Tags), partial cDNA sequences which uniquely characterize the gene transcripts and hence the genes from which they come. In any case, it is unlikely that the human genome will be sequenced indiscriminately: some bits are much more interesting than others.

The genome project has been a major motivator behind the development of new methods of sequencing and mapping DNA, and the sheer volume of data involved has driven the development of much of bioinformatics. However, the rewards in terms of drug discovery have been disappointing, and most companies working in the area are turning to 'functional genomics' as the next step in the programme.

Human growth hormone

Human growth hormone (hGH) is one of the earliest proteins to be made by genetic engineering that has gained approval for use as a drug: Genentech sold $150 million worth in 1990. Mammalian growth hormones are produced naturally by the pituitary gland in young animals before and during adolescence. They increase the rate of growth and stimulate the body to put on muscle mass. After the age of 30 production of growth hormone falls: injections after this age cause muscle to build up and fat to decrease.

Human growth hormone is used medically for rare children's diseases where the body does not produce its own growth hormone. It can also be used to treat several diseases where extremely short stature is part of the disease, although not because of a shortage of growth hormone, such as the chromosomal disorder Turner's Syndrome.

Recent work suggests that HGH may reduce, or even reverse, the reduction in muscle mass that occurs with ageing, and also improve skin elasticity and muscle tone. It could therefore be an anti-ageing drug. Recent trials have suggested that the effects are not significant, and it may even have an adverse effect, so the jury is out. However, even if it only reduces the effects of ageing, without lengthening life span, it could still be enormously attractive. Against this must be set the near certainty that the drug will have some side-effects: whether they will be trivial or life-threatening remains to be seen. In practice, any trial of hGH as an anti-ageing drug must focus on symptoms, as it would take 20 years to prove a genuine anti-ageing affect. A related area of use could be as an anti-wasting agent in diseases such as AIDS.

A third area of use of hGH is simply illegal, but may go on anyway. This is in drug abuse in sport (*see* **Sports and biotechnology**).

Hybridization

Hybridization has several meanings in biotechnology and molecular biology.

DNA hybridization. This is the formation of a double helix of DNA from two DNA strands. The two separate strands of DNA will come together to make a double helix if their bases are complementary, so that wherever there is an A in one strand there is a T in the other, a G in one strand and a C in the other. (In fact there is a slight degree of laxity about this, and, depending on how long the DNA strands are, up to 10% wrong or 'mismatched' bases can be tolerated.) DNA hybridization is used as a method for using one bit of DNA (the probe) to find out if a complementary bit of DNA is present in some mixture of DNA species.

It is used in 'blot' techniques, PCR, gene library screening, DNA fingerprinting, and a range of other techniques.

DNA usually comes in a double helix. To make a hybrid molecule, the original DNA helix must be separated into two strands, a process called 'melting'. The temperature at which a particular helix 'melts' depends on:

- how long it is (for helices shorter than about 50 base pairs, the shorter they are the lower the melting temperature),

- how many G and C bases it has (more $G+C$ means a higher melting temperature),

- whether the helix is perfectly matched, or whether one or more of the bases in it is mismatched, i.e. is not opposite its complementary base in the helix,

- a range of chemical conditions, like how much salt there is in the solution.

The same 'melting temperature' governs the temperature at which a new hybrid helix will form. Thus it is often used as a measure of how stable a new helix would be.

Original DNA

Dissociate the double helix into single DNA strands (denature or 'melt' the DNA)

Single stranded DNA

DNA probe

Add DNA 'probe'

Hybrid molecule (heterodulpex)

Those parts of the two DNA molecules that have complementary base sequences form a double helix

Molecular hybridization. This is forming a new molecule that has functional parts from two different molecules. It could therefore have a combination of properties from its two 'parents'. Examples of this approach are the new antibiotics that can be made by combining the enzymes that make two old antibiotics in one cell, and making fusion proteins by joining two functional domains of other proteins together.

Cellular hybridization. This is essentially another term for cell fusion.

Species hybridization. This is forming a hybrid between two species. It can be done between closely related species by suitable breeding programmes: however, species do not hybridize readily, and, apart from a few closely related species such as donkey and horse, animals rarely form hybrids in this way. Alternative methods include making **chimeras**, cell fusion (for plants; this rarely works for animals) to produce a new species with all the genes of both the original, or using bacterial plasmids to shuttle genes between bacterial species.

Hydrophobicity

A hydrophobic molecule is a molecule that dissolves very poorly, if at all, in water, but dissolves quite well in solvents such as butanol or toluene. They are non-polar molecules, i.e. they are essentially electrically neutral all over. The opposite is the hydrophilic molecule, which dissolves well in water or DMSO (dimethyl sulfoxide) but poorly, if at all, in toluene or long-chain alcohols. These molecules usually have partly charged groups on their surface, and often form ions when dissolved in water. Most biological molecules are to some degree hydrophilic, a major exception being the triglycerides ('fats'), which are hydrophobic. Hydrophobic molecules are therefore sometimes called 'lipophilic'.

When given a choice of environments (for example, a mixture of water and oil to dissolve in) hydrophobic molecules will chose a hydrophobic environment (in this case oil) and hydrophilic molecules a hydrophilic environment (water).

Hydrophobicity and hydrophilicity are measures of how well molecules dissolve in non-polar solvents and water, respectively. There are degrees of hydrophobicity and hydrophilicity. Thus, among the amino acids, glutamic acid or lysine are highly hydrophilic, because they form ions easily and dissolve readily in water, while tryptophan has a large, uncharged side chain and is more hydrophobic in character. These differences in hydrophobicity can be used to separate the molecules. Hydrophobic chromatography uses this phenomenon: a mix of molecules is passed over a solid material that is mostly hydrophobic in character. The molecules that are more hydrophobic will stick to the material more strongly, and so will not be washed across the solid support as fast as the hydrophilic ones.

Many biological molecules are sufficiently large to have distinct hydrophobic and hydrophilic bits. These molecules are called amphipathic. If the two regions of the molecule are at opposite ends, then the result is a surface-active material: it will tend to congregate at the junction between a hydrophobic and a hydrophilic solvent. Phospholi-

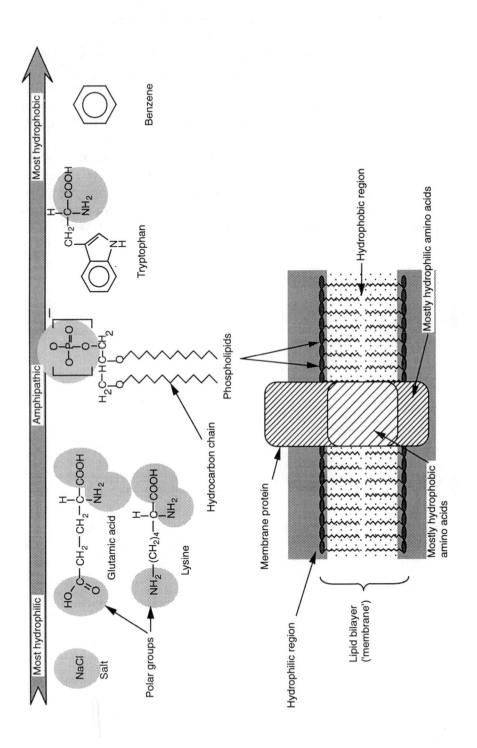

pids are of this type. Phospholipid membranes are arranged so that the 'tails' of the phospholipids form a layer of hydrophobic 'liquid' that dissolves quite different chemicals to the watery phase around it. Proteins also almost invariably have a mixture of 'hydrophobic' and 'hydrophilic' amino acids. The protein folds so that most hydrophilic molecules are exposed to the waterery solution they are dissolved in, and the most hydrophobic ones tucked away inside the protein. Thus the distribution of hydrophobic and hydrophilic molecules along a protein (sometimes called a 'hydrophobicity plot') can be a clue to how the protein will fold up. In particular, proteins with a large region of hydrophobic amino acids in the middle of their sequence are often associated with membranes, with the hydrophobic amino acids embedded in the hydrophobic layer in the middle of the lipid layer.

Imaging agents

A range of proteins are being developed as imaging agents or contrast agents. This means that they are for use with the various types of body scanners. The proteins (usually antibodies) are linked to a chemical group, which allows the scanner to see them very easily. The proteins bind to specific tissue types, usually tumours, and so allow the scanner to distinguish those tissues from the surrounding tissue: in the absence of a contrast agent the target cells could look exactly like the surrounding tissue.

Imaging agents can be made for any of the main imaging systems:

CAT scanning (computerized axial tomography). This techniques uses X-rays, and consequently the label tagged on to the antibody is an X-ray-opaque substance, usually something made from a 'heavy' metal like gold.

PET scanning (positron emission tomography). This technique injects a very small amount of a radioisotope into the body, and then tracks where it goes by following the path of the radioactive particles. The favoured element to tag on to your antibody for this is technetium.

NMR (nuclear magnetic resonance). This uses the way that the body absorbs microwaves when it is in a strong magnetic field. Chemical groups absorb microwaves differently depending on what sort of field they are in and what the group is. A wide range of materials can be used as 'contrast agents' for NMR scanning.

ESR (electron spin resonance). Little used, but of substantial potential, ESR detects 'unpaired' electrons that turn up in some types of compounds, usually those involved in energy metabolism. NMR and ESR use no radioactivity, and so are gaining favour as diagnostic systems because of widespread nucleophobia, which is especially prominent in the USA.

Immobilized cell bioreactors

Many of the plant and animal cells grown by biotechnologists are most effectively handled when the cells are immobilized on to a solid support. This helps to shield them against the stirring forces necessary to mix a bioreactor's contents, and makes them easier to move around and separate from the substrate.

There are a wide range of immobilized bioreactors. They fall into three classes:

Membrane bioreactors. These grow the cells on or behind a permeable membrane, which lets the nutrients for the cell through but does not let the cells themselves out. Variations on this theme are the **hollow fibre**

reactor, a common way of growing hybridoma cells for making monoclonal antibodies.

Filter or mesh bioreactors. Here the cells are grown on an open mesh of
an inert material, which allows the culture medium to flow past it but
retains the cells. This is similar to membrane and hollow fibre reactors,
but can be much easier to set up, being similar to conventional tower
bioreactors with the meshwork replacing the central reactor space.

Carrier particle systems. The term 'immobilized cells' is often taken to
mean that the cells are immobilized on to something not all that much
larger than they are, such as small nylon or gelatin beads. The reactor
can then handle the beads in the same way as granular catalysts are
handled in chemical reactions. There are several ways of doing this.
Normal bioreactors of all sorts can be adapted to handle the larger
particles. This works best if the particles are of neutral density, so that
they do not sink or float. Alternatively, if the particles settle out fast, the
bioreactor can be a fluidized bed or a solid bed bioreactor. In the former
the particles are kept suspended in a dense, fluid mass by liquid pumped
through them from the base. The mass behaves like a liquid, even though
it is made of solid particles. In the latter the fluid flow is not fast enough
to push the particles apart, and so they sit in a bed at the base of the
bioreactor with the fluid flowing past them. Packed bed reactors come in
various shapes (conical—'tapered bed reactor'; disc-shaped—'radial
flow packed bed') to help fluid flow through them evenly.

Immobilized cell biosensor

These are biosensors, i.e. detector devices, that use a biological bit to
allow them to detect only one thing at once, and use living cells as their
detector system. They are often called microbial biosensors, since the
cells are usually bacterial ones.

As with any biosensor, there are two parts to immobilized cell sensors:
the immobilized cell, which does the sensing and produces a very weak
signal of some sort, and a detection and amplification system, which
takes that very weak signal and amplifies it to something detectable. The
cell used depends on what you want to detect. Some typical examples of
analytes ('things to be analysed') are:

- amino acids (using bacteria that metabolize them)

- glucose (using almost any cell)

- toxic chemicals (using any bacterium that is sensitive to the chemicals
  to be detected)

- carcinogens (using bacteria that are defective in DNA repair genes)

- BOD (biological oxygen demand, i.e. how much organic matter there is in waste waters)

- heavy metals (using metal-resistant bacteria)

- herbicides (using plant cells or blue-green algae)

- toxicity (using cultured animal cells)

Only a few of these have been transformed into realistic sensor systems.

The readout methods can be equally diverse.

Gas generation/depletion. A favourite method where the amount of oxygen burnt up or carbon dioxide produced by the bacteria is measured. This is rather non-specific since almost anything the bacterium does burns up oxygen and generates CO_2.

Light production. This uses luminescent bacteria, either ones that are naturally luminescent or ones that have the relevant genes (for luciferase, the light-generating enzyme) genetically engineered into them. Light production is either a measure of general bacterial well being (for toxicity sensors) or is coupled to the presence of a specific chemical.

Direct electrochemical coupling. Some groups are working on hooking the electrode directly into the bacterium's own electron transport system. This is a more sophisticated version of measuring oxygen uptake.

Bacterial biosensors are usually much less specific than other biosensors since bacteria are very diverse and complex things. However, they have the substantial advantage that they are very active, and so make a 'signal' that is easier to detect than that produced by antibodies or DNA probes.

Of the few commercial 'biosensor' systems, several are bacterial biosensors: two luminescence-based bacterial biosensors (for toxicity and BOD measurement) are in use in the water industry, for example.

Immortalization

Immortalization of a cell type is its genetic change into a cell line that can proliferate indefinitely. Cells taken from a mammal, called primary cells, will divide in culture for 20–60 divisions, but then stop dividing. This is not because they have run out of room or nutrients to grow, but rather because they have become incapable of dividing any more. They usually show characteristic changes in their structure, and reduce the amount of biotechnologically useful product they produce, be it metabolite or protein. This is called cell senescence, and severely limits the manipulations that primary cells can be put through.

To avoid this cells are 'immortalized': put through a treatment that allows them to overcome senescence and divide indefinitely, keeping whatever differentiated features they had to start with. This is done in a number of ways. Several oncogenes, when transfected into a cell, will immortalize the cell. Some genes from oncogene (tumour-causing) viruses can also immortalize cells, notably the T antigen gene from the SV40 virus. A third route is to look for a spontaneous mutant in the cells you want to immortalize. This is done by growing a large number of the primary cells in culture and simply looking for any that keep on growing when the others have become senescent. The rate at which this happens varies between different organisms (mice seem to generate immortal cell lines much more readily than humans, for example).

The last common route is via **cell fusion**. If a mortal, primary cell is fused with an immortalized cell line, the result is usually an immortalized cell. This is how the technology of making monoclonal antibodies immortalizes the one lymphocyte that is to make the antibody characteristic of a hybridoma. All the lymphocytes in a sample are fused with a suitable immortalized cell, so they are all then immortalized: the experimenter can then grow them indefinitely in the search for the hybridoma that is producing the antibody they want.

A related concept is transformation of cells. This is their immortalization and conversion into a more aggressively growing form of the cell. Cells taken from advanced cancers are not merely immortalized, they are transformed. Typically, such cells will grow to form clumps of cells (colonies) in soft agar media, whereas normal, untransformed cells will not: they need a hard, charged plastic surface to grow on. You can measure the extent to which a chemical or a virus can transform cells by measuring its clonogenic potential in a clonogenic assay, i.e. exposing cells to the material and then seeing how many can form colonies in soft agar.

Immune system

The immune system is the system of cells and molecules that gives us protection against invading organisms. The system cannot be preprogrammed to recognize pathogens, because the pathogen would simply alter its appearance at a molecular level to evade recognition, so the immune system has to 'learn' what each invader looks like, and then mount an appropriate response to it. Doing this takes a lot of different cell types and molecular interactions. These are important to biotechnology in two ways:

- as a source of materials. Many materials made by the immune system are valuable biochemicals, and biotechnology has invested a lot of effort in making them efficiently. Leading the list are antibodies and cytokines.

- as a target for therapy. The immune system can go wrong in many ways, and biotechnology has expended much ingenuity in trying to cure those diseases without destroying the beneficial functions of the system.

How the immune system works. The immune system recognizes molecules on potential invaders. The molecules are called antigens, and the specific part of the molecule recognized by a specific cell is called an epitope (*see* **Antibodies**). The immune system makes several families of molecules that fit these epitopes, binding to them very tightly. It does this by generating many random variants on a common structural theme and selecting the ones that bind. There are two classes of cells making such molecules: B cells (which make antibodies) and T cells (which make T cell receptors; TCRs).

How do the B and T cells know which antibodies and TCRs to make? The key to the system are a group of T cells called helper T cells, or CD4+ cells. Potential antigens are continuously scavenged from the blood, broken down, and 'presented' to the T cells by antigen-presenting cells. If the fragments of antigen bind well to the TCR, then the T cell becomes 'activated' and passes on the news to other T cells and B cells. The B cells become activated in turn to make the corresponding antibody. Other types of T cells (CD8+ cells, called killer cells) become activated to identify that antigen when it is part of a cell, and attack the cell directly. This latter type of immune response is called cellular immunity or cell-mediated immunity, and is important for fighting off viral diseases and immune attack on cancers. The antibody-type response is called humoural immunity, and is important for attack of bacteria. When the immune system experiences an antigen (usually called 'challenging' the immune system), it responds with antibodies and/or a T cell response, depending on the 'arm' of the immune system that has been stimulated the most.

Antibodies can kill their targets in several ways. Antibodies stuck to a cell are a signal for macrophages to engulf the cell and digest it. Antibodies can also trigger 'complement', a complex set of proteins that puncture holes in cell membranes.

Immune modulators. Also immunomodulators. Cytokines come into this as part of the signalling system that collects all these cells together at the site of a potential infection. There are over 18 cytokines known, each with a slightly different message to carry. There are also lots of other cell-surface molecules involved. Here we will only mention the MHC (major histocompatability complex) which acts as a 'carrier wave' for the information that passes between cells about the epitope they are 'seeing'. Antigens that do not bind to the MHC properly are often not 'seen' by the immune system.

What can go wrong? If the immune system is constantly scavenging molecules, why does it not scavenge the body's own molecules, present them to T cells, and start of a process of self-destruction? Sometimes it does, and the resulting disease is autoimmunity: the molecules it recognizes as targets are called autoantigens. Autoimmune disease can range from minor ones such as eczema, where the body attacks itself to a small degree as a side-effect, through more severe attacks on particular tissues, as happens in rheumatoid arthritis, to the complete destruction of one tissue, as happens in insulin-dependent diabetes. It is also possible for the self control to break down completely, and the body to self-destruct massively: one example of such a disease is system lupus erythromatosis (SLE).

Usually, however, the body prevents T cells that think that the body's own molecules are dangerous from ever reaching the blood. They are eliminated in the thymus, where T cells are made ready for action. T cells can also be made permanently unresponsive to a stimulus outside the thymus. They are then said to be in a state of anergy, and the immune system is 'tolerant' of that antigen. One of the aims of biotechnological intervention in the immune system is to induce tolerance to selected foreign antigens, such as the foreign MHC of transplanted tissues. The immune system would then 'see' them as part of the body, and not as a potential invader. The only other route is to knock out those sections of the immune system with drugs that paralyse or kill the immune cells. This means that the body is now no longer able to fight off infection properly.

The other type of disease of the immune system is inappropriate response. Some people's immune systems react far more violently to, for example, peanuts than to rabies virus, even though no one every died of an infection of peanuts. These people are allergic to peanuts: the immune system appears to be presensitized to reacting to some of the molecules in peanuts. Hypersensitivity of this sort is very common, and lies behind asthma, eczema, and probably many other diseases like arthritis. Asthma, in particular, is thought to be triggered by an allergy to the mites that breed in the warm fabrics of modern homes, especially mattresses.

Disease like these are inflammatory diseases, where the immune system's actions cause inflammation that is not associated with fighting off an invader, so tissues get red, swollen, and ultimately attacked and even destroyed. Cytokines play a dominant role in this effect. As mentioned above, such diseases often have an autoimmune component, the body attacking not only the invading peanuts but also any other bystander molecule, which includes the body's own molecules.

One hundred years ago, such immune diseases were rare. Asthma, for example, was considered a mild disease that would be not be seen in most doctor's practices at all. Now between 2 and 5% of all children have asthma in the West, and the disease would be frequently fatal without controlling medication. Why this massive increase? Modern life styles may be a factor: only in the last 50 years have houses been warm, dry, and almost totally enclosed, favouring a population explosion in dust mites. Pollution may also be relevant. The immune system uses nitric oxide (NO) as a signalling chemical, and the NO in pollutants, particularly car exhaust, may be confusing the signals. Basically, however, no one knows.

Immunization

Immunization is the process by which we get an animal to make an antibody against something. The animal could be a human or a farm animal, in which case the purpose of immunization is to provide that animal with the ability to make that antibody so that they are protected against a disease. Or the animal could be immunized so that we could collect its blood and extract the antibody from it, thus providing us with a source of that antibody. There are a number of steps involved.

The animal in injected with the antigen, i.e. the material that we want the antibody to react to. If that is a very small molecule (such as a steroid hormone or a short peptide) then it is usually chemically linked ('conjugated') to a much larger molecule, such as a protein. Favourite proteins are bovine serum albumin (BSA) and keyhole limpet haemocyanin (KLH).

There are several terms for subsequent stages of the immunization process.

Adjuvants. The antigen can be injected together with an adjuvant, a material that increases the immune response. Common adjuvants are mineral oils and similar complex mixtures that cause inflammation. A common brew is 'Freunds complete adjuvant'. For human use, aluminium phosphate or hydroxide gels are used.

Boosting. The first injection will give rise to a primary immune response, the production of a relatively small amount of antibody. The antibody will be mostly IgM (*see* **Antibody structure**) and its K_a will be low (*see* **Binding**). If the same antigen is then injected again, a secondary response will occur, producing much more antibody, this time mostly IgG with a higher affinity. This subsequent injection is called boosting, and is usually done several times.

Titres. To test how the immunization is going, a small sample of blood is removed and the ability of the antibodies in it to bind to the antigen is tested. The blood is diluted until the antibodies in it are so dilute that they can no longer bind to the antigen to any detectable degree. This dilution is called the antibody's 'titre'. Therefore, when measuring the strength of an antibody preparation, people often cite a dilution figure (1 in 100 000 being pretty good, 1 in 1000 being fairly hopeless). As immunization proceeds with additional boosters, the titre of the antibody should go up as the amount of antibody and its affinity goes up.

Immunoconjugate

A compound that is a combination of an antibody molecule (or part of one) and another molecule. There are several types.

Immunotoxins. See separate entry.

Antibody contrast agents and tracers. These are used in conjunction with 'scanners': CAT (computerized axial tomography, an X-ray technique), PET (positron emission tomography, a radioactive tracer system), or NMR (nuclear magnetic resonance) diagnostic devices. They all produce images of the insides of a patient, but these images can be much improved (in the case of CAT and NMR) or are only possible (in the case of PET) if some chemical is injected into the patient which the scanner can detect. If the chemical is linked to an antibody, then the scanner becomes a very sensitive way of tracking where the antibody ends up. Contrast agents are chemicals that increase the 'darkness' of the scanner image, and apply to NMR and CAT scanners (and to conventional X-rays, too). Tracers are materials that do something unique, so they 'light up' in the scan: some NMR reagents and PET scanner chemicals fall into this category (*see* **Imaging agents**).

Antibody–enzyme conjugates. These are complexes where the antibody has been chemically linked to an enzyme. They are widely used in immunoassays, where the enzyme acts as a 'flag' to mark the presence of the antibody. A few nanograms of antibody can easily be detected if a suitable enzyme is attached. Common ones are horseradish peroxidase (HRP) and alkaline phosphatase (AP).

Immunodiagnostics/immunoassays

One of the success stories of biotechnology, these are medical diagnostic methods that use antibodies. The antibody is used to detect the presence of something in a sample. The antibody latches on to its target very specifically, so is a very precise reagent. It can also latch on to the antigen at very low concentrations, and so can produce a very sensitive test. This combination has meant that monoclonal antibodies are used in about

35% of all medical diagnostic procedures. Exactly the same test technology can be used in other, non-medical applications, which are called immunoassays.

The problem with immunodiagnostics is that the antibody does not do anything obvious when it latches on to its target, so we have to arrange the assay so that some other process detects that binding has occurred. There are various facets to this.

The label. Antibodies can be labelled in various ways. As well as labels used for **imaging agents**, immunodiagnostics can use a variety of labels in '*in vitro*' assays. These usually have different names:

- ELISA: enzyme-linked immunosorbent assay. Uses an enzyme label on the antibody.

- RIA: radioimmuno assay. Uses a radioactive label on the antibody or antigen.

- FIA: fluorescent immunoassay. Uses a fluorescent tag on the antibody or antigen.

- CLIA: chemiluminescent immunoassay. There are a variety of luminescent (light-generating) labels that can be used (*see* **Luminescence**).

- immunogold: labelling the antibody with colloidal gold particles. This can be used as a label in a macroscopic assay (where is acts as a very intense dye) or in immunohistology (staining tissue sections with antibodies), where the gold particles show up very clearly in the electron microscope.

The second facet is the physical format of the assay; what reagent is attached to what object. Common aspects of assay formats are:

Sandwich assay. Here, two antibodies are used that bind to different parts of the antigen. One is trapped on a solid surface (e.g. the bottom of the wells on a 96-well plate), the other has a label attached to it. If the antigen is present it links the two, and so the label stays in the plate.

Competitive assay (competition assay). This is like a sandwich assay, but the analyte is a small molecule that competes with the binding of an enzyme chemically linked to the analyte hapten (the hapten–enzyme conjugate) to the antibody. This is virtually the only way of making an immunoassay that can detect a small molecule.

Latex. 'Latex' particles are very small particles of plastic, usually coated with the antibody: typically they are polystyrene spheres 100 nm to 1 μm across. In the presence of antigen the particles stick together into larger lumps, held together by the antibodies that coat them, hence the name latex agglutination assay.

	If antigen is there	If there is no antigen
Sandwich assay	Antigen links the labelled antibody onto a solid material	If no antigen present, the label is not bound to the solid support
Competitive assay	Antigen binds to the antibody in solution	If no antigen is present, then the antibody is free to bind onto the solid support

Type of assay	When antigen is present	When antigen is absent
Latex agglutination assay	Microspheres held together by antigen — Latex forms clumps	Microspheres not held together — Latex forms uniform suspension

The third aspect is the physical format of the assay. Is it:

Homogeneous, i.e. an assay that simply gives a result when the sample is added (together with suitable reagents), just as a colour pH indicator does. The chemistry of this can be very complicated.

Microtitre plate format, i.e. done on microtitre plates (which have to undergo a series of washes between each reaction). Performing the assay on other surfaces, such as glass plates or silicon chips, can be essentially the same, but usually results in smaller tests.

Microparticle based, i.e. the antibody is bound to tiny beads and these are moved between solutions by centrifugation, filtration, or other methods. (This is different from a latex agglutination assay, where the particles are the readout system as well.)

There are a range of semi-official trade names for more complex immunoassays (the competition for a good acronym for your immunoassay is intense). Among the more common are:

- ARIS. This uses a complex reaction in which the binding of an antibody to a synthetic target prevents glucose oxidase from working. This type of assay is now almost out of its patented period. It is a homogeneous assay (i.e. there are no washing or separation steps involved). It works for small molecule analytes.

- EMIT. This is another small molecule, homogeneous immunoassay, but one that is more sensitive than ARIS.

Other immunoassay formats fall into the biosensor category, which is considered to be more in the mainstream of biotechnology.

Immunosensors

Biosensors have a biological part and a detection part. The biological part confers selectivity on the sensor, the detection part detects the effect of whatever the biological part does and turns it into a recognizable 'signal' (usually an electrical signal). In immunosensors the biological part is an antibody. The physical part is usually a physical mass detection system or an optical device. The optical devices are discussed under **optical biosensors**.

There are two groups of immunosensors based on mass detection. Both use extremely small mass detectors, usually manufactured on a silicon 'chip' (and hence sometimes called 'microchip biosensors'), to detect the tiny changes in mass found when an antibody binds to an antigen. They are all resonance devices that measure the binding of the analyte to the probe. The simplest type is based on the tuning fork principle. The note a tuning fork sounds depends on the mass of the tines. If the mass increases, the note goes down. The sensors have the equivalent of a microscopic tuning fork with the antibody coated on the tines. The silicon surface from which the tines are made detects the frequency with which they vibrate. When something binds to the antibody, the note falls and the circuit picks this up. Surface acoustic wave (SAW) devices are a variation on this theme. Because the tuning fork is made of piezoelectric material, these are sometimes called piezoelectric sensors.

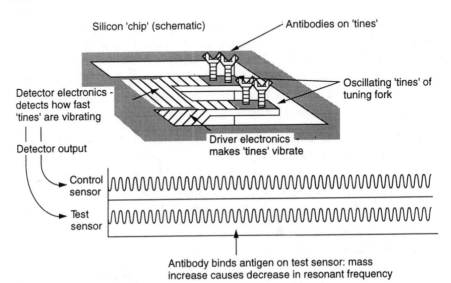

Antibody binds antigen on test sensor: mass
increase causes decrease in resonant frequency

The problem with such sensors is that anything landing them gives a signal. Thus, despite having a very specific antibody as a biological element, they are very prone to interference. So, while 'tuning fork' microdevices are well known in mechanical applications such as strain gauges and gas sensors, they have not provided a reliable biosensor yet.

Immunotherapeutics

These are drugs, usually biopharmaceuticals, that act on the immune system. Because the immune system regulates itself with a vast array of proteins that signal between the cells (including cytokines), most immunotherapeutics are proteins made by genetic engineering to boost some aspect of the immune system: the way the white blood cells grow, differentiate, or act. The cells of the immune system produce only minute amounts of these proteins, so the biotechnologist has to clone and express them in another organism to make them as drugs. Many of them have only been discovered by cloning their genes and then seeing what the resultant protein does.

Among the ones being developed as drugs are:

- interferons: second oldest biotechnological protein, being used as an immune system booster for many diseases. There are three different interferons (alpha, beta and gamma) with different actions.

- interleukins: especially interleukin-2 (IL-2).

- CSFs (colony stimulating factors). These stimulate the growth of the cells that make the white blood cells, which are responsible for the immune response.

See also **Cytokines**.

Immunotherapy

This is the use of antibodies or antibody-derived proteins in treating disease. The use of antibodies as targeting agents (in immunoconjugates or immunotoxins, for example) is usually not considered to be immunotherapy. Rather, immunotherapy means giving the patient an antibody that they cannot make for themselves, because their immune system cannot work fast enough, because their immune system is not working at all because of disease, or because the antibody is against an antigen that the body would not normally recognize as 'foreign'.

Antibody immunotherapy has been fairly spectacularly unsuccessful, costing much more to develop and being much less successful that originally hoped. It is now understood that you need to chose the target as well as the antibody very carefully. The molecular target must be highly specific to the disease you want to cure, and must be accessible to the antibody: most antibodies do not enter solid tumours, for example, so an antitumour antibody is not likely to have any effect. In addition, the antibody must be highly specific, and have the effect you want it to have.

Antibodies can be used 'naked', using the body's immune system to attack antibody-coated cells. They can also be conjugated to a 'warhead', such as a radioactive atom, a toxin (*see* **Immunotoxins**), or an enzyme (*see* **ADEPT**). Bispecific antibodies can also be used, which bind the target cell with one arm and lock on to another cell with the other, bringing a killer cell to bear on your target.

Immunotherapy can also mean using whole cells from the immune system as a therapy. This latter approach has been tried under the title of adoptive immunotherapy, where NK cells ('natural killer' lymphocytes, some of the white cells in the blood able to destroy other cells) were taken from terminal cancer patients, stimulated using cytokines to become more energetic, and then injected them back into the patient. The therapy had some effect, but severe side-effects. A variant of this is to use another class of white cells [tumour infiltrating lymphocytes (TILs)], which can target a cancer more specifically. Again, they must be obtained from the patient first. TILs have also been tagged with foreign genes in a first approach to gene therapy in treating terminal cancer. The initial gene experiments put a 'useless' gene into the cells: the ultimate idea is to put a gene into TILs that will increase their effectiveness at killing tumours.

Immunotoxins

Immunotoxins are protein drugs. They consist of an antibody joined on to a toxin molecule. The toxins used (toxins from bacteria *Diphtheria*, *Pseudomonas*, or *Shigella*, or the castor bean toxin ricin) are extremely poisonous. Probably only a few molecules of ricin inside a cell can kill the cell. Thus they are of no use as systemic drugs. However, if they can be placed at a specific site, then they can be used to kill off one type of cell with very high efficiency. This is the idea behind immunotoxins. The toxin is joined to an antibody molecule that can bind specifically to one type of target cell. The resulting 'conjugate' is injected into the blood at extremely low concentration. When it encounters its target cell, the conjugate binds to it, concentrating the toxin there. The toxin then has a much higher chance of killing the cell. ImmunoGen has a ricin-based immunotoxin of this sort in clinical trials as a treatment for leukaemia.

Refinements use parts of the toxin molecule, not all of it. Most toxins consist of a part that enables the toxin protein to enter the cell (the A chain) and a part that kills the cell (the B chain). The A chain is not toxic and the B chain on its own cannot get inside the cell to work, and so is much less toxic. Conjugating the B chain to an antibody makes a much less dangerous material: however, it can still kill cells if the antibody binds to them since the local concentration of B chains around that cell is so high that a few B chains get inside anyway.

Immunotoxins have some limitations. As large molecules they cannot get inside solid tumours easily. They are also rapidly mopped up by the patient's own immune system unless the patient is on immunosuppressive drugs. Also, there are some cells that bind antibodies non-specifically as part of the normal immune reaction. They will bind the immunotoxin, and so be killed.

Immunotoxins can be made by linking the toxin and antibody molecules chemically. They can also be made by fusing the genes for the toxin and the antibody: the resulting fusion protein is more stable, and can be smaller and less prone to binding to other tissues than a chemical conjugate. The antibody can also be 'humanized', reducing other complications.

A related idea is to use the toxins themselves as biotherapeutics (*see* **Toxins**). A reverse version of an immunotoxin is to couple a non-antibody binding site to the constant region of an antibody, producing an immunoadhesin (*see* **Dabs and other engineered antibodies**).

Induction

In biotechnological terms this means getting an organism to make a protein, usually an enzyme, by exposing it to some stimulus, usually a substrate for growth that requires the organism to have that enzyme in order to metabolize it. Induction involves the control of gene expression, but it is not a strictly genetic phenomenon since no new genes or gene rearrangements are involved. It is only the expression of genes already there.

In general, an inducible gene, i.e. one that is capable of induction, can be induced by one or a few compounds. These are called inducers. These compounds (or sometimes their metabolites) affect how a protein binds to the promoter region of the gene concerned, and so affect the control of that gene. The exact mechanisms involve are as varied as they can be (as is the case for nearly all biology). Thus, in order to be inducible, a gene needs to have the right promoter region. Some expression vectors (*see* **Expression systems**) have inducible promoters in them. They must also carry the genes for any proteins involved, of course: the inducer does not bind to naked DNA on its own.

A related term is repression, in which a compound has the opposite effect, i.e. reducing gene activity and so making the cell lose an enzyme activity. Such genes are called repressible, and the phenomenon repression. This can be very important in biotechnology since many genes for useful enzymes, such as those that make antibiotics and other secondary metabolites, are repressed by common substances such as glucose. A common mechanism is catabolite repression, where the presence of common metabolites in the growth medium of a microorganism represses the enzyme pathways that would otherwise make them. There are specific genes involved in catabolite repression, which can be removed by appropriate genetic manipulation in some cases.

Induction also means a form of logic that reasons from specific examples of something to general rules about that something. This is something biochemists do a lot (despite the fact that it is logically indefensible, it seems to work), but that is rarely what they mean by induction. Induction is also a way of having babies—despite many women's pleas, there is no equivalent repression in this case.

Inoculation

Inoculation (apart from in the sense of immunizing someone) is introducing a small culture of a microorganism into a new environment with the intention that it should grow there. Thus, fermentors are inoculated at the start of a run with a batch of organisms that have been grown to a state where they are ready to grow rapidly in the conditions provided by the fermentor. This may take some skill to achieve since the conditions under which the inoculant were grown are probably not the same as those inside the final fermentor, and so the organisms could be adapted to a rather different culture condition. The small dose of organisms (typically between 1 and 10% of the number of organisms expected in the final fermentation) is called the inoculant.

This is inoculating in the laboratory or production plant. Bacteria can also be inoculated into soil (to help bioremediation or to colonize the roots of plants), or on to plant roots or seeds directly. Again, the aim is to get them to grow in their new environment.

In vivo, etc.

There are a number of latinisms in biotechnology.

In vitro vs. *in vivo*. These are widely used when scientists are talking about doing something 'simple' in the laboratory and then taking the result and applying it to a more complicated, living system. *In vivo*. Literally means 'in the living', and means in a living system, such as a complete animal. It is contrasted with *in vitro*, literally meaning 'in glass', which is translated by every English newspaper to mean 'in the test-tube'. It means 'in the laboratory', and is taken as the opposite of '*in vivo*'.

There is no clear ruling about whether cells are *in vivo* or *in vitro*: it depends on what you are talking about. The terms are usually used to contrast one experiment with another, rather than as absolute definitions.

In situ. This means doing something in place, rather than taking it out of its usual context. Sterilization *in situ* is sterilizing some piece of equipment in place in the overall machine, rather than taking it out, cleaning it, and then putting it back.

There are a number of *in situ* techniques in cytology, the study of cells and their structure. They mean that you do an analytical technique on a cell that has been sliced and stuck to a microscope slide, rather than on chemical components extracted from a cell. This produces an image of the distribution of whatever you are testing for in the cell, for example, *in situ* hybridization. Usually, DNA hybridization is done on DNA that has been chemically extracted from the cell. *In situ* hybridization uses a DNA probe to hybridize to DNA in the chromosomes in a cell as it sits on a slide. This tells us where in the chromosome that piece of DNA lies. (Of course, the cell has to be dead to do this.) A common *in situ* hybridization technique is fluorescent *in situ* hybridization (FISH) which produces very attractive colour photographs of chromosomes with brightly glowing spots on them where the target genes are.

Ex vivo. This is used in reference to medical therapy strategies, and means that you are going to treat something outside the body rather than inside it, where it normally resides. Usually this term is applied to cells, especially blood or bone marrow cells, that are taken out of the body, treated *ex vivo*, and then put back in. The advantage is that only the target cells are treated, not the whole patient.

In silico, in machina. These mean 'done in a computer, not in real life'. Most biology cannot be done in a computer; it has to be done using the much more expensive, much messier, and much slower techniques of laboratory experiment. Some aspects of genome analysis can be done *in silico*, now that computers are powerful enough and there are enough data known about genes and their sequence. Thus Incyte Pharmaceuticals, a Californian company specializing in genome sequencing, has coined the phrase 'digital Northern' to describe the idea of comparing the amount of specific RNAs in two tissues through computer analysis rather than by doing a Northern blot. The same ideas are also called 'virtual biology'.

ISFET

Ion-sensitive field effect transistor. A field effect transistor (FET) is a semiconducting device in which the electric field over an n–p–n or p–n–p junction is used to modulate the current flowing through that junction. (The electric field attracts electrons or holes into the junction region, so providing charge carriers to carry current through the depletion layer.) It is a standard component of integrated circuits. Closely related in terms of its electronic effect is MOSFET (metal oxide semiconductor FET).

The transistor may be made into a sensor device by allowing ions to accumulate above the junction region. If the material above this region absorbs some ions specifically, then they will accumulate there and build up a charge. This will create an electric field, and so the FET will 'switch on' and a current will flow. Thus this device (an ion-sensitive FET) will allow a current to flow that is dependent on the amount of a specific ion present.

These devices have been suggested as a method for monitoring ion concentration in a range of biotechnological processes. However, they have also been turned into biosensors by replacing the ion-selective layer with an enzyme that generates ions by its action. Urease is a common example since it takes uncharged urea molecules and splits them into ammonia and carbon dioxide: the ammonia picks up a proton to become charged ammonium ions, which the electrode detects. This sort of device is also called an enzyme FET (EnzFET or ENFET).

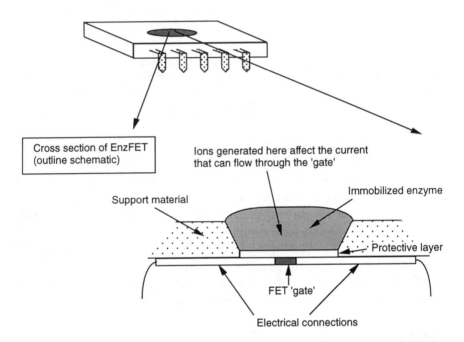

Cross section of EnzFET (outline schematic)

Ions generated here affect the current that can flow through the 'gate'

Support material

Immobilized enzyme

Protective layer

FET 'gate'

Electrical connections

The attraction of ENFETs is that they could, in principle, be manufactured by the large-scale manufacturing processes used by the semiconductor industry. The drawback is that they are too unreliable and too difficult to manufacture to be usable in most cases. A few exceptions use a FET as a detector for urease, the enzyme being used as a 'tag' to track the presence of some other molecule such as DNA or an antibody.

Claimed advantages of ISFET-based sensors include:

- They can be mass produced by the techniques of silicon chip manufacture.

- Several sensors can be put on one chip, together with control and reference electrodes.

- The extremely small size of the device means that it can measure extremely small charge changes, and consequently is very sensitive.

While all these are true for the semiconductor base of the sensor, they have not yet proven true of the whole device, except in some research laboratories.

Kinetics

Kinetics is the study of the rate of change of something. In biotechnology, it can mean a variety of things.

Cell growth kinetics. This is the rate at which cells grow, and is discussed in more detail under **cell growth**.

Reaction kinetics. The speed with which chemical reactions can be made to happen is essential to many aspects of biotechnology. Among the commonly mentioned types of kinetics are:

Binding kinetics. How fast a molecule such as an antibody binds to its target. A broad use of the term includes considerations of how fast you stir the reaction, whether the molecules can physically get at each other, and so on. Usually the molecule doing the binding is called the ligand, and so these are ligand-binding kinetics. Related to this is ligand-binding thermodynamics, measuring how much energy is gained or lost when a ligand binds to its target, and hence how tightly it will stay bound. Values for this binding energy are usually expressed as ΔG, also called the 'free energy of binding'. Basically, the more negative this number is, the tighter the ligand will bind to its target.

Solid phase reaction kinetics. See **Solid phases**.

Enzyme kinetics. This is a rigorously defined area of science, and studies how fast enzyme-catalysed reactions can go. Key terms are the Michaelis–Menten equation, and its constants (K_m and V_{max}). The latter is a measure of how fast an enzyme works, the former of how dilute a solution of its substrates it works on. (Of course, any enzyme can work as fast as you want it to, simply by adding more enzyme to the reaction. V_{max} measures how fast each unit of enzyme can work, when compared with other enzymes.) Another kinetic measure is turnover number, which is the number of substrate molecules each enzyme molecule processes each second. Yet another is Q_{10}, which is the amount the rate of reaction is increased by increasing the temperature. You can guess from the Q_{10} whether a reaction is limited by how fast the molecules are getting to the enzyme ('diffusion limited') or whether it is limited by how fast the enzyme works.

Also part of the study of enzyme kinetics is the study of inhibitors. These are compounds that slow down the catalytic activity of the enzyme. They can act in many ways. Competitive inhibitors bind to the active site of the enzyme, and so stop the substrate getting in. Allosteric inhibitors bind to some other part of the molecule, and specifically alter the shape of the enzyme so that it is no longer as effective a catalyst. Both these types of molecule are reversible inhibitors: if they are removed, the enzyme can work again. Irreversible inhibitors chemically modify the enzyme so that it will not act as a catalyst ever again.

227

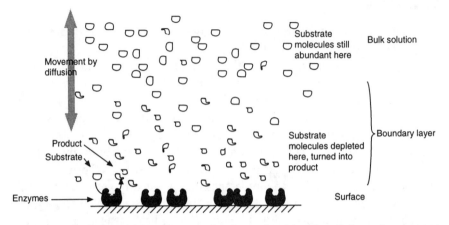

Diffusion limitation

End-product inhibition is a result of enzyme kinetics. It is an important effect controlling how much product an enzyme can make. In this case the end-product of a chain of enzyme reactions (a metabolic pathway) acts as an inhibitor of the first enzyme in the pathway. So when the pathway operates and the end-product builds up, the first enzyme is automatically 'turned off' and the pathway stops again until whatever it is making is used up. This is a form of feedback control, also called feedback inhibition. End-product inhibition relies on one of the end-products of a pathway being an allosteric inhibitor of one of the enzymes.

Knock-outs and mutants

Knock-out organisms are ones where a specific gene has been inactivated, or knocked out, by recombinant DNA techniques. They are widely used as tools to find out what a gene does: knock it out, and see what effect this has. Knock-outs are made by using homologous recombination between the target gene and a piece of cloned DNA to insert a piece of 'junk' DNA into the gene you need to disrupt. If the organism is haploid (i.e. only has one chromosome set) then this technique will result in that organism's only copy of the gene being knocked out. If it is diploid, then only one of the two alleles will be knocked out, and you will have to do some conventional breeding to make an animal that has two copies of the gene knocked out.

Knock-out organisms, especially knock-out mice and yeast, have been used extensively to try to find out what the genes discovered by genome projects are actually for. Many of the genes discovered in even such a simple organism as yeast have no obvious function, so this is the only way to know what a new gene is doing.

Knock-out mice can also be used as models of human disease (*see* **Transgenic disease models**).

Labels and substrates

A range of standard methods for labelling or tagging macromolecules are used in the life sciences.

Enzyme labels

Many enzymes are used as tags in immunoassays. They are chemically tied on to the antibody, so that their catalytic properties can be used to track the antibody. Usually they are used to turn a colourless chemical (the substrate) into a coloured one, so they can, literally, be seen.

Some common enzymes used as labels are:

Enzyme	Abbreviation	Substrate	Colour
Horseradish peroxidase	HRP	ABTS, NBT	Blue
Alkaline phosphatase	AP, AlkPhos	DNPP	Yellow
Beta-galactosidase	β-gal	X-gal	Blue

Chemical labels

A number of groups are useful labels in their own right. These include:

Fluorescent labels. The most common is fluorescein, which glows yellow-green, and rhodamine, which glows red, when illuminated with UV light. Fluorescein is usually linked on to a molecule by using the reactive derivative fluorescein isothiocyanate (FITC). An issue of importance for all fluorescent labels is quenching, the reduction in the fluorescence output by other chemicals. Many fluorescent labels show self-quenching: when many fluorescent molecules are tagged on to one protein, they quench each other so that the whole does not fluoresce very well. Fluorescent proteins are becoming popular, because they are extremely efficient fluorophores and hence can be detected at very low concentrations. The most widely used are phycocyanins (from seaweed) and green fluorescent protein (GFP) (from the jellyfish *Acquorea victoria*). A variety of GFP mutants have been isolated, so you can get GFP labels that glow different colours.

Chemiluminescent labels. These are reacted with another chemical to generate light. There are two approaches. The luminescent chemical can itself be linked to what you want to detect. Alternatively, an enzyme can be linked to the target, and used to generate light in an enzyme-catalysed reaction (*see* **Bioluminescence**).

Radioactive labels

The ultimate label is a radioactive atom in your molecule, because its presence does not affect the chemistry of the molecule significantly. Radioactivity is used far less now for two reasons. Concerns over worker safety have made people use far less radiation generally. Additionally, more scientists use laboratory robots, and these are extremely difficult to clean up if they accidentally get contaminated with a long half-life isotope. Nevertheless, radioisotopes are still used in many applications. They are characterized by half-life (the time for the amount of radioactivity to decay to half its initial level, and not, as the press believes, the time taken for them to 'become safe'), and the energy of the radiation they emit. Low energy labels like tritium (^3H) give radiation that does not penetrate far into solid matter, so they give finer 'pictures' and are safer, but you cannot detect them so easily (or at all inside, for example, a human body). Higher energy labels (like ^{131}I or ^{99}Tc) are therefore used when you want to detect only tiny amounts of label, or label in an inaccessible site, such as inside the body.

Langmuir–Blodgett films

These are thin films of molecules formed on the surface of water. The original Langmuir–Blodgett film was a lipid layer on top of the water, but the term is often used to describe lipid films in which both sides are in water, or these films when they are transferred to a solid surface.

Lipids have a polar, water loving (hydrophilic or lipophobic) 'head' and a water hating (hydrophobic or lipophilic) 'tail'. Thus, half of the molecule is soluble in water, half is not. The most stable arrangement of such molecules is to have them in clusters with the tails on the inside, away from the water, and the head on the outside. One such cluster arrangement is a flat sheet, with the tails in the middle and the heads on either side. This is the Langmuir–Blodgett film, or lipid bilayer. It is the basis of the membranes that surround living cells and some of the organelles inside cells.

Lipid bilayer films or membranes are only one example of 'liquid membranes', in which a thin layer of liquid is stabilized so that it can last for a long time in water. They all have to be stabilized by some chemical means, otherwise they collapse into little globules of liquid, or dissolve in the water.

Lipid bilayer membranes have applications in drug delivery systems (as liposomes), in biosensors, in separation processes, and in some bioreactors. Nearly all of these applications are still only laboratory demonstrations. **Liposome** and **liquid membrane** separations are described separately.

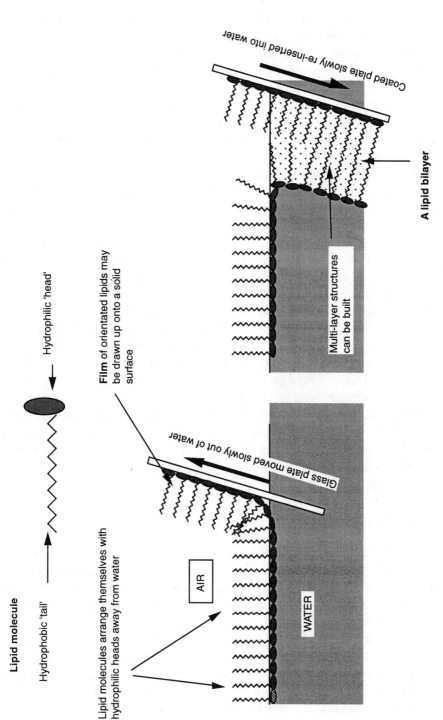

Lipid molecule

Hydrophilic 'head'

Hydrophobic 'tail'

Lipid molecules arrange themselves with hydrophilic heads away from water

AIR

WATER

Glass plate moved slowly out of water

Film of orientated lipids may be drawn up onto a solid surface

Coated plate slowly re-inserted into water

Multi-layer structures can be built

A lipid bilayer

Biosensor applications rely on the high electrical resistance of a Langmuir–Blodgett film, or on its optical properties.

Electrical sensors are based on the ability of some proteins to carry ions across a lipid membrane. Some antibiotics, proteins from nerve cell membranes, and a variety of 'transport' proteins that allow cells to get materials from outside the cell into the cell without making holes in the membrane, can all be inserted into the lipid membrane. The protein can allow one material or type of material (an amino acid, a metal ion, or maybe simply protons) to cross the membrane: in the presence of that material the membrane will conduct electricity. In its absence, the membrane will have a much higher resistance, because there will be no path for any other charged species to cross it. Thus the membrane could be a very sensitive detection system. The problem with this is that the membranes are mechanically and chemically unstable, as are most of the proteins we want to put in them. Thus, while a sensor system may be improvised in the laboratory, none have worked 'in the field'.

A related use of Langmuir–Blodgett films is as switching elements in computer-like circuits (*see* **Molecular electronics**).

An alternative sensor system based on Langmuir–Blodgett films is an optical sensor system. Because the films are extremely thin, they cause interference effects when light is shone through them or reflected off them, and these effects depend critically on how thick the film is. If antibodies are immobilized on the surface of the film, then when they bind to their antigen the total thickness of the assembly will change (from being film + antibody to film + antibody + antigen), and so the colour of reflected light will change. Again, this can be demonstrated for some very simple, model systems in the laboratory, but not for any realistic sensor application.

Leaching

Microbial leaching, or bioleaching, is the use of microorganisms, usually bacteria, to isolate metals from mineral ores by solubilizing them and allowing them to be washed ('leached') out of the ore. Thus it is a method of mining, and is a major component of microbial mining (**biohydrometallurgy**) technology.

Many ores cannot be processed economically because the concentration of metal in them is too low. Some of these ores are low-grade ores which are discarded as waste during mining operations aimed at higher grade ores. (The 'grade' of an ore depends mainly on how much metal there is in it, but also on how accessible that metal is. Clay has a very high content of aluminium, but it is extremely expensive to extract aluminium from clay.) However, if the metal can be released as a soluble salt, then it can be washed out and collected without the ore having to be mined, crushed, and smelted, as in a normal mining operation.

Leaching is also used to extract gold and uranium from normal ores (*see* **Gold and uranium extraction**).

Leaching can be done in three physical arrangements. Slope or dump leaching is where a pile of the metal ore on the side of a hill is sprayed with the bacterial culture at the top and the eluant with the metal is collected at the bottom. Heap leaching is similar, but the material is in an isolated heap, which is more common in most mining sites. *In situ* leaching pumps the bacterial culture into the centre of the ore body along pipes or tunnels and then allows it to filter down to the base, where it is collected.

Leaching is a chemical process. In some cases bacteria such as *Thiobacillus ferrooxidans* or *Thiobacillus thiooxidans* oxidize sulfur compounds in the mineral to sulfuric acid, deriving metabolic energy. The sulfuric acid solubilizes the metal (copper sulfate is soluble, while the sulfide is not, for example), and so the metals wash out in an acid solution. In others the bacteria act directly on the metal ion, for example oxidizing uranium(IV) (insoluble) to uranium(VI) (soluble). The ore to be leached is sprayed with the bacteria in a suitable nutrient mix, which supplies all the other chemicals it needs for growth. Thus the bacterium is limited by the energy it can obtain from digesting the mineral, and thus digests the ore as fast as it can. Optimizing the nutrient mix is a critical factor in making a bioleaching process work at a commercially useful rate.

Ligand

A ligand is anything that binds to another molecule. Usually it means something that binds to a receptor or enzyme molecule, such as a hormone or protein growth factor. Thus the antigen molecule that binds to an antibody is an example of a ligand. Enzyme substrates can be considered as ligands, although they are not usually called this because, soon after they bind, they are converted to something else.

Discovering the natural ligands for receptors is a substantial part of the research into what receptors are doing. Often a ligand will bind to more than one receptor, and a receptor will bind more than one ligand. These networks are important in the control pathways in many living things.

Ligands that bind receptors on cell surfaces can be agonists or antagonists. The former trigger the receptor to perform its function, i.e. signalling to the cell, opening an ion channel, or whatever. Antagonists block that function.

The strength of the ligands' binding is measured by its affinity constant (*see* **Binding**), or the reciprocal of the affinity constant, the dissociation constant.

LIMS

Laboratory information management systems (LIMS) are the computer systems used to manage the information that a laboratory needs and uses. They have become topical in biotechnology because of the large amounts of experimental results produced by high throughput, automated laboratories in genome projects and high throughput screening.

LIMS systems need to solve several related problems.

1. How to get lots of different machines to communicate with one system. This sounds trivial, and is not.
2. Capturing data directly off instruments, so a scientist does not have to read it off a screen or printout and then type it in to the system. This is particularly important for high throughput discovery systems such as DNA sequencers, mass spectrometers, and automated microtitre plate readers that are part of HTS systems (*see* **High throughput screening**). DNA sequencer LIMS systems, for example, will read the gel outputs and turn them directly into DNA sequence data suitable for input into analytical programs.
3. Tracking samples consistently, so that each sample, analysis, or run is identified, its results linked to it, and the protocol used in that run is known. All this data should be linked up, not stored on separate (often incompatible) systems.
4. Appropriate data storage and presentation. Data should be summarized to a suitable level. A chemist may want to see detailed NMR spectra, but a molecular biologist will not, so the details must be summarized in the presentation software.

In theory, a good LIMS system can result in a completely paperless laboratory, but that ideal is rarely achieved.

Lipases

Lipases are enzymes that break down lipids into their component fatty acid and 'head group' moieties. The lipases used in biotechnology are almost invariably digestive lipases, meant to break down the fats in food. However, they can be turned to a number of different uses.

They can be used to break down complex fats into their components, which are then used to make other materials. This, however, is a relatively prosaic use. Two more valuable uses are lipid synthesis and transesterification. Lipid synthesis aims to make the lipase put a lipid back together from a fatty acid and an alcohol. This replaces conventional acid-catalysed chemistry, which, as well as using high temperature and pressure, produces a product contaminated with side products, which must then be purified. Unilever produce isopropyl myristate (an ingredient of cosmetics such as moisturizing creams) using a lipase-based biocatalytic synthesis. In order to reverse the normal reaction of lipases (which is to use water to split lipids), you have to be able to remove the water from the reaction. This means that the reaction must be carried out in non-aqueous solvents, such as benzene (*see* **Organic phase catalysis**).

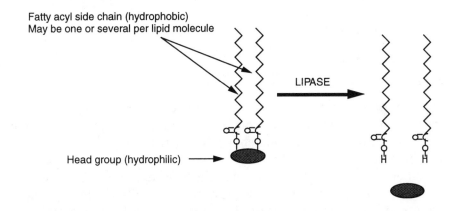

Fatty acyl side chain (hydrophobic)
May be one or several per lipid molecule

LIPASE

Head group (hydrophilic)

A related reaction is transesterification, a process where a lipase is used to swap the fatty acid chains between lipids without ever releasing significant amounts of fatty acid. This is a useful thing to do since it enables the biotechnologist to take a saturated fat (with a high melting point) and an unsaturated one (with a low melting point) and produce mixed molecules with intermediate properties: depending on how the ingredients are mixed, the properties can be determined fairly accurately. This application also requires that the lipase work in organic solvents.

Transesterification of triacylglycerol fats (the normal 'fat' in animal tissue) which is specific to the 1 and 3 fatty acids is a relatively specific and widely used transesterification, and is called Interesterification.

Lipid

A group of molecules containing esters of fatty acids. Typically, lipids contain long chain aliphatic 'tails' linked to water-soluble 'heads'. Lipids can be broken down into their component water-soluble heads and sparingly soluble tails by alkali (a process called saponification) or by lipase enzymes. Common tails are fatty acids, of which the more common saturated ones (fatty acids with no carbon–carbon double bonds) are:

- lauric acid 12 C
- myristic acid 14 C
- palmitic acid 16 C
- stearic acid 18 C

and the more common unsaturated ones are:

- oleic acid $18:1^9$
- linoleic acid $18:2^{9,12}$
- linolenic acid $18:3^{9,12,15}$
- arachidonic acid $20:4^{5,8,11,14}$

(The numbers after the acids are the number of carbon atoms in them, followed by the number of double bonds, and then the positions in the chain of the double bonds are shown as superscripts.)

Common head groups include several derivatives of phosphoglycerol, including phosphatidylcholine and phosphatidyl inositol. However, there are many other chemical variants in lipids, including sugar groups, amine compounds as head groups, many different hydrophobic moieties as 'tails', and linkage to proteins and other molecules in complex macromolecules.

Lipids are important to biotechnology in many ways. They are major components of living things, and as such are raw materials for food production. They have a wide range of physical and chemical properties which are exploited in as wide a range of products as biosensors, drug delivery systems, and cosmetics. Understanding how lipids function is also critical to understanding how living things work.

Liposome

A liposome is a small capsule made of lipids. Lipids can form stable structures in solution, in which the polar 'heads' point outwards into the watery solution and the apolar 'tails' stick together into the middle. If this is a droplet of lipid, it is called a miscelle. If it is a sheet of two layers of lipid molecules, then it is a **Langmuir–Blodgett film**. If such a film closes up into a ball, the result is a sphere with watery solution outside and inside separated by a lipid 'bilayer'. This is a liposome. Liposomes can have multiple layers stacked inside each other, but are often considered as if they were single (unilammellar) bags. Some lipids will spontaneously form miscelles or liposomes if shaken up in water at their 'critical miscelle concentration'. (At higher concentration the miscelles just lump together into a bulk of lipid, and at lower concentration the lipid dissolves to form a true solution.)

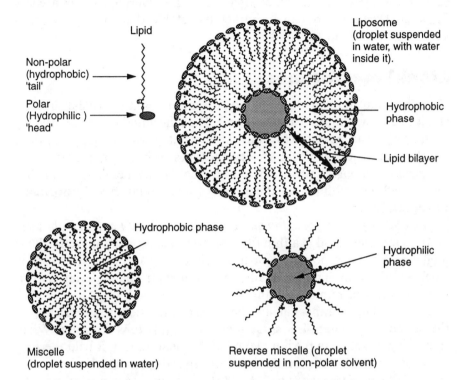

Lipid

Non-polar (hydrophobic) 'tail'

Polar (Hydrophilic) 'head'

Liposome (droplet suspended in water, with water inside it).

Hydrophobic phase

Lipid bilayer

Hydrophobic phase

Hydrophilic phase

Miscelle (droplet suspended in water)

Reverse miscelle (droplet suspended in non-polar solvent)

Liposomes have been suggested as the basis of several methods of drug delivery, especially for delivering peptide drugs. This is because they could protect their contents from digestion in the stomach and so deliver them to the intestine, where they would be absorbed, or they could allow them to be injected into the bloodstream and be carried around to a specific organ. Here the organ would recognize the lipids and absorb them specifically (this would work well for the liver, which tends to absorb liposomes from the blood spontaneously). Alternatively, antibodies linked on to the outside of the liposome would bind it to the specific cells that the drug is meant to affect (a combination sometimes called an immunoliposome). Liposomes tend to accumulate at sites of inflammation, and in some tumours, and so they are potential delivery vehicles for anti-inflammatory and antitumour drugs.

Lipsomes are also being as adjuvants in immunization: they can be used to carry the proteins of a synthetic vaccine to an antigen-presenting cell in the immune system and stimulate it to produce an effective immune response. Liposomes are considered particularly useful for this sort of application as they are made of the same materials (lipids) as the outside of cells, and so are less 'alien' to the body.

Trapping things inside liposomes is a form of **encapsulation**, and as such can be used in many other areas. Here, however, liposomes are less favoured because they can be rather less stable than other, polymer-based encapsulation methods.

Liquid membranes

These are thin films made up of liquids (as opposed to solids) that are stable in another liquid (usually water). Thus the liquid must not dissolve in the water, but nevertheless must be prevented from collapsing into a lot of small droplets. There are several types of liquid membrane.

Langmuir–Blodgett films. These are 'true' liquid membranes, in that there need be nothing in them except the liquid (*see* **Langmuir–Blodgett films**).

Immobilized or supported membranes (immobilized liquid membranes; ILM). Here the liquid is a thin film trapped in some solid material. This might be a porous polymer (such as scintered glass) or a fibrous one (such as cellulose). The liquid fills up the pores in the material, and so forms a series of minimembranes.

Supported membranes can also be ion exchange membranes (IEMs), if the supporting material is something that binds ions strongly. When something dissolves in the liquid part of the membrane, it is held there by the solid part. This can be the basis of separation methods.

Emulsion liquid membranes (ELMs). Here the watery part and the non-watery liquid are mixed up with a detergent. This makes small droplets of water in the other liquid (or the other liquid in water) stable. The result is a mix of water inside liquid droplets, themselves inside water. This is a membrane since it is a liquid barrier between two volumes of water.

Liquid membranes can be used in a number of applications. (The specific applications of **Langmuir–Blodgett films** are described separately.) Their main potential is as the basis of separation systems (*see* **Liquid membrane separations**).

Liquid membrane separations

Liquid membranes are thin layers of a liquid that does not mix with the water on either side. (In principle, they could also be thin layers of water with some other liquid on either side.) If something can dissolve in the liquid, then it can pass through the membrane. This can be the basis for separating materials that do dissolve in the liquid from those that do not, by putting the mixture on one side of the membrane and pure water on the other: the soluble component diffuses across the membrane, the contaminants cannot.

More sophisticated separation mechanisms can be based around this idea. The membrane can be impregnated with a carrier molecule which can carry across one type of molecule but not others. These usually bind to target molecule, making it soluble in the lipid (as a 'complex') whereas it usually would not be. Chemicals that can do this could include some peptide antibiotics, clathrins, crown ethers, or cyclodextrins. The transport of the molecule you want can also be linked to the transport of another molecule (for example, a proton): this is called 'coupled transport', and is the way that living cells concentrate many molecules inside themselves.

Ion exchange systems can also be used with a supported liquid membrane, in an ion exchange membrane (IEM).

Live vaccines

Live vaccines are vaccines containing living organisms or intact viruses, rather than inactivated (killed) organisms or extracts of them. They can cause better immunity in patients, but have the potential drawback that, unless they are thoroughly 'crippled' in some way, they may cause disease. They can also be harder to transport and store, because you have to avoid killing the organism

Biotechnology has generated ideas and research products for live vaccine development in a number of areas. **Viral vaccines** are dealt with elsewhere. Bacterial live vaccines can be developed in a number of ways.

Attenuation. Bacteria need a number of specific genes (virulence genes) to be able to cause disease, but these genes are not essential for growth in the 'test-tube'. When pathogenic bacteria are grown outside their host, they tend to lose their virulence genes by mutation. The result is an attenuated bacterium, which will cause an immune response similar to the original but which is harmless. Usually, several mutations are needed to make sure that a strain is really attenuated. If the nature of the virulence genes is known, then conventional and 'molecular' genetics can be used to select for mutations in or loss of those genes.

Gene cloning. An alternative is to place some key genes from the pathogenic bacterium into another, harmless organism. These may be the genes for surface parts of the pathogenic bacterium, such as pili or transport proteins, that are seen by the immune system. The degree to which an antigen, or a particular part (epitope) of an antigen is detected by the immune system, and hence the amount of antibody response that the immune system makes against that antigen, is called its immunogenicity. A key part of designing a good vaccine is deciding how to make the vaccine highly immunogenic, so that it is 'seen' by the immune system very clearly.

On vaccination with such a material, the immune system 'learns' what those parts of the pathogen look like without ever encountering the whole organism. This is similar to cloning the protein as a vaccine, but has the advantage that, as part of a living organism, it can stimulate the immune systems to greater prodigies of invention in generating good antibodies against it.

Live bacterial vaccines are generally suggested being used for enteric infections, including tooth decay, and some parasitic diseases.

Loop bioreactors

Also loop fermentors, these are bioreactors in which the fermenting material is cycled between a bulk tank and a smaller tank or loop of pipes. The circulation helps to mix the materials and to ensure that gas injected into the fermentor (usually oxygen or air) is well distributed amongst the liquid. The reactors are also very useful for photosynthetic fermentations, where they allow for the photosynthesizing organism to be passed along a large number of small pipes, where the light can get to them easily, rather than inside a single volume, where only the organisms near the edges get much light.

There are lots of types of loop bioreactors, but they break down into those that have an internal loop (e.g. a stirred tank bioreactor with an internal draft tube) and those that have an external loop. Some airlift fermentors are of the first type, as are pressure cycle reactors: reactors where air or oxygen is injected into the riser half of the reactor and this drives the liquid in that half up, so pushing the flow round the vessel. A variant that is found in all of them is the jet loop reactor, in which the recycled liquid is injected with considerable force back into the main tank. This means that it does not merely circulate the re-injected liquid around, it stirs the rest of the tank contents up as well. This has the advantage that the recycling mechanism is also the stirring system, removing the need for stirrers and baffle plates.

One popular type of loop bioreactor is the **airlift fermentor** (or bioreactor). See also **tank bioreactor, gas transfer**.

Jet loop fermentor

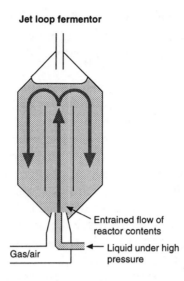

Entrained flow of
reactor contents

Gas/air — Liquid under high
pressure

Luminescence

Luminescence, the production of light by chemicals, is gaining increasing use as a labelling system for antibody- and DNA-based tests. Luminescent tests are of interest because, if carried out in a rigorously light-proof box, they can be extremely sensitive: a photomultiplier tube can detect when only a handful of photons have been given out by a reaction, so offering the potential of detecting only a handful of DNA or antibody molecules.

There are two ways of generating light using chemicals:

Chemiluminescence. This uses specific chemical groups which, when reacted, give out light. They can be attached to many other chemicals (e.g. proteins, DNA). There are also chemiluminescent groups that are 'blocked' (they do not themselves react to give out light). However, when the blocking group is chopped off, they react rapidly to generate light. A phosphate group is a common blocking group, which allows a chemiluminescent reaction to be used to detect an enzyme that cleaves phosphate groups, such as the widely used alkaline phosphatase (AP). Adding chemiluminescence to such an assay enhances its sensitivity greatly.

Bioluminescence. Some specialized enzyme systems can also generate light, using the energy of ATP (adenosine triphosphate) to do so. The most commonly used is luciferase, an enzyme from bacteria. Enzymes from fireflies have also been used. As well as being used as a label, these enzymes can be used as exquisitely sensitive detectors for ATP: a few hundred molecules of ATP, the amount in a few bacterial cells, can produce a detectable glow. This makes the enzymes useful for rapidly detecting bacteria for hygiene monitoring. (However, the enzymes are also subject to interference from many of the products they are meant to test.)

Luminescent labels are widely replacing radioactive ones in some applications. Thus the scientist will chemically link a luminescent group on to an antibody or a DNA probe where before they would have used a radioactive group: luminescent labels can be as sensitive as radioactive ones. The light can be detected by a sufficiently sensitive camera, or by exposing the gel or blot directly to photographic film, as is done in an autoradiograph. (Confusingly, this is sometimes called an autoradiograph as well, although it has nothing to do with radioactivity.)

This approach has the advantage that the active luminescence of living cells can be measured (it is usually too weak to be seen by the naked eye). This has been used to monitor cellular processes as cells metabolize. A commercial example is the Microtox environmental monitoring system, which uses living luminescent bacteria to detect if an environmental sample contains any toxic material. If it does, the bacterial glow fades.

MACS, BACS, etc.

MACS are mammalian artificial chromosomes. BACS are bacterial artificial chromosomes. Both are complex pieces of DNA that can be used as cloning vectors, to allow a foreign piece of DNA to replicate in a cell or organism. Together with **YACs** they can be used to clone very large pieces of foreign DNA, in theory tens of millions of bases long.

Most cloning vectors are based on plasmids or viruses; they allow our cloned DNA to grow rapidly in their host cells. However, these are not genetically stable; the cells will lose the cloned DNA easily if we do not continuously select for it to be there. The DNA will either be rearranged by the cell, usually removing part of the gene we want to clone, or will be lost completely. Chromosomes, by contrast, fit in with the normal gene-building machinery of the cell which ensures that every new cell gets an exact copy of the genetic material of the old cell. BACs and MACs are synthetic versions of chromosomes that use this machinery.

All chromosomes need to include an origin of replication, the signal that DNA polymersase uses in the cell to start replicating the DNA before the cell divides. Thus MACs and BACs must provide these. Chromosomes also need genetic elements that make sure that each cell contains one copy of the chromosome, and that when they are duplicated during cell division that exactly one copy goes to each new cell. Some BACs use the F1 plasmid to provide origins and other necessary sequence; others (sometimes called PACs) use the bacteriophage P1. Mammalian chromosomes need several origins. They also need a centromere somewhere in the middle (this attaches to the spindle during cell division, so the two new copies of the chromosome end up in the two daughter cells, not both in one), and telomeres at the ends. Telomeres round off the ends of the chromosome and make sure that they are genetically stable.

In theory, many cloning vectors can handle very large bits of DNA. In practice, YACs, MACs, etc. are the only convenient way of handling very long pieces of DNA, i.e. over 100 000 bases long. MACs are being explored as ways of putting very large gene clusters into transgenic animals, and as vectors for gene therapy. BACs are used in cloning large stretches of bacterial DNA into other bacteria. Both are also used to study the control and function of genes, and of telomeres, centromeres, and origins of replication.

Marine biotechnology

Approximately 70% of the world is covered by sea, but only about 1% of biotechnology uses this resource. Among the potential applications of marine resources to biotechnology are:

Natural products from marine organisms. Marine organisms live in a very diverse range of chemical habitats, and have to prevent a huge range of different animals from eating them. They produce a range of chemicals that can be used as candidates for drug development (*see* **Natural products**).

Enzymes. As well as making a range of novel enzymes associated with their specific metabolisms, marine microorganisms are adapted to a wide range of temperature regimes, from $-2°C$ in the lake discovered in 1996 beneath the Antarctic ice to $+115°C$ in black smoker vents in the mid-Atlantic ridge. Their enzymes are correspondingly adapted to work at temperatures that are of great potential value to industry (*see* **Extremophiles**).

Algae. Algae produce a range of products directly valuable in their own right (*see* **Algae: commercial uses**).

The sea is also a potential source of minerals, which biotechnology has been aiming to extract for some years, but without success (*see* **Sea water**).

Markers

One of the most flexible words in biotechnology, this can mean many things in different contexts.

Assay markers. The system that allows you to see the result of a chemical assay or test. Also called labels; examples include enzymes, radioactive atoms, and luminescent groups (*see* **Labels and substrates**, **Luminescence**).

Genetic markers. Landmarks in a genetic map (*see* **Genetic map**).

Biological markers. These are chemical signposts that something is happening in a cell or tissue. Often they can be produced by **reporter genes**, genes we introduce into a cell to show when something has happened. Markers can also be chemicals produced by the cell, even things such as cell shape. 'A marker for' an event is usually shorthand for 'something that is convenient to measure to show that that event has happened'.

Disease markers. These can mean genetic markers in the sense above for genetic diseases. Usually this means a chemical 'signpost' that someone has a disease. Thus the presence of some heart-specific enzymes in the blood (such as CK-MB) is a marker for heart attack: it shows that the patient has recently had a heart attack. Biotechnology contributes to these by finding them, and by developing tests to measure them.

Mass spectrometry

Mass spectrometry (MS) has been adopted by a wide range of biological researchers in the last five years, as the instruments and methods have made it more routine. A mass spectrometer measures the exact molecular mass of a molecule by measuring its flight path through a set of magnetic and electric fields. More exactly, it measures the mass/charge ratio. To allow the molecules to fly uninterrupted, the operating chamber of the mass spectrometer contains a vacuum. In vacuum nearly all biological molecules have no charge, but the method relies of them being charged, so that the magnetic and electric fields can affect them. So, mass spectrometers include a method of bombarding the target molecules with radiation to charge them up. This can fragment the molecule. This is a benefit because then you can measure the mass/charge ratio of bits of the molecule as well as the whole thing in one experiment. Piecing together these data allows the full structure of the original molecule to be deduced.

Ionization methods include bombarding the molecule with X-rays or high-speed electrons (which breaks them down a lot), atom beams (called fast atom bombardment, or FABMS, which only breaks them down a little), or spraying them out of a charged nozzle (called electrospray, which hardly breaks them down at all). An increasingly widely used technique for biological molecules is MALDI (matrix-assisted laser desorption ionization) where the sample is embedded in a volatile matrix and a laser vaporizes it. The nature of the mass of material your sample is embedded in is crucial for this. A common combination is MALDI–TOF (TOF is time-of-flight spectrometry, a fairly simple way of identifying the molecular mass of ions). Two mass spectrometers can also be linked together in tandem MS (also called MS/MS), so that the first MS machine finds molecules with one particular size, and the second smashes them up and so gives information about what their chemical structure is.

Mass spectrometry is being used to identify proteins directly from mixes, and to sequence proteins. It can be used to sequence DNA, but conventional techniques are so fast this is not a common application yet. If we already have a database of known protein sequences, we can identify which of them an unknown protein is by its atomic weight if we can measure the weight accurately enough: mass spectrometry allows this accuracy. The technique also only requires subpicogram amounts of chemical to identify, so it is increasingly popular with scientists tracing which genes are active in cells and tissues. MS can also be used to track post-translational modifications in proteins, which genetic methods cannot do.

MS is also being used as a general analytical method, for example, to trace the metabolites of a drug in clinical studies. It is potentially better than conventional methods like HLPC because it can find the size of any molecule; so, if you have some idea of what molecules you might expect, the MS can tell you if they are there (you do not have to develop a new analytical method to check for them). In practice it is used in conjunction with HPLC or CE (capillary electrophoresis): the HPLC separates chemicals, and the MS then identified what they are. Waters and Applied Biosystems provide systems that integrate MS and HPLC for biological uses.

Maxicells

Maxicells are bacterial cells that have a mutation in the genes that control how they divide. Under the 'right' conditions (usually when the temperature of the medium is raised) they simply stop dividing. However, they do not stop growing, so the result is a huge bacterial cell. This can be helpful, in that such large cells are much easier to separate from their medium than the relatively small normal cells: they will, for example, settle out of solution under their own weight in a fairly short time.

A related idea is the minicell. This is another cell division mutant. Here, under the 'right' conditions the cells divide, but they do not divide in the middle. Rather, a small 'minicell' splits off from one end. All the bacterial DNA stays in the main cell, so the minicell has no DNA of its own. Therefore it cannot make any new RNA, and as soon as the RNA that happened to be in the cell has broken down it cannot make any new protein either. However, this rule is broken if the cell contains some types of plasmids: a few plasmid molecules can end up in the minicell. Thus, when all the trapped RNA has broken down, the only proteins being made by the minicell are those coded by genes on the plasmid. This is very useful for studies of gene expression, since, by isolating minicells, the proteins being made by the plasmid can be studied without having to purify them from all the other proteins being made by a normal bacterial cell.

Metabolic engineering

The manipulation of cells to make them produce more of a product is sometimes called metabolic engineering. This focuses on altering the enzymes present in a cell, rather that just altering their growth conditions. Typically, a cell will be genetically engineered to produce more of a specific enzyme, or to replace one enzyme by another with more desirable properties.

The difficulty in metabolic engineering is that just replacing one enzyme by another usually does not have much effect, because other control systems in the cell stop the new enzyme from working effectively. It used to be thought that the amount of chemical being processed by a specific pathway (the 'metabolic flux') was determined by flow through one key 'rate-determining' step in the pathway. If you doubled the amount of the enzyme that catalysed this step (the rate-determining enzyme), then you would double the amount of stuff made by that pathway. Many experiments have shown that this is not so: all the enzymes of a pathway contribute to controlling how fast that pathway works to a greater or lesser degree, and if you double the amount of one enzyme it has little effect overall. Thus, metabolic engineering is about predicting changes in the whole of metabolism, and altering a number of enzymes and physiological conditions to bring about those changes.

Applications are generally in fermentation, where metabolic engineering seeks to improve the yield of primary or secondary metabolites. However, it can also be applied to plants that produce valuable chemicals, and even to human cells where the same analysis can be used to discover what goes wrong with metabolism in diseases such as diabetes and obesity. Transgenic plants have been suggested as practical ways of making a variety of biomolecules such as modified starch and oils through manipulating their metabolism.

The degree to which a single enzyme affects the flux through a pathway is called its control coefficient, and is given as a number between 0 (has no effect at all) and 1 (is the only control on that pathway).

Metabolic pathway

The chemistry that turns an organism's food into more organism is collectively known as 'metabolism'. The whole of metabolism is a very complex network, but specific series of reactions can be found in which chemicals ('metabolites') are converted one into another to get from a starting material to an end material in a series of relatively small steps. Such series of reactions are called metabolic pathways.

Metabolism is usually divided into anabolic pathways [pathways that build up molecules for use by the organism (e.g. those pathways that make amino acids)] and catabolic pathways [ones that break down molecules, either for energy or simply to get rid of undesired materials (e.g. breaking down hydrocarbons to obtain energy)]. Some pathways, especially those at the centre of metabolism (e.g. those that break down glucose) perform both functions, and are called amphibolic. Pathways can also be divided into 'primary' and 'secondary'. Primary metabolism is making the basics of life that are common to nearly all living things.

Secondary metabolism is making specialist chemicals such as hormones or toxins. Mutant organisms that lack enzymes of the anabolic pathways of primary metabolism are called auxotrophs: they require that the end-product of that pathway be provided as food since they cannot make it themselves. Mutants of the pathways that make secondary metabolites are called idiotrophs.

Many anabolic pathways are end-product inhibited. This means that the end-product of a pathway is an inhibitor of the enzymes in the pathway, usually the first enzyme that is specific to that pathway. So, if a lot of the end-product builds up, the pathway automatically shuts down the synthesis of more of that product. This can be very irritating for the biotechnologist, who therefore has to remove the end-product of the pathway very fast, so that it does not build up, or find some way of knocking out the inhibition.

Popular press articles about 'your metabolism' almost always mean energy metabolism: the metabolic reactions specifically concerned with handling molecules whose main function is providing energy, and 'burning' them to generate that energy. Someone with a 'slow metabolism' burns off less energy. In measurable fact, very few people have a 'slow' metabolism, and, specifically, people who put on weight easily have no slower metabolisms than anyone else.

Microbial mining

This is the use of microorganisms to remove minerals, and particularly metals, from rocks. It is a specific application of **biohydrometallurgy**. Microbial mining is related to the use of microbes in **desulfurization**, and for **bioremediation**.

Microbial mining falls into two areas:

Leaching. This is using bacteria to process ores to make the metals in them more accessible. Usually it involves using bacteria to release the metals as soluble salts, to be washed out for subsequent recovery. However, it can also involve preprocessing of the ores which, while not releasing metals directly, allows them to be separated more easily by washing out, flotation, or other 'traditional' methods in a further processing step (*see* **Leaching**).

Purification. Using microorganisms or microorganism components to separate and concentrate metals from very dilute solutions. This is also called **biosorption**.

Biohydrometallurgy is used commercially to recover copper and uranium from low-grade ores, especially chalcopyrite ($CuFeS_2$), covellite (CuS), chalcocite (Cu_2S), and uraninite (UO_2). A number of other metals (antimony, arsenic, molybdenum, zinc, cadmium, cobalt, nickel, and gold) can be extracted using bacteria, but these are not used to a significant extent.

Plants can also be used to extract metals, although this is not used as an extraction technology commercially to any significant extent. Phytoextraction is the concentration of metals from the soil into the plant, usually the leaves. Some soils can produce quite toxic crops because plants concentrate elements such as selenium and cadmium in their leaves. The roots of water-growing plants such as the water hyacinth can also concentrate metals from their water supply, a process called rhizofiltration, which is used in reed bed biofilter technology. The only commercial use of such plants is to use metal-tolerant plants to stabilize waste tips (and to make then less ugly).

Microbial physiology

Physiology is the description of the chemical processes that happen in a living organism. The physiology of microorganisms is important because many biotechnological processes such as fermentation, biosorption, and even bulk growth, depend on the right chemistry happening under the right conditions. So microbial physiology is critical to controlling a range of processes.

Aspects of microbial physiology that scientists seek to control include:

- overall thermodynamics (where the energy goes to) and mass balance (where the mass goes to) in the organism.

- specific substrate utilization and growth kinetics: how fast microorganisms grow on a particular substrate, and how much of it they convert to the product you want.

- growth energetics (where the specific energy sources go to in the bacteria, and how energy is used in different parts of metabolism).

- redox state or redox balance: the balance of oxidizing and reducing molecules in the bacterial cell. This is important for the energy balance, and also for what chemicals the cells can produce. A related concept is proton motive force (pmf) which is the chemical 'push' behind the central reactions of energy metabolism.

- transport mechanisms: how materials get to the cell, and then into or out of it.

- heterologous protein production: how the chemical conditions inside the cell affect how much of a 'foreign' protein is produced, and whether it is produced as a soluble material, as an inclusion body, or as denatured, useless material.

Microcarriers

In biotechnology, microcarriers are generally small particles used as a support material for cells, and particularly mammalian cells, in a large-scale culture. Mammalian cells are too fragile to be pumped and stirred in the same way that bacterial cells are, but still need to be fed with oxygen and nutrients and must still be separated from their culture medium when the time comes to collect the product.

In mammalian cell culture, microcarriers are particularly useful for culturing cells that would normally grow attached to a solid surface (attached or surface culture, as opposed to suspension culture). Rather than have to have acres of flat plastic surface, the cells are grown over the surface of small polymer spheres, especially, polystyrene, gelatin, collagen, or polysaccharides like dextran or cellulose. The surface area for growth is huge, and the spheres can be treated like bacterial cells for filtering and for (gentle) centrifugation, and they protect the cells from the shear forces involved in pumping and aeration. Some microcarriers are simply solid spheres, some are porous. Porous ones have a larger surface area for cells to grow on, since the cells can grow inside them as well as on the outside, so giving them greater protection. However, it is harder to see the cells on the carriers, which can be important if you want to know how well your culture is growing.

An alternative to growing cells on microcarriers is to grow the cells as aggregates. Cell aggregates have some of the mechanical robustness of cells on microcarriers, and have much higher cell content for a given amount of solid matter. However, getting cells to grow in aggregates can be much harder than getting them to grow on suitably treated polymer surfaces.

Microorganisms

There are a very wide range of microorganisms used in biotechnology. A few are mentioned elsewhere in this book under specific headings. Some are used for more than one thing, and so crop up in several biotechnological contexts.

Microorganisms, indeed all life, are classified into prokaryotes (organisms without a cell nucleus) and eukaryotes (organisms with a cell nucleus). Animals, plants, and fungi are all eukaryotes, bacteria and archebacteria are prokaryotes.

Bacteria are classified into Gram-positive and Gram-negative. These names reflect whether their cell walls will absorb Gram's stain, but the division they represent is quite a substantial one, and Gram-positive and Gram-negative organisms are biochemically and genetically quite different. However, they may look much the same under the microscope. Microorganisms may be ball-shaped (cocci), rod-shaped, or made of very long 'strings' called hyphae. Hyphae may be branched or unbranched: in either case they are often harder to grow in bulk because the stirring needed to get nutrient to all the hyphae can break them. Organisms that grow as long strings or filaments are called 'filamentous'.

Microorganisms are also classified into aerobes (grow in the presence of oxygen) and anaerobes (do not use oxygen). These can be facultative or obligative: facultative aerobes can use oxygen or not: obligative aerobes must have it. Obligative anaerobes are killed by oxygen.

Some of the more commonly mentioned organisms are:

Aspergillus. A type of filamentous fungus that has been used for genetic engineering in a few cases, and which is also used to produce citric acid by fermentation.

Bacillus subtilis. This Gram-positive bacterium is also widely used as a cloning host, especially for the expression of secreted proteins. Strains that lack any protease activity have been developed, which therefore do not break down the product protein when it is secreted into the fermentation medium.

Candida utilis. A yeast, this organism is used in fermentations to produce chemicals.

Clostridium acetobutylicum. A bacterium used in the past to produce acetone and butanol by fermentation, and now used as a source of enzymes. (Acetone and butanol are now made chemically.)

Corynebacterium glutamicum. This is widely used in fermentation processes producing amino acids for food supplements.

Escherichia coli. Usually abbreviated to *E. coli* in print (and almost always in conversation), this very versatile Gram-negative bacterium is used in many biotechnological processes. Its genetics are the best known of any organism, with the majority of its genes known and about 90% sequenced. It is by far the most common host cell for recombinant DNA work. It is also used in fermentations to make many amino acids and other products since it grows on many very cheap fermentation substrates, grows fast, and can be manipulated genetically to accumulate many different chemicals. It is also very chemically versatile and quite non-pathogenic (with the exception of a few strains which, obviously, are not used for biotechnology).

Penicillium. A group of filamentous fungi used primarily to produce penicillin antibiotics.

Pseudomonas. A group of soil bacteria that contain some extremely diverse chemical abilities, which biotechnology has harnessed in bioremediation

Saccharomyces. A group of yeasts. *Saccharomyces cerevisiae* is brewers' and bakers' yeast, and as such is probably the most widely exploited microorganism. *Saccharomyces* are also used in recombinant DNA work because they are eukaryotes and hence have the same sort of genetic structure as humans, secrete proteins in a similar way, and so on, but are almost as easy to ferment in bulk as bacteria.

Streptomycetes. Gram-positive bacteria that are used to produce a range of chemicals, especially antibiotics. They have also been used as the hosts for genetic engineering, in part to manipulate their antibiotic synthetic pathways.

Also mentioned elsewhere are *Agrobacterium tumefaciens*, *Thiobacillus* and *Ferrobacillus* (**microbial mining**), and *Methanococcus* (**SCP**).

Microorganism safety classification

One major concern about biotechnology is whether it is safe. As much of biotechnology involves the genetic manipulation, selection, or physiological manipulation of microorganisms and their subsequent production in large amounts, some of this concern translates into a concern about the safety of industrial-scale microbiology.

Most guidelines and codes of practice in handling microorganisms are aimed at medical microbiologists and microbiologists handling pathogens to produce vaccines. Thus most of the guidelines on how microorganisms should be handled in biotechnology are derived from these medical examples. The World Health Organization has no evidence that genetically manipulated organisms are inherently more dangerous than any others, and has found no case where a laboratory or industrial worker has contracted an infection as a result of contact with a genetically engineered organism.

The principle of classifying the danger from a microorganism, and so of deciding how to contain that danger, is to classify the organism according to how likely it is to escape, how likely it is to survive if it does escape, and how much damage it could do if it does survive. Different countries have different rules about how this is done: the table summarizes a few of these.

If an organism is outside the Class 1/Group 1 area, then it can be contained by a variety of physical or biological methods (*see* **Clean room**, **Physical containment**).

A range of national safety committees ensure that appropriate containment is used for organisms in each class (even if, in other industries, no containment is needed for those same organisms at all).

Institution:	Risk: Minimal	Normal microbiological	High risk to individual only	High risk to individual and risks to community
ACDP[1], ACGM[2]	—Group 1—	—Group 2—	—Group 3—	—Group 4—
EFB[3]	Class 1	Class 2	Class 3	Class 4
WHO	Group I	Group II	Group III	Group IV

[1]Advisory Committee on Dangerous Pathogens (UK).
[2]Advisory Committee on Genetic Modification (UK).
[3]European Federation for Biotechnology, which has the same groupings as the US Public Health Service (PHS).

Micropropagation

This is a term used in plant production for the use of biotechnological methods to grow large numbers of plants using tissue culture methods. In essence, a desirable plant is cut up into many very small pieces (sometimes single cells, sometimes clusters of a few to thousands of cells) and cultured. The culture conditions are tuned so that the cells grow into a callus, a mass of cells that looks like a small mould. The conditions are then switched so that the callus starts to develop into a small plant 'embryo' (*see* **Embryogenesis**). Once this embryo has grown sufficiently it can be planted out as a small plant. In some techniques the embryo is encapsulated in a protective sheath so that when it is sown it has a 'shell' similar to the seed produced by more conventional breeding.

The advantages of micropropagation are that it can produce a very large number of plants in a relatively small space of time, and that all the plants are usually genetically identical. The drawbacks are that it is skill intensive, and hence much more expensive than conventional breeding, and that it can only be done on plants for which the right cell culture conditions have been worked out.

Microscopy

Microscopy is a standard method of examining biological systems, both to determine their structure and to examine their chemistry. Two more specialist techniques are:

Fluorescence microscopy. This version detects the light given out by fluorescent molecules in a sample. Almost always the fluorescence comes from a fluorescent label attached to an antibody or DNA probe, so we can tell what part of the cell the fluorescent reagent has bound to. Several colours of label can be used, and the results are very pretty. Techniques using fluorescence microscopy include FISH and immunocytochemistry.

Confocal microscopy. Here the microscope effectively focuses on only a very thin slice of the sample at a time. So a picture is built up, not as if you were looking down through a thick sample, but as if you had sliced it into layers. Often, confocal microscope systems are connected to computers which can reassemble the slices into a 3D picture of the original object.

See also **Scanning tunnelling microscopy**.

Model organisms

Many of the experiments that biotechnologists might conceive of doing to develop new products or processes cannot be done for ethical, moral, legal, financial, or practical reasons. Breeding experiments on people or trees, studies of individual cells in a whole cow or a pig, such experiments may give valuable knowledge to agriculture, healthcare, or environmental studies, but we cannot sensibly do them. So, instead we do a model experiment, one that seeks to duplicate some critical aspect of the experiment we want to do, but avoids the impraticalities.

Some organisms have become standards for such model experiments, and so are called model organisms. They are models for some aspect of other organisms. Among the more popular are:

Caenorhabditis elegans. A microscopic nematode worm with only a few hundred cells and a very small genome. The fate map of the cells (exactly where every cell in the organism comes from and fits in the organism as it grows) has been completely worked out. This means that any new mutants of C. *elegans* can be related to the effects they have on specific populations of cells, and thus to specific biochemical effects. Thus C. *elegans* has been a model organism for the study of the role of genes in development, a role enhanced by the sequencing of over 95% of its genome.

Arabidopsis thaliana. A small weed with 'only' 80–100 million bases in its genome and a short life cycle, which has become a staple of plant genetics research. It is a model for more complicated and slower growing plants like wheat and apple trees.

Drosophila melanogaster. The fruit fly, which has been a staple for animal genetics, and particularly the genetics of growth and development, since the 1920s. *Drosophila* geneticists call their mutant flies by more entertaining names than most, such as *dunce*, *sevenup*, *bride of sevenless*.

Zebra fish. A small fish which is being used as another genetic model. It is the most human-like animal that can be cloned easily (mice can be cloned, but only in small clones of eight individuals.) There is also a zebra fish genome project.

Mouse. The mouse is the most common laboratory animal in biomedical research, and is used for all sorts of aspects of the discovery and development of drugs as well as fundamental research. There are many inbred strains of mice, which have particular and very well known peculiarities.

Rat, rabbit. Rats and rabbits are also used in pharmaceutical research and development, but are less popular for fundamental research, in part because they are not as extensively researched as mice, and in part because there are not as many inbred lines as there are of mice, but mainly because mice are smaller and hence cheaper to rear and keep. Also, if a mouse bites you, it does not do as much damage. Rats can bite quite severely, and other animals (such as dogs or baboons) can take off your hand.

Dictyostelium discoideum. A soil amoeba that was used a lot for research into how simple cells differentiated and signalled to each other, but which has become less popular over the last decade.

Biomedical research has a slightly different problem from other research, being concerned with human disease. The aim here is to find a good disease model, an animal disease which exactly mimics the human disease, so that you can try the drugs out on animals instead of humans. This is not only because we do not want to test completely untried chemicals on people. It is also because the model organism's genetics are much better known that of people (so there is less variability in the results), and often the disease progresses faster in the model than in humans (so we can do an experiment in a sensible time). Of course, if a good animal model is not available, there is no point using a bad one. In that case, you have to make a leap of faith and 'go straight to humans'. AZT, the first AIDS drugs, was tested directly on people because there was no animal model of AIDS available.

A popular journalistic fantasy is that scientists do experiment on animals out of sheer perverse pleasure in 'vivisection'. There are too many refutations of this argument to list exhaustively here: I will mention only that scientists have spouses, children, and pets too, and a lower record of animal or child abuse than the general population. The journalistic attitude reflects the '**yuk factor**'—diabetic journalists are happy to inject insulin, but not willing to accept than Banting and Best killed many dogs to discover it.

Molecular biology

Much of biotechnology is based at least in part on molecular biology. Molecular biology, and its twin science molecular genetics, started in the late 1940s around a group of biologists, and physicists turned biologists, who were looking for a new way to attack the fundamental problems of life. Traditional biochemistry at that time (and much of biochemistry today) sought to attack complex systems by taking them apart and analysing all the bits as carefully as possible in terms of chemicals and biochemicals. The 'new' science instead chose the simplest system they could and sought to analyse it using genetics as their primary tool. The system they chose was the bacteriophage, and hence many of the founders of molecular genetics were members of a semi-official 'phage group'.

The genetic approach paid off handsomely in three ways. First, it opened up whole new areas of genetics (genetics at a molecular level rather than the whole organism genetics that had characterized previous work on *Drosophila*, mice, and plants, or the biochemical genetics of bacteria and yeast). This in turn allowed the investigators to start to decipher the genetic code and deduce some of the mechanisms of protein synthesis. But probably more importantly, it gave credibility to a new way of thinking about biology. This is now entrenched in the normal way of thought, and envisages the molecular basis of biology as being made up of quite understandable building blocks that bump into each other and latch on to and off each other in defined ways. Whereas in 1950 an enzyme was a squiggle in an equation, in 1990 it is a coloured 'blob' on a computer graphics display. The molecules at the basis of life became much more real and much more important. Life became a discrete machine, and the instructions for that machine lay in DNA. Hence the centrality of DNA to much of biology today. This approach to living systems, as discrete blocks labelled 'protein' and 'gene', has been called 'molecular Lego'.

As well as this philosophical change, which enabled us to imagine a 'biotechnology' of engineered genes and molecules, the phage group work gave us the basic tools of recombinant DNA technology. Restriction enzymes, DNA ligase, and many cloning vectors all come directly from bacteriophage genetics.

Thus, molecular biology is not a science in the sense that it studies molecules or biology—biochemistry, physiology, pathology, microbiology do that too. It is more a way of doing biology, both a way of thinking about it and a way of getting the tools to do experiments. It is, in Thomas Kuhn's terminology, a paradigm.

Molecular electronics

This speculative idea means making electronic or computing devices out of single molecules, or small groups of molecules. Talk about switches that are made of a single protein molecule leads to computers with greater-than-human powers, which could fit in a matchbox.

Fairly obviously this is speculative, but may not be as speculative as nanotechnology in general. First, conventional computer chips are made by layering materials on silicon, and it takes only a leap of imagination to suggest that layering proteins could achieve similar things. Secondly, many proteins do have charge-transfer and charge-switching properties, which could, with much greater understanding of the properties of proteins in general, be harnessed to provide some aspects of the information processing capability of a semiconductor device. Thirdly, Langmuir–Blodgett films (thin films of lipids) are known to be an essential part of the electrical properties of nerve cells, and can be made quite readily in the laboratory. Nerve cell proteins are inserted into the lipid film and alter the film's ability to let ions pass depending on what other ions are present or the electric field that they are exposed to. This has got to the stage of building films, putting proteins into them, and demonstrating the electrical characteristics of the protein, rather like where we were with transistors in 1930.

A lot of work has focused on using bacteriorhodopsin, a pigment from *Halobacterium halobium* (a purple, salt-loving bacterium), which has electrical and light-absorbing properties as well as being able to self-assemble into membranes. Membranes of bacteriorhodipsin assembled in the right way can store light images when exposed to electric fields, a first step to a holographic computer memory system.

Molecular computing was a trend term a few years ago, but has now been partly supplanted by **nanotechnology**.

Molecular graphics

This is the display of a molecule's shape, usually on a computer. It has gained a lot of publicity because of its application to **rational drug design**. Molecular graphics takes the description of how the atoms of a molecule are arranged in space from a database and draws a picture of what the molecule would look like if it were made of solid balls (the atoms) or thin sticks (the bonds between atoms). Some examples are shown in the figure. Usually, molecular graphics does not calculate the structure of the compound (*see* **Computational chemistry**).

Because the human brain is extremely good at perceiving patterns in complex pictures, but rather poor at seeing patterns in large collections of numbers, molecular graphics is a good way of allowing people to see the similarities in structures between molecules, and also of seeing if two molecules fit together well. This in turn is useful when, as part of a rational drug design programme, a scientist wishes to find a molecule that will fit into the known structure of the active site of an enzyme, or the hormone-binding site of a receptor.

Different representations of the same molecule. All these are views of the peptide hormone oxytocin, viewed from the same angle

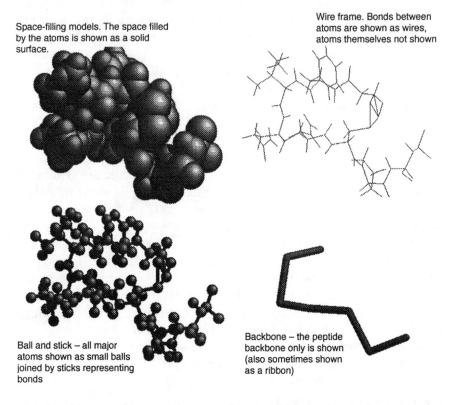

Space-filling models. The space filled by the atoms is shown as a solid surface.

Wire frame. Bonds between atoms are shown as wires, atoms themselves not shown

Ball and stick – all major atoms shown as small balls joined by sticks representing bonds

Backbone – the peptide backbone only is shown (also sometimes shown as a ribbon)

Molecular graphics packages often produce extremely attractive pictures as part of their output, which is another reason why they have a high profile in the public relations material of pharmaceutical and biotechnology companies. The more sophisticated display methods can produce 3D pictures that the user can manipulate as if he were in a room full of bits of molecule that he can move around with his hands, a type of computer interaction called 'virtual reality'. However, these pictures can be misleading: molecules are not 'really' like this, and all the pictures here are only approximations of one aspect of reality.

Molecular imprinting

This is an approach to synthesizing a non-biological chemical that imitates some of the properties of an enzyme or antibody. A polymer material, often polyacrylamide, is 'imprinted' with gaps that exactly fit one, and only one, species of small molecule, in the same way that the binding site of an antibody exactly fits its antigen. This is done by forming the polymer matrix in the presence of the small molecule, so that the chains fold around those molecules. When the polymer has set solid the small molecule is washed out using suitable solvents, leaving 'holes' in the polymer. These can have quite a high affinity for the molecule that has been washed out, and so could be used to purify those molecules away from many others.

In addition, just as antibodies that are raised against a transition state analogue can have catalytic activity (*see* **Catalytic antibodies**), so an imprinted polymer, which has gaps that are 'shaped' to fit a transition state analogue, could be a catalyst. So far, molecular imprinting techniques have been used experimentally for purification, but not for catalysis. Molecular imprinted matrices have been used to separate a wide range of small and large molecules, including separating chiral isomers from each other. Their binding capacity can also be used to form the binding moiety of a sensor.

Molecular modelling

This is the use of computers to draw realistic models of what molecules look like. At one end of this range of techniques is **molecular graphics**, which is simply drawing 3D representations of what a molecule would look like (if the atoms were solid balls). At the other end it shades into **computational chemistry** (the calculation of what the physical and chemical properties of a molecule are). Usually, molecular modelling falls more into the graphics end of the spectrum.

Using molecular modelling, rational drug design programmes can look at a range of different molecular structures of drugs that may fit into the active site of an enzyme, and by moving them on the computer screen it can be decided which structures actually fit the site well. Molecular modelling can add sophistication to picture drawing by calculating the hydration (degree to which individual bits of the molecule bond neighbouring water molecules) and charge distribution across the molecule. These also affect how molecules bind together.

Monoclonal antibodies

Antibodies produced in the blood are made by a large number of different lymphocytes (B cells). Each B cell makes a unique antibody, so the antibodies that recognize any particular antigen are a mixture of molecules. This mixture is called 'a' polyclonal antibody: an antibody preparation that reacts with only one antigen, but which nevertheless derives from many different 'clones' of B cells. While this is useful for the body (covering the antigen with antibodies) it is a problem for the biotechnologist who wants defined materials to work with. Monoclonal antibodies are a way round this. They are antibodies made from a single clone of B cells which has been isolated and immortalized for growth *in vitro*. The invention of the methods for producing monoclonal antibodies won Cesar Milstein a Nobel Prize and, characteristically, very little money.

Monoclonal antibodies are generated as follows:

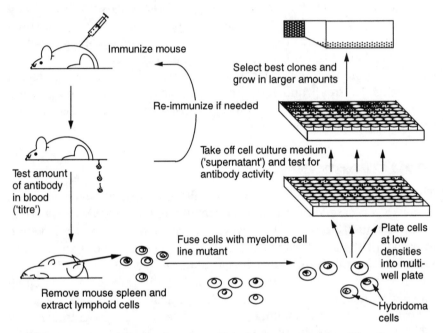

- Immunization. A mouse (usually) is immunized with the target antigen. This is done by injecting the antigen, sometimes with another material (an adjuvant) to stimulate the immune response (*see* **Immunization**).

- Splenectomy. The spleen (a useful concentrated source of B cells) is removed from the mouse.

- Fusion. The lymphocytes are fused with an immortal cell line (myeloma cell). This immortalizes them, i.e. means that they will grow for ever in culture.

- Selection of clones. The fused cells are grown in HAT (hyopxanthine/ aminopterin/thymine) medium. Because of the genetic defects in the myeloma cell line and the metabolic peculiarities of normal B cells, neither myeloma nor B cell can grow in this medium, but the combined, fused cell can. Therfore a small proportion of fused cells is selected.

- Cloning. The fused cells are put, at very low concentrations, into the wells of a multiwell plate. On average each well with a cell in will only have one cell in it, so that, on average, the cells that grow in each well will be a clone, i.e. derived from a single cell. This ensures that you have a pure cell line. This cell line is termed a hybridoma.

- Selection. The clones are screened by whatever method is to hand to find the one producing a good antibody against the antigen we want. Most of the cells will *not* be making the antibody we want.

A 'good' antibody is one that binds tightly to the antigen (in chemical terms, has an affinity of 10^9 or better), does not bind significantly to anything else, and is the right class and subclass (IgG, IgM, etc.; *see* **Antibody**).

If the target molecule is a very small one (like a drug molecule) then injecting it into a mouse will rarely produce an antibody response. In this case the molecule is chemically linked on to a larger molecule, usually a protein and often bovine serum albumin (BSA) or keyhole limpet haemocyanin (KLH), so that the immune system can 'see' it. The small molecule is called a 'hapten' in this case.

Monoclonal antibody production

Monoclonal antibodies can be produced commercially in a number of ways, depending on the scale of production.

As mouse ascites fluid. Mice can be injected with the hybridoma cell line that makes the monoclonal antibody. The cells form a tumour (an ascitic tumour) which secretes the antibody into their ascitic fluid (the fluid that surrounds the lungs) and the blood plasma. Either can be collected and the antibody purified from it. This is simple, and does not require sterile culture conditions. However, it does require animal facilities, and only produces around 50 mg/mouse. Ascites is generally not used in the West very much, partly because of controls on the use of animals, and partly because of the expense involved.

Tissue culture methods, which are used to make the hybridoma in the first place, can be used to make antibody: the tissue culture 'supernatant', i.e. what is left of the medium once you have removed the cells, is a source of antibody. However, this is rarely effective at producing more than 10 mg of antibody.

Suspended cell fermentors. The traditional biotechnology bioreactor technology has been used to grow hybridoma cells in bulk. Celltech (now Lonza), for example, has a 1000 litre airlift fermentor which can produce 100 g of antibody in a two-week fermentation with a hybridoma. This is a similar technology to medium-scale microbial fermentation, although, because mammalian cells are very sensitive to chemicals, temperature changes, shear (squashing), and other environmental disruptions it can be more difficult to get it to work reliably, and also costs a lot more in expensive culture media.

Immobilized cell reactors. Several types of immobilized cell reactor have been used to make monoclonal antibodies on a scale of a few grams. The most popular is probably the hollow fibre reactor. A few grams of antibody is enough for several million tests to be used for medical diagnosis, for example, and so suffices most commercial needs.

A rising new technology is to use bacteria to produce antibodies. The genes for the heavy and light chains must be spliced into one bacterium, but when this is done the bug is very much easier to grow than mammalian cells. This also makes genetic engineering of chimeric or humanized antibodies easier since the cloning technology necessary to do this is done in *E. coli*. Production in bacteria is preferred because it is cheaper and faster than in cell culture, and the systems are usually quicker to set up on a large scale. Against this, the antibody gene has to be engineered into a suitable bacterium. This is a minor problem if the antibody has already been created by recombinant DNA techniques such as phage display or antibody fragment engineering (*see* **Dabs and other engineered antibodies**).

Motifs

Neither protein nor DNA sequences are random. If nature 'wishes' to evolve a protein to do something, she starts out with proteins that already exist to do something else and, usually, shuffles bits of the relevant genes around to make the new entity. Thus certain strings of bases or amino acids crop up time and again in different genes and proteins. These are often called motifs. Usually they are significant because they denote some bit of the molecule has a particular function. Thus 'zinc finger motifs' in proteins suggest that the protein has a section that binds to DNA. Similarly in DNA the 'TATAA' motif is suggestive of a promoter sequence in eukaryotic cells.

'Motifs' are similar to signal sequences in proteins. However, signal sequences are meant to be 'read' by the cell. Motifs are only read by the biotechnologist as clues to what a particular part of a gene or protein does. Amongst the signal sequences known are the leader sequences that lead to secretion (*see* **Secretion**), another that tags the protein as headed for the endoplasmic reticulum and lysosomes, one that sends the protein to the cell nucleus, the 'stop transfer' sequence which anchors a protein in the cell membrane, and so on. Being able to read signal sequences is also helpful to the biotechnologist since it gives clues as to where in the cell a particular protein is meant to end up, and hence what its function might be. Signal sequences are only of relevance to proteins (although, of course, they are coded in the DNA), whereas sequence motifs can be found in DNA or protein.

Mutagenicity tests

There are a range of tests using biological systems to see whether compounds can cause mutations. It is argued that chemicals that cause mutations are also likely to cause cancer in humans, a correlation that has generally been found to be true. The main single-cell test systems used are:

Ames test. Named after Bruce Ames, this test exposes *Salmonella typhimurium* to the chemical. New mutants are detected as bacteria that can grow without being provided with histidine ('back mutations'). The test is one of the standard tests required for mutagenicity testing of products.

SOS-Chromotest. This is an alternative bacterial test which detects when an *E. coli* has had its DNA repair enzymes activated. Mutagens activate specific enzymes that repair damage in DNA, and the test uses a side-effect of these enzymes to detect their activity. The test is not generally accepted.

Micronucleus test. This looks for a characteristic aberration of the chromosomes (formation of small fragments of genetic material outside the nucleus, which are called 'micronuclei') in cultured mammalian cells, usually Chinese hamster ovary (CHO) cells.

Ames himself has said that most mutagenicity testing, including his own test system, is largely irrelevant to human health since 99% of the mutagens and carcinogens we are exposed to come from the 'natural' environment and not from man-made sources.

Mutation

A mutation is a change in the genes. The original gene is usually called the 'wild type' or, in human medical genetics, the 'normal' gene. Strictly, the 'mutation' is an identifiable change in one allele of the gene, i.e. one of the two copies in a cell. However, some scientists assume that if a gene is different from its form in the majority of individuals, or if the variant of the gene does not work, then it is a mutant.

Mutations come in a variety of forms, depending on exactly how the DNA sequence of the gene is altered. These are:

● point mutation: a single base is altered, usually to another base.

- deletion: a region of DNA is missing. This can be anything from a few bases to millions of bases.

- insertion: a new bit of DNA is inserted into a gene. Usually this is a result of an insertion element, a piece of DNA that can put copies of itself in new places round the genome, such as a **transposon**. A variation on the deletion/insertion theme is mismatch recombination, which alters the number of copies of a sequence in a repetitive array, like a VNTR (*see* **DNA fingerprinting**).

- rearrangements: everything else is a rearrangement. These are usually rearrangements of large regions of DNA rather than of a few bases. Chromosomal rearrangements shuffle whole chromosomal regions, switching them end-to-end (inversion) or swapping regions between chromosomes (transposition). Sometimes transpositions are straight swaps (balanced), sometimes some DNA gets lost in the process (unbalanced).

These are natural mutations. The other sort are artificial mutations. These can be induced mutations, where we bombard an organism with radiation or mutagenic chemicals to accelerate the normal rate of mutation—any of the above mutation types can result. Alternatively, we can manipulate the DNA itself, a process called *in vitro* mutagenesis or **site-directed mutagenesis**.

Mutations in genes are detected by a wide range of **DNA probe** or **PCR** technologies. If you do not know what the mutation is, then it can be detected by sequencing the DNA and looking for differences. There are also gel electrophoresis techniques that will show if a target piece of DNA is different from one you have already cloned. The most common are denaturing gradient gel electrophoresis (DGGE) and single strand conformational polymorphism analysis (SSCP). These will rapidly tell you if a difference exists, but not what the difference is. Gene 'chips' are also being developed to detect mutations.

Mythogenesis

Biotechnology has been extraordinarily successful in attracting scientists and investment. This has occurred despite that fact that few biotechnology companies even today are breaking even, and there are very few truly biotechnological products on the market that were not there 10 years ago. One entirely reasonable explanation for this is that much of the new wave of biotechnology is aimed at medical problems, and these take a long time to solve, are great intellectual and social challenges, and could earn enormous profits for their conqueror. Another explanation, favoured by a few arts-trained observers, is that this is a *post hoc* rationalization for something much deeper, and that biotechnology is attractive because it promises the realization of very ancient dreams, in Jungian terms the physical embodiment of mythic archtypes.

Thus some believe biotechnology promises life extension via pharmaceuticals that are specific and 'natural' (both secondary metabolite products and biotherapeutics), the creation of plausible supermen (*see* **Sports and biotechnology**), reproduction without sex, human cloning (and thus both a kind of immortality and the potential for children who are an extension of their parents), wild new animals such as chimeras and giants, and so on.

In literal terms this is nonsense—chimeric animals look like any other animals, 'giant' mice are only 30% longer than normal mice, and human reproduction has never been so carefully legislated. However, this need not matter. If biotechnology is perceived subconsciously as opening the doors to such archetypical dream worlds, then it will attract and repel much more strongly than if it is perceived just as a lot of scientists making money out of clever brewing. Serious press reports of a 1995 British meeting on biotechnology glossed over all the achievements of serious science in favour of reports of a scientist who claimed that he would be able to produce a cheese-flavoured cauliflower. (The non-serious press did not report the meeting at all, of course.) Why this focus, when it was only meant as a light-hearted example of what might be possible using plant genetic engineering? Because the 'all-food', the single food that is all you need to eat, has strong mythic roots going back to Greek ambrosia and Biblical manna, and anything that suggests that scientists are working on such an all-food is more attention-grabbing, even if it is nonsense, than the number of people dying of AIDS.

This could be important for the science and industry of biotechnology since it suggests that much of the campaign that is waged to get the public to accept biotechnology is aimed at the wrong motivations, and consequently will not convince many people. Indeed, this has been the case: in general, a study on European attitudes to biotechnology carried out in 1991 tentatively showed that the more the people of a country knew about biotechnology from the education that the government and industry put out, the more they were against it.

Names

One of the most fiercely competitive areas for biotechnology start-ups is finding a good name. As well as the obvious ones ('Monoclonal Antibodies Inc', 'Affinity Chromatography Ltd.') biotechnology companies' names are assembled from a wide range of standard units. Start with one of:

- bio-: almost an essential part. Means, to do with life.

- immu- or immuno-: to do with the immune system, usually to do with antibodies.

- hyb- or hybri-: usually to do with DNA hybridization. Can also be related to making hybrid species. Hybritech is the maverick here, being concerned with monoclonal antibodies.

- trans-: across, suggesting multidisciplinarity. 'Transgenics' is a special case.

- eco-: now needs no introduction—anything that can be considered 'ecological'.

- agro- or Agri-: agricultural.

- myco-: to do with fungi.

- onco-: to do with cancer.

- cyto-: to do with cells (usually means mammalian cells).

- gen-: to do with genes, and hence with recombinant DNA.

- enz- or enzo-: to do with enzymes.

and finish with one of:

- -gene or -gen: anything to do with genes.

- -zyme or -ase: to do with enzymes.

- -med or -medix or -medic or -medics: implies an application in the heathcare industry.

- -tech: superfluous, and obvious.

- -probe: either something to do with DNA probes, or something to do with medical diagnostics. Ideally both.

- -clone: suggests recombinant DNA technology.
  The name can also have 'Sciences', 'Systems', or 'Technology' added to the end. If the name has several words in it, a memorable acronym can be helpful: hence DNAX, ABC, etc.

Nanotechnology

Nanotechnology is the production of machines of nanometre size components, popularized by K. Eric Drexler. Enthusiasts say that biotechnology can provide the methods necessary to build such small machines. This is because many biological molecules assemble themselves chemically into large complexes and the nanotechnologists hope to use the same self-assembly processes to get the nanomachines to build themselves. A much quoted example is that of a tiny submarine which can be injected into a patient to clear out arteries clogged by atherosclerosis. Biology could provide some of the elements of this (for example, the world's smallest screw propeller is the flagellum of a bacterium). However, this is definitely 21st century stuff, and is unlikely to be realised in any of the forms being discussed today. Just think how Charles Babbage saw his calculating engine, and what computers actually looked like when they became real.

Micromechanics, building engineering structures on silicon chips, works on a scale of tens of micrometres rather than the 100-nanometre scale needed of nanotechnology, and has focused on a few well-defined products such as pressure and strain gauges. The success of micromechanics in a few fields does not imply that molecular electronics or nanotechnology are realistic for the next few years.

Natural products

Natural products are any chemical or material derived from any organism, as opposed to ones that are synthesized chemically. In biotechnological terms this often means drugs or therapeutics derived from living things, usually fungi or bacteria. There are a range of examples.

Microorganisms produce a wide range of chemicals which are used directly as drugs or as the starting materials for semi-synthesis of drugs, for example, penicillin and cephalosporin antibiotics. Major research groups and several specialist companies are dedicated to discovering new natural product drugs of this type by screening bacteria and fungi.

Plants make a dazzling array of chemicals to deter animals from eating them. They include flavour and fragrance compounds such as menthol from mint, limonene from lemons, caffeine from tea, and opium from poppies. Many of these have been used as the starting point of drug synthesis (such as steroids) or as drugs in their own right (such as taxol).

Animals also make a variety of potential therapeutic products. 'Natural products' usually means animal products from non-mammals, such as the antithrombin protein hirudin and the anticancer protein drug antistatin, both from leeches.

Marine organisms are also a potentially diverse source of natural products, in part because of the range of chemical environments they inhabit, and in part because of the range of organisms involved: bacteria, coral, sponges, seaweed, and cephalopods have all been examined for natural product drug leads, as well as fish. This is a fairly new and specialized activity, needing skills with the aqualung and Zodiac boat as well as with biochemistry. It is also unusually difficult to get enough material to do trials: it takes hundreds of kilograms of deep sea sponges to get a few grams of halichondrin B (undergoing trials as an anticancer agent).

A wide range of materials such as wood and straw (traditional, natural building materials) and PHB (the biodegradable polymer made from microorganisms) are also products of living organisms. However, neither the traditional biomaterials nor other traditional products such as food are usually called 'natural products' in this sense. They are often labelled 'natural' to encourage the consumer to distinguish them from the products of industrial civilization.

Nerve cells (neurones)

The nervous system is one of the most complicated systems in the human body, and one whose failure causes severe disease. Biotechnology has consequently put a lot of effort into understanding and treating nerve-related diseases.

The nervous system is a network of nerve cells (called neurones) and their supporting cells. Nerve cells consist of a cell body (with the cell nucleus); short, branching 'arms' called dendrites; and usually a long connecter called an axon. Axons can be up to a metre long in humans.

Nerves are electric cells. They pass information along in a wave of electric activity. When 'at rest' there is voltage across their cell membrane (the 'resting potential'). When they 'fire', an 'action potential' wave passes down them: as the voltage changes in one bit of the nerve, it triggers the next bit of nerve to alters its voltage. Voltage changes are a result of ions flowing across the membrane, so ion channels are critical to how the nerves work. If you plot the voltage difference between the inside and outside of a nerve cell as a graph against time, an action potential appears as a 'spike' on the graph. The action potential 'spike' can travel down a nerve at up to 200 metres per second.

Conduction between nerve cells is different. When the action potential spike reaches the end of a nerve (a synapse), it makes the cell release specialized chemicals, which travel across the small gap between that cell and the next cell (or the muscle, if this nerve is a motor nerve), and triggers the second nerve cell to start its own action potential. The chemicals are called neurotransmitters, and there are a lot of different ones depending on the part of the nervous system involved. Neurotransmitters are detected by specific receptors in the receiving nerve cell.

Myelinated nerve cell

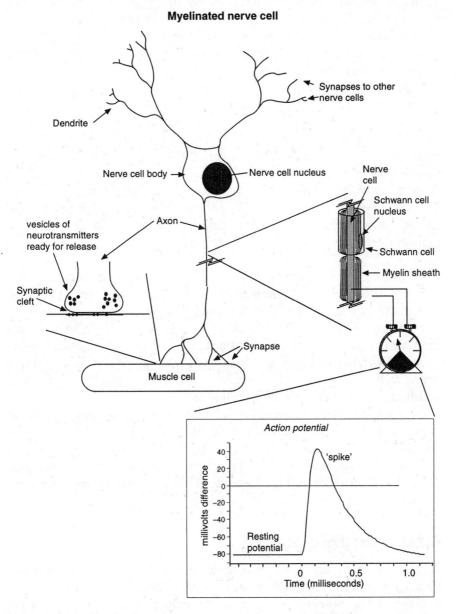

Nerve cells do not exist in isolation. In the peripheral nervous system (i.e. all of it except the brain and spinal cord), most cells are surrounded by Schwann cells, which nourish the nerve fibres and coat some axons in myelin, consisting of many layers of phospholipid insulator. Myelin provides electrical insulation for the nerve cell, and thus speeds the transmission of action potential waves down the axon. Axons with myelin on them are (not surprisingly) called myelinated nerves. Schwann cells also help peripheral nerves to regrow their axons if they are damaged, laying down a 'path' along which the axon grows.

In the central nervous system glial cells play the role of Schwann cells, expect that they are not so effective at allowing regeneration of damaged nerves, so CNS tissue does not regenerate efficiently.

In the peripheral nervous system nerves are conveniently divided into motor nerves (which drive muscles) and sensory nerves (which send sense signals to the brain). Motor nerves end in a special synapse that connects to an 'end plate', a dedicated bit of the surface of a muscle cell that receives the nerve impulse and triggers the ion changes that makes the muscle contract. Sensory nerve axons usually connect with specialized sensory nerve cells, which detect pressure, heat, and other sensations. The same system links to special photoreceptors in the eye ('rods' and 'cones', the specific nerve cells that detect black-and-white and coloured light patterns, respectively) and in the nose and tongue for smell and taste.

Biotechnological interests in nerve cell-related diseases include:

Nerve cell repair and regeneration. Nerve cells in the central nervous system are not readily replaced, so diseases such as stroke and multiple sclerosis (MS) are effectively incurable. MS, for example, is an auto-immune disease where the body attacks its own myelin sheaths, leaving spherical patches of the brain ('plaques') without myelin. Encouraging nerve cells to regrow into damaged areas, or replacing them with cultured cells, would be hugely valuable.

Neurological drugs. Psychological and neurological diseases, from depression to Parkinson's disease, could be cured if we knew how to correct the malfunction of the nerve cells. A lot of work on this relates to finding out how neurotransmitter receptors and ion channel proteins work. Genomic discovery has uncovered large families of genes for both types of protein, so biotechnology has a lot of work to do here.

Neurotrophic factor

A general name for a nerve-specific growth factor, i.e. a molecule (usually a protein) that will encourage nerve cells to grow or to repair damage. Their potential is as drugs to help patients overcome nerve damage caused by spinal or head injuries, degenerative diseases like multiple sclerosis or Alzheimer's disease, or ageing. Among the neuro-trophic factors are:

- nerve growth factor (NGF), the first neurotrophic factor to be discovered.

- neurotropin-3 (NT-3), which is generating particular interest because it may have potential as a therapeutic for degenerative neural diseases such as multiple sclerosis or Alzheimer's disease.

- ciliary neurotrophic factor (CTNF), which is believed to enhance nerve growth in embryos, and hence may be valuable for helping nerves to regenerate. Synergen are cloning this protein for eventual therapeutic use.

- brain-derived neurotrophic factor (BDNF), which is similar to NGF but targets brain cells.

- basic fibroblast growth factor (bFGF), which, in combination with NGF can help regeneration of central nervous system nerves in some animal studies.

New diseases

Because it has a panoply of new and powerful techniques at its disposal, biotechnology is always looking for new ways of using them. One way is to identify a disease that has not been identified before, or one that is now thought to be much more serious than it was previously, and develop a treatment for it. For other diseases, of course, treatments already exist, which makes it harder for a new one to be accepted. Some of the 'hot' diseases discussed as targets for biotechnological solutions are:

Any viral disease (since there are so few antiviral drugs). Particularly AIDS, but also:

- hepatitis, which is degenerative liver disease. Hepatitis A, B, C, and D viruses are recognized, as well as environmental causes such as alcohol or solvent abuse.

- herpes simplex, and particularly genital herpes, which can be life-threatening to new babies if caught from their mothers. Cytomegalovirus (CMV) is another of the herpes group for which better treatment would be useful.

Other diseases in the news are:

Lyme disease. A debilitating bacterial disease caused by the spirochete *Borrelia burgdorfei* first recognized in 1982 and now affecting thousands of people. A vaccine is wanted.

'New' viruses, such as Lassa, Ebola, Marburg, and Hanta viruses.

These generally are not natural pathogens for humans, but if humans get infected the viruses are almost always lethal, usually very quickly. Despite the film *Outbreak*, starring Dustin Hoffman, these viruses are rare in the West and seldom spread beyond the immediately infected individuals.

Helicobacter pylorii, a bacterium that scientists now accept is a major cause of gastric ulcers.

Nitrogen fixation

Nitrogen is an essential macronutrient (something we need a lot of in our diet) for all living things. Air comprises 80% of nitrogen gas: however, plants and animals cannot convert this into protein. They rely, instead, on other forms of nitrogen: ammonia and nitrates for plants, proteins and amino acids for animals. Only a few organisms can convert atmospheric nitrogen into these forms of nitrogen, which can be assimilated readily, a process called 'fixing'. The rate at which fixed nitrogen can be supplied to growing crop plants is one of the limiting factors in their growth and yield

Nitrogen-fixing organisms are bacteria. Some live free in soil and some in symbiosis with plants. The symbiotic ones are the most interesting to the biotechnologist, although the free-living organisms like *Azobacter* and *Klebsiella* are easier to handle in the laboratory, and so most researchers prefer to use them. Symbiotic nitrogen-fixing organisms live in nodules in the roots of a few plants, and convert atmospheric nitrogen to ammonia for the plants in return for a supply of C_4 acids, made by the plant from carbon dioxide. The genes that code for the enzymes that fix nitrogen (the nif genes) have been cloned and characterized in some detail. *Klebsiella pneumoniae*'s nif genes have been studied the most: there are at least 17 of them, showing how complex this process is. The nod genes, which induce the plant to make nodules in which the bacteria can live, are less well characterized, but the subject of intense study. Biotechnology has tried several routes to fixing nitrogen for agriculture more efficiently.

Only a few types of crop plants (legumes, clover, rice, lupins) fix nitrogen via symbiotic bacteria of the genera *Rhizobium* and *Bradyrhizobium*, which live in their root nodules. Some other non-legumes fix nitrogen, but are not widely used as crops. One route is to force the *Rhizobia* to live in other plants, by introducing the bacteria to the plants in tissue culture or by engineering the cell-surface receptors of the plant root cells so that the bacteria absorb on to these roots in the same way as they do on to beans and clover. This route has been moderately successful on a laboratory scale. Another route, much touted 10 years ago, was to transfect the nif genes into the plants themselves so that they did not need the bacteria at all. This is now thought to be a route that is not likely to work. The bacteria provide a lot more enzymatic machinery than just the nif genes to fix nitrogen, and the roots also provide specific proteins (such as the plant haemoglobin protein, leghaemoglobin) which are an essential part of the nitrogen-fixing process: the nodules are not just passive containers for the bacteria.

The simplest use of biotechnology is in producing legume inoculants, to increase the soil population of rhizobia around a growing legume. Since each plant has to pick up the bacteria from the soil (there are no bacteria in the seed), nitrogen fixation can be limited by the rate of infection of the growing roots. Thus, dosing the soil with suitable bacteria can give a better rate of fixation. (It is controversial as to whether this is an effective economic measure.)

A variant on this approach is to improve the efficiency of the bacteria in fixing nitrogen. BioTechnica tried an engineered *Rhizobium meliloti* in 1988, in which there were several copies of the gene for nitrogenase instead of the usual single copy. Nitrogenase is the enzyme that actually takes nitrogen molecules from the air and splits them open. The engineered bacterium was used to infect alfalfa, but it did not result in any increased yield from the plant and so the trial was stopped.

If fixing nitrogen frees a plant from dependence on soil nitrates, why don't all plants fix their own nitrogen? The reason is that fixing nitrogen takes a great deal of metabolic energy, so if there is any other way to get nitrogen for the plant (or indeed for bacteria), then they will take it as the more energy efficient course. It is not clear, therefore, whether getting plants that do not normally fix nitrogen to do so would actually decrease crop yields because they would divert a lot of energy away from producing the edible portions of the plant and into nitrogen fixation.

Nutraceuticals

Also called functional foods, these are foods that have some specific health benefit (other than preventing starvation). They are distinguished from 'health food', which is generally believed to be good for you but which does not claim to any more for you other than not making you ill. They are also differentiated from food supplements, which are meant to make up for a lack of an essential nutrient in your diet.

Examples of the claimed benefits of food components are:

- carnitine (fights heart disease)
- calcium (helps prevent osteoporosis)
- folic acid (helps prevent neural tube defects in babies)

- cranberry juice (helps prevent urinary tract infection)
- pectin (lowers cholesterol)
- fibre (reduces risk of bowel cancer)
- vitamin E (fights cardiovascular disease)

Some nutraceuticals are scientifically accepted, others are still controversial because of a lack of medical-style trials of their effect. If trials are done, their results can also be controversial. For example, the first 'double blind' study of the effect of vitamin E on heart disease, published in 1996, showed that large doses of the vitamin actually seemed to increase the death rate slightly. However, there is substantial sympathy for nutraceuticals and proponents of nutraceuticals say that they are not drugs, and should be tested in other ways.

Generally, biotechnology contributes to nutracteuticals by providing foods and food ingredients that have increased levels of the 'healthy' components.

Oil

Biotechnology has tried to produce or use oil for fuel applications. None of these applications have proven economic since the oil industry has had over 100 years and trillions of dollars of turnover to optimize its processes.

Two major applications are the use of oil as a feedstock for bacterial growth (for **single-cell protein** production), and the use of organic material to make oil-like fuels or oil substitute (**biofuel** production).

Biotechnological products are also used in conventional oil extraction. Oilmen use a material called 'drilling mud' which is pumped down an oil well in order to carry bits of rock drilled off the base of the hole to the surface. It is made from water, fine clay, and polymers. Xanthan gum is a favoured component because it is viscoelastic, i.e. its viscosity is much lower at high shear forces than at low ones, so it behaves like water near the drilling head (does not slow up the drill) but like jello when it is flowing back up the drill shaft, and so supports the rock fragments.

Mud is also used to force oil to the surface. In a new well, the pressure of oil may force it to the surface. However, after a time the pressure drops, and a pump is needed to get the oil to flow upwards, often pumping water into the rock to displace the oil. When this 'secondary' recovery cannot produce oil economically, usually when the less viscous water can no longer effectively displace the thick oil, 'tertiary' recovery systems can be tried. These include pumping viscous 'mud' into the well, including mud with biotechnological ingredients. Archaeus Technologies goes one step further and pumps bacterial cultures in cheap substrates such as molasses down oil wells to make the polymers and surfactants underground. The bacterial suspension has low viscosity, and so is easy to pump.

Oligonucleotides

Oligonucleotides are short DNA (or, rarely, RNA) molecules, usually defined as 100 bases long or less. This is the length of DNA that an automated DNA synthesis machine (a DNA synthesizer, oligonucleotide synthesizer, or 'gene machine') can make in one go and still have a significant yield of product. Oligonucleotides are usually defined by their origin: if it is made chemically it is an oligonucleotide, if it is cloned then it is a gene or a gene probe.

Oligonucleotides are usually named for their length. The naming follows the monomer–dimer–trimer scheme, up to decamer (10 bases). Beyond that, the name of an oligonucleotide is generally its length as a number followed by '-mer'. Thus a 17-base oligonucleotide is called a '17-mer'.

Automated DNA synthesizers, often called gene machines, use a series of chemical reactions to build up the DNA chain one base at a time. Each reaction consists of four steps since the chemistry has to make sure that only one base is added each time, so building up a 50-base oligonucleotide (a '50-mer') requires 200 reaction steps. Clearly, if one of those steps is slightly inefficient, then the overall efficiency will be very poor: this is why synthesizing oligonucleotides of greater than 100 bases becomes quite difficult. Most gene machines are completely automated, so all the biotechnologist has to do is type in the DNA sequence required and collect the DNA.

Oligonucleotides have become critical to biotechnology for three reasons.

1. They can be linked together to form larger lengths of DNA which can function as completely synthetic genes (*see* **Gene synthesis**).
2. They can be used as DNA probes for a variety of genetic studies. In this they are particularly useful since they can distinguish between versions (alleles) of a gene that differ by only one base. Such oligonucleotides are called allele-specific oligonucleotides (ASOs).
3. They are the primers for the widely used PCR technique.

Oncogene

Oncogenes are genes that are believed to be necessary for cancers to develop. There are a large number of them and, as would be expected from the variety of cancer types, they act in many different ways. Most are present in normal cells as protooncogenes, i.e. versions of the gene that are benign, and indeed are essential to the body's normal development. A mutation turns them into the malign oncogene.

Oncogenes are of interest to biotechnology because of the importance of cancer as a cause of morbidity and mortality in Western societies. Many biomedical R&D programmes are aimed at curing or palliating cancer, and hence are interested in directly or indirectly preventing the effect of oncogenes. The approach depends on the oncogene involved. Some oncogenes make proteins that are detectable on the outside of cells or in the blood: these proteins can be 'tumour markers', i.e. markers to show where a tumour is developing. In turn they may be used to diagnose cancer or to target a biotherapeutic to the cancer cell and so destroy it specifically. Oncogenes that act only inside the cells cannot be used as tumour markers in this way.

Among the more talked about oncogenes or oncogene proteins are:

- erb, a family of proteins, of which erbB-2 is associated with breast cancer.

2. The substrate may be more soluble in organic solvents (or indeed, only soluble in them).

3. The enzyme may be more stable or have an altered specificity in the new solvent.

4. There will be no side reactions involving water.

5. Products may be easier to recover from a solvent (e.g. by evaporation or extraction with water).

Thus, for some reactions, and especially those involving materials that are very poorly soluble in water or are very easily hydrolysed, getting an enzyme to work in a non-aqueous solvent could be a very good thing. Examples are the synthesis of peptide by proteases (in water, proteases only break up peptides), and the transesterification of lipids by lipases (in water, lipases overwhelmingly break lipids up rather than put them together). Using lipases in organic solvents has been one of the more successful applications of this technology.

The problem is that, as usually prepared, enzymes rarely dissolve in anything except water, and even if they do dissolve they do not work. This is in part because enzymes are prepared as aqueous solutions, and so a mixture of the enzyme with an organic solvent is just that, a mixture of two immiscible liquids. If the enzyme is dried thoroughly so that an absolutely minimal number of water molecules stick to it, some enzymes can be made to work in organic solvents such as octanol. Variations are the use of **supercritical fluids** for the enzyme reaction, **reversed phase biocatalysis** or emulsion systems, or **bioconversion** in organic solvents.

An alternative approach is to engineer the protein genetically to be more stable or more active in the solvent concerned, and this is attracting some interest.

Ornamental plants

Plant biotechnology has made major advances in developing new ornamental plants. Plants are inherently easier to 'engineer' through cloning or genetic engineering than animals. Crop plants can be hard to engineer because a change in them usually reduces the efficiency with which they produce their crop: seeds, leaves, tubers, or other valuable parts. However, this does not apply to ornamental plants, which just have to survive and look pretty.

The world market for ornamental plants is over $30 billion, making this a huge area for biotechnology. Top producers are the USA and the Netherlands (the fields of tulips are one of the Netherlands' major industries).

There are two major areas of applications.

Cell culture and cloning technologies. Rare variants of plants can be propagated very quickly using cloning technologies, and the same technologies sometimes generate new variants through **somaclonal variation**. Thus 'new' plants such as tropical species or ornamental cabbages, onions, and other plants can be sold in garden centres very fast.

Genetic manipulation. Putting new genes into ornamental plants is a growing area of interest. Among the targets are:

- blue rose. This is the 'holy grail' of rose breeders, and several groups have tried to put the genes for blue pigments into roses. The colour of the blue pigments is dependent on the pH inside the flower, and so far no one has been able to get both pigment and pH right (*see* **Flowers**).

- Pest-resistant transgenics. Plants that have inherent pest resistance are created through incorporating genes for pest-toxic proteins, such as B.t. toxin (*see* **Pest resistance in plants**). This is attractive for garden plants, some of which have very poor abilities to fight off pests

- ethylene binding and senescence control. Cut flowers 'die' largely through cell senescence (the flowers go through the same changes as they would at the end of their natural flowering season). Most of these changes are triggered by the gas ethylene (ethene), which acts as a hormone in plants. The effects can be blocked either by blocking the receptors for ethylene or blocking one of the enzymes that makes it. This has been tried for the cut flower market, but could also be done to block ripening in some fruit (such as bananas), which is also triggered by ethylene. The grower would then ripen the fruit when it arrived at its destination with added ethylene.

Orphan Drug Act

A US law which gives an incentive to a company developing a drug for a comparatively rare disease. For drugs that provide novel treatments for diseases that are suffered by only a small number of people, the Orphan Drug Act gives the developer of the first drug of any one type seven years exclusive right to market that drug. This is meant as an incentive to develop drugs for which the market would otherwise be marginal, given the intense competition in the pharmaceutical industry. It has been invoked quite widely by the biotechnology industry since many bio-pharmaceuticals are so specific in their effects that they can only be used for a very narrow group of patients.

Osmotolerance in plants

Osmotolerance is a measure of a plant's ability to stand up to drought or to large amounts of salt in its water supply. Salt tolerance is sometimes called halotolerance. Because a reliable supply of fresh water is a limiting factor to agriculture in some places, osmotolerance is an important characteristic for plant breeders to achieve. Biotechnology has been suggested as a way of achieving it.

Plants survive water stress (i.e. environmental effects that tend to dehydrate them, such as drought or high salt) in a number of ways. These are by structural adaptation (e.g. thickening cell walls to stop water loss, making leaves round to reduce the surface area), physiological adaptation (e.g. developing molecular pumping mechanisms to pump water into the cells or salt out), or metabolic adaptation, by producing internal chemicals that counteract the effect of drought or salt. These are metabolically neutral chemicals that counterbalance the osmotic potential of the salt in salt water. They are often sugars or sugar alcohols like mannitol, and are collectively called osmolytes. Metabolic adaptations tend to involve only a few genes, whereas the other two adaptations involve many (dozens to hundreds). So metabolic adaptations are the favourite target of biotechnological efforts to transfer osmotolerance to crop plants.

Metabolic approaches to osmotolerance usually involve filling the plant cell up with something that is innocuous, that the plant can make easily, and that 'attracts' water through its osmotic potential (i.e. just by being there, not because it expends any energy). A range of such compounds are known, and the enzymes to make them have been characterized to some degree. Consequently they could be genetically engineered into crop plants to make them capable of withstanding greater water stress. There are the usually problems of plant genetic engineering (e.g. will it work? will the resulting plant produce commercial levels of crop?), together with the additional problem that the osmoprotectant material must end up in the right part of the cell to be effective.

Oversight

In US regulatory contexts, this means 'having regulatory responsibility over'. Thus the definition of which organisms are subject to 'regulatory oversight' is a critical feature of the regulation of biotechnology: it defines which organisms must be approved by which authorities before they can be used for industrial biotechnology.

Panning

In a discovery context, also called biopanning. This is a technique for finding which of a library of molecules (often generated by combinatorial chemistry or by phage display technology) will bind to a target molecule. One of the partner molecules is immobilized to a surface, usually plastic but sometimes glass, and the other is swirled over it like a gold-rush 'panner' looking for gold in river silt. If any of the mobile molecules bind to the surface, then you know that a match has been found. This has been used extensively in receptor binding studies, to find which of many millions of chemicals will bind to a receptor. It is also used to select from phage display libraries of antibodies to find the antibody that binds best to an antigen (*see* **Phage display**).

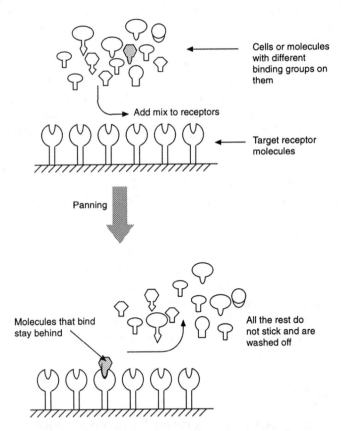

Cells or molecules with different binding groups on them

Add mix to receptors

Target receptor molecules

Panning

Molecules that bind stay behind

All the rest do not stick and are washed off

Panning can also be used to purify cells from a population. It has been used to purify stem cells from blood. (Stem cells are cells with the capacity to proliferate into many types of blood cells, usually found only in the bone marrow, which can be used to regenerate the blood system after massive irradiation as part of cancer therapy.) Here the surface is coated with a ligand which binds to a receptor on the cell you are looking for, and then blood is passed over it. This is, in fact, a whole-cell version of **affinity chromatography**.

Patents

Whether a biotechnological process or product can be patented, and if so how, has been one of the biggest legal stumbling blocks to the application of biotechnology since the early days of genetic engineering. The first patent for an organism altered by recombinant DNA technology was granted to Chakrabarty in 1980, for a hydrocarbon-degrading *Pseudomonas* strain to be used for oil clean-up. Since that patent established that recombinant DNA, as well as some more conventional biotechnology techniques and products, can be patented, legal battles have been fought generally over: (i) whether something is a patentable invention or not, and (ii) if it is, who invented it.

In principle, patents are based on claims of novelty, utility and enablement: a patent must describe something new, something useful, and must describe it in a way that enables someone else to duplicate it. Perpetual motion machines fail at this last hurdle, for example, which is why most patent offices now require a working model. Living things are not usually patentable, but biological materials obtained by non-biological means (one in which 'the hand of man' has had a part) are patentable.

Nearly 23.5% of all patents granted in OECD (Organization of Economic Cooperation and Development) countries in 1987 were granted in Japan. 30.5% in the USA, 8.8% in the FRG, and < 6% in the rest for any one country. However, Japan has a tradition of patenting everything (nearly 50% of all applications are Japanese). Patents often form a type of trade barrier between countries, making it difficult for non-residents to obtain protection, and hence to use their invention in that country. For example, the US patent office claimed that the Japanese patent system puts any foreign language applicant at a disadvantage.

The material that is patentable varies from country to country.

In addition to patents covering things ('composition of matter' patents), patents covering processes for making or using microbes are allowed in all areas, although methods for breeding are not allowed by the EPO.

Apart from the differences and ambiguities in patent law, biotechnology companies face a substantially longer time between filing their patent and getting it granted than companies in most other fields, especially in the USA. This means that they cannot defend their patent in the courts for several years after it has been made public.

Area	Macromolecules or viruses[1]	Unenegineered microorganisms	Plant varieties	Animal varieties	Genetically engineered organisms
USA	yes	yes	yes	yes	yes
Canada	yes	yes	no	no	yes
EPO[2]	yes	yes	no	no	yes
Japan	yes	yes	no	yes	yes

[1]However, there is some uncertainty about what is the difference between a recombinant protein and its (presumably) identical natural counterpart, for example.
[2]The situation with the European Patent Office (EPO) is not clear, as the draft directive on patenting was rejected by the European Parliament in 1995, and so there are no effective overall rules. Thus, by the end of 1995 the EPO had rejected a plant patent application (by Plant Genetic Systems) on the basis that it was an attempt to patent a variety, but accepted an animal patent (for the Oncomouse) which is subject to a still-ongoing appeal.

Biotechnology companies have discovered that a patent is only as good as its last court case. While obtaining world-wide patent protection is complex and costly, the patentee has then to be willing and financially able to defend the patent against infringement in the courts, which can take years and cost millions of dollars.

Key organizations in patenting are: EPO (European Patent Office), PTO (US Patent and Trademark Office), and the various European national patent offices.

The most high profile patent cases in biotechnology are:

PCR. There is no doubt that Cetus publicized and developed polymerase chain reaction. But did they invent it? Hoffmann LaRoche claims that they did not, and that it was described in 1973.

EPO. Amgen and Genetics Institute have been working on genetically engineered erythropoietin (EPO) roughly at the same time, and both tried to claim patent protection. In April 1991 the US Court of Appeal effectively gave complete rights to Amgen, because the supporting technical information provided with the Genetics Institute patent did not (the court said) enable someone else to reproduce what they did. ('Enablement' is a key part of a patent—the patent has to describe something new in a way that enables someone else to reproduce it.) This decision was a big surprise for industry observers, who had expected a 'mutual infringement' ruling, i.e. they both infringed each other's patents.

Factor VIII. Factor VIII is used to treat haemophilia, and Genentech, Scripps Clinic, and Chiron had developed methods for purifying it from blood, and had claimed a patent on the product. The US Appeals Court decided that they could not claim rights to the product (although their specific ways of making it are patentable).

DNA sequence. See **DNA sequence patents**. Can you patent a DNA sequence, even if you do not know what it can do, and so do not know what you could 'invent' with it? If you know what the gene does, should you be able to patent anything so fundamental to life? Especially when we are all carrying copies? Can you patent single base variants on genes? There are lots of arguments, so in short, no one knows.

Transgenic animals. The Oncomouse was granted a US patent in 1988, leading to further patents in the USA for transgenic animals. By the start of 1996 the PTO had granted nine patents on transgenic non-human animals, including three that claimed all such animals. In Europe the EPO still does not allow animals to be patented.

Most of the high-profile cases are fought in US jury trials, which means, like the O. J. Simpson trial of 1995, the jury must be educated on some extremely complex technical issues before they can even begin to understand the arguments. As in the O. J. trial, therefore, the jury often ends up putting the technical arguments aside and judging by 'gut feel'.

PCR (polymerase chain reaction)

Polymerase chain reaction is a method for amplifying DNA that was invented by Kary Mullis of Cetus (but *see* **Patents**). It takes a single copy of a DNA molecule and uses it to create millions or billions of copies of itself. Because of the specificity and accuracy of the reaction, this is an ultimately sensitive detection system.

The diagram outlines how PCR works. The key ingredients are Taq polymerase (DNA polymerase (an enzyme that makes new DNA isolated from the bacterium *Thermus acquaticus*) or some other, equivalently heat-stable DNA polymerase, and two primers (short DNA molecules) which are complementary to two sites either side of the piece of DNA you want to amplify. The Primers are usually oligonucleotides that you have synthesized. Given these ingredients PCR will amplify almost any piece of DNA.

Many uses of PCR have been developed since its invention in 1985. Most obviously it has been used for detecting DNA sequences, for the diagnosis of genetic disease, for **DNA fingerprinting**, for detecting bacteria or viruses, and for research (especially such arcane matters as cloning the DNA out of Egyptian mummies, and from dodos). Its use in genetic diagnosis is widespread, while the use in bacteriology is much less common. This is, in part, because of the problem of contamination. If PCR can amplify a single molecule of DNA, then a single molecule 'escaping' from the amplified product, if it gets back into the starting materials, can start the PCR reaction. Several workers have had to abandon research into a particular gene because their laboratories have become saturated with contaminating PCR products, and some genetic diagnoses detecting male-specific genes in fetuses have to be done exclusively by female workers since the skin cells falling of male workers are enough to contaminate the test. Controlling contamination was a major problem for developing machines to do PCR as a routine diagnostic test, and one of the reasons that it was 10 years before Hoffmann La Roche launched their automated PCR system. (Abbott Diagnostics launched an LCR-based system a couple of years earlier.)

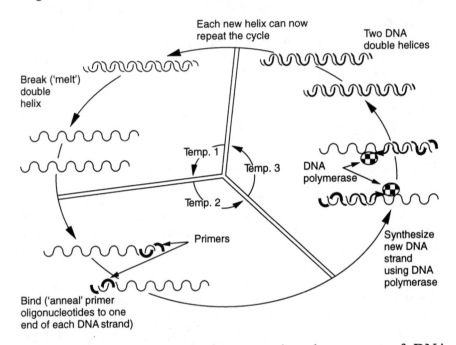

It is difficult to use PCR for measuring the amount of DNA (quantification) (*see* **DNA amplification**).

PCR can also be used to clone genes, if two suitable primers can be made, and to select the correct gene construct out of a mixture of constructs when making a synthetic gene: the use of PCR in cloning is very widespread. If the primers are chosen correctly, PCR can be used to make deletions, insertions in genes, and recombinants between them. It sometimes does this accidentally anyway, generating artefacts.

Variants on PCR, such as single-sided PCR and inverse PCR (which rearrange the DNA before amplification so that only one primer is needed), and random PCR (which stitches synthetic DNA on to the ends of a segment to be amplified so that no new primers are needed), have been developed. A common variant is RT–PCR, which stands for reverse transcriptase–PCR and starts with an RNA target. Reverse transcriptase makes a DNA copy from the RNA, and then the PCR amplifies that

DNA copy. This is widely used to clone and characterize mRNAs.

PCR was the subject of a bitter patent dispute between Cetus, who claim they invented it, and Hoffman La Roche, who said that it was essentially invented 15 years before. Partly because of this, and partly because the Cetus patent covers all applications of PCR, there are a number of other amplification systems that do similar things but which work via a different mechanism (*see* **DNA amplification**). Roche has now bought rights to PCR from Cetus.

PEG (polythylene glycol)

Polyethylene glycol (PEG) is a water-soluble polymer used in many guises in biotechnology. It is the basic structure of a family of polymers that can have PEG units linked to other types of polymers. PEG polymers are typically soluble in many solvents, and are surfactants, i.e. they can promote mixing of water with other solvents. Biotechnological uses include:

- Large-scale purification of proteins and viruses, through two-phase systems or precipitation (*see* **Purification methods: large scale**).

- As a fusogen to make mammalian cells fuse (*see* **Cell fusion**).

- As a molecule to conjugate on to proteins (a process sometimes called PEGylation). PEGylated proteins are effectively covered with a 'shell' of PEG, which can protect them from recognition by the immune system and slow their breakdown by the body when injected. Over 40 PEG-modified proteins are now in use as therapeutics. PEG-conjugated enzymes can also be more soluble in organic solvents, or be more resistant to breakdown in industrial applications. This has been applied successfully to stabilizing lipases, which can then be used to build up lipids from fatty acids and alcohols in organic solvents such as trichloromethane and benzene.

Peptides

Peptides are short protein molecules, but are usually produced using different methods from other proteins. In general, something is a peptide if it contains 20 amino acids or less, a protein if it is 50 amino acids long or more: in between, it depends on who you talk to.

Peptides were very popular in the 1980s since it was discovered that a large number of hormones and neurotransmitters (the hormones that carry signals between nerve cells) are peptides. They can be produced by chemical, biochemical, or genetic means, unlike larger proteins, which are usually produced solely by genetic or cell biological methods.

Chemical synthesis adds amino acids one at a time to a growing chain using a cycle of reactions (*see* **Peptide synthesis**).

Peptides that have been made commercially include calcitonin (for osteoporosis), glucagon (for hypoglycaemia), and thyrotropin-releasing hormone (TRH, for thyroid disease). Aspartame, the artificial sweetener marketed under the Nutrasweet label, is a two-amino acid peptide, and is produced in amounts that dwarf the other, pharmaceutical products.

Peptide synthesis

Peptides are very short strings of amino acids, usually 10–20 amino acids long, but sometimes only two or three. They are made by different routes from proteins, for two reasons. First, peptides are usually broken down rapidly by bacterial cells, so it is difficult to make them by recombinant DNA methods. Secondly, because they are relatively small molecules, it is feasible to make them by chemical or enzymatic methods.

There are three general routes to making peptides. The first is by genetic engineering. The peptide is usually produced as a fusion protein, the peptide itself being joined on to a much larger protein. It then has to be cut ('cleaved') off the larger entity after that has been purified when the bacterium or yeast has made it. This can be difficult to achieve effectively since you need a chemical reagent (such as cyanogen bromide, which cuts at methionine residues) or an enzyme that cuts the fusion protein at exactly the junction between the peptide and the larger protein, but not within the peptide itself.

The second route is *in vitro* enzymology. Many proteases that break peptide bonds are known. By altering their reaction conditions they can be made to work in reverse and synthesize peptide bonds. Conditions can include making them work in organic solvents (*see* **Organic phase catalysis**), under extremely high pressure, or modifying the amino acids so that the peptide is removed from the reaction (by precipitation or because it dissolves in a second, organic solvent phase) as soon as it is formed.

In order that the protease does not assemble a whole string of amino acids, but rather adds them one at a time, the amino acids are 'protected' by adding groups on to them that prevent uncontrolled polymerization. A cycle of reactions adds an amino acid, then removes its protecting group, then adds another, and so on.

The third route is chemical synthesis. This performs the same sort of reaction cycle as enzymatic synthesis, but using traditional organic chemical reactions. The reactions can be carried out on a solid material (in a reaction series called the Merrifield synthesis) so that the peptide chain 'grows' while attached to a support structure, or in solution (which is usually easier for large amounts but cannot make long peptides). The efficiency of each step is high, but because it is not 100%, the yield is usually low after a couple of dozen amino acids have been added. Biochemical methods use enzymes to do the same chemistry. Recombinant DNA can be used to make a short gene for a peptide, but short peptides are usually broken down very fast by bacterial and yeast cells, and so very little product survives. Thus, the usual route to making a genetically engineered peptide is to make a larger fusion protein, manufacture that in bulk, and then split it up chemically or biochemically into smaller fragments, one of which is the peptide you want.

Chemical routes usually require more reaction steps than enzymatic ones, but the materials are usually cheaper. Either enzymatic or chemical synthesis can produce kilograms of a peptide, and there are fully automated 'peptide synthesizers' that can perform the chemistry to synthesize grams of a peptide in a few hours.

Permeabilization of cells

Cells are normally surrounded by a thin membrane of lipids and proteins, the plasma membrane. This is meant to keep out anything not essential for the cell's survival (and, for cells from animals and plants, they function as part of the whole). However, they can also keep out materials that biotechnologists want to get in. To get round this, cells can be 'permeabilized'. This effectively makes small holes in the plasma membrane, so that we can get material into the cells but all their contents do not leak out: they remain capable of doing what we want.

Permeabilization can be done by treating the cells with organic solvents (which dissolve out small patches of the lipid membranes), detergents such as bile salts, some special purpose ionophores (molecules that introduce channels of molecular size into the membrane, and which usually only let in a limited number of types of molecule), or physical treatments like freeze-drying or sonication. Many types of cell also become rather more permeable to some chemicals after they are immobilized on to solid supports.

Permeabilized cells can have other advantages over intact cells for use in a bioreactor. They are rarely viable, so they do not waste metabolic energy (and hence your valuable starting materials) building more cell mass. They will also not grow inside a bioreactor, blocking it up.

Pest resistance in plants

Genetic engineering has sought to engineer genes conferring resistance to pests in plants by several means. (This is a potential alternative to using conventional pesticides.)

The first is to identify genes in plants that confer resistance to pests, and transfer them to crop plants that are more valuable but are susceptible to that pest. This route is favoured in searches for resistance to bacterial and fungal pathogens. The plant genes often show 'gene-for-gene' matching with genes in the pathogen called 'virulence genes': the virulence genes have a role in causing the disease, and the corresponding plant genes have evolved to stop them. The difficulty is that exactly what the genes are doing is often unknown.

The other route is to add a completely new gene to the plant. This is a way to fight off pests that will not respond to changes in plant biochemistry, usually pests that do gross damage to plants by eating them. The approaches currently used are:

1. To include the gene for *Bacillus thuringiensis* toxin in the plant. The toxin stops gut function in some insects, so if they nibble the leaf it kills them. Calgene have successfully done this with tobacco, and Monsanto with tomato (the latter was a dramatic success as far as the plant's resistance to insect pests was concerned) (*see* **Biopesticide**). Plant Genetic Systems have a number of field trials underway of plants engineered with B.t.k. toxin in Europe and the USA, including potatoes and tomatoes, and Sandoz Pharmaceuticals is commercializing their transgenic B.t.k. toxin tobacco in the USA. Since tobacco is grown to be burnt, not eaten, there are fewer concerns about the health safety of genetically engineered tobacco than about almost any other crop.
2. To include an enzyme that attacks insects in the plant. DNA Plant Technologies is working on this, using chitinase as the enzyme: chitin is a major component of insect's skeletons, and chitinase is an enzyme that breaks it down.
3. To include a protein that blocks a pest's usual method of attack or digestion of the plant. This has been used with good effect: the gene for cowpea trypsin inhibitor, a protein that inhibits the protease trypsin (and related enzymes), has been engineered into tobacco . This blocked the action of digestive enzymes in insects' gut, and so killed them. Chitinase also works through this route to some extent, breaking down the wall of the gut.
4. To include a gene from the pathogen in the plant. This is called pathogen-derived resistance (PDR). The idea is that making an excess of one protein from a virus in the plant cells disrupts the control of how the virus is assembled when it infects the plant, and so viruses cannot propagate in the plant.

Phage display

Phage display is a method of making and testing proteins, usually newly invented proteins. The gene for the protein is cloned into a bacteriophage genome in such a way that the protein appears on the surface of the phage. You can then select the phage you want by selecting the protein directly, often using **panning**. The protein you have selected is still bound to the phage, so you can grow the phage up to identify the gene sequence, and hence the protein sequence, and to manipulate it further. Typically, the M13 phage is used for this, the protein being inserted into the Gene III. Libraries of up to 10^{12} phage can be made with 'random' variants of a protein in them, and the half a dozen variants that we want identified by panning.

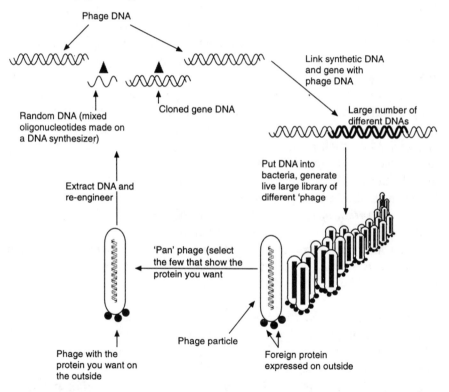

Phage display has been used mainly for generating antibodies by cloning. Antibody genes are either cloned from human lymphocytes or are synthesized on a gene machine. In either case, they are then spliced into an appropriate phage DNA and put into bacteria, where they make antibody fragments (such as SCAs) on the surface of the phage. If you select the phage that bind most strongly to the antigen, not only will they have the appropriate antibody on their surface but they will also have the gene for that antibody inside. This is an alternative to monoclonal antibody technology, and one that can generate human antibodies (as monoclonal technology cannot).

Phage display can also be used for a variety of protein engineering projects for other proteins, screening a large number of variants efficiently, providing that some sort of binding technique can be used to find the phage we need.

Pharmaceutical proteins

Pharmaceutical proteins, also often called biopharmaceuticals and sometimes also 'biologics' (in regulatory contexts), are proteins made for use as drugs. The first biopharmaceutical made by recombinant DNA technology was insulin, approved for human use in 1982, followed by human growth hormone in 1985 (both proteins had been used as drugs before). The first truly new recombinant DNA biopharmaceutical was alpha-interferon, approved for treatment of Karposi's sarcoma in 1985.

Usually biopharmaceuticals, which have to be human proteins to be fully effective in humans, are made from genetically engineered bacteria, as the only other source is cadavers or live human tissue. The genetic engineering of such products is covered elsewhere (*see* cytokine, growth factor, stem cell factor). Issues peculiar to biopharmaceuticals are usually the result of the stringent regulation that any drug must pass before it is allowed to go into general use. They include:

Demonstration of efficacy. A curiosity of the regulations is that each biopharmaceutical must be demonstrated to be effective on its own, whereas many of them are meant to be adjuncts to therapy with other drugs and not to have an effect of their own. Demonstration that the production version does what it is meant to do is usually performed by bioassay.

Demonstration that the product is free of contaminants. This is particularly true of bacterial proteins and cell wall material (endotoxins), which could act as a 'pyrogen', i.e. a material that could cause a feverish immune response in someone injected with it.

Demonstration of purity and stability. There can be other materials than the biopharmaceutical in the preparation—indeed, some are so powerful that a single dose is only a few milligrams, invisible to the naked eye, so something else has to go in there just to make it easier to handle. However, the 'something else' must be exactly characterized. The whole preparation must also be demonstrated to be stable. This is often achieved by freeze-drying it.

Freedom from side-effects. Apart from those caused by impurities or extremely high doses, these include principally the body's ability to recognize the protein as foreign, and so mount an immune response against it. Differences as small as the removal of an N-terminal methionine from a protein can alter the body's immune response to it. Pharmaceutical proteins include:

- insulin. The first major seller, now produced as a recombinant DNA product by Novo Nordisk and Eli Lilly. Insulin is used to treat diabetes.

- erythropoietin (EPO). Used as an adjunct to cancer therapy, it stimulates the bone marrow to produce more blood. It is one of Amgen's big sellers.

- human growth hormone (HGH). Produced by Genentech and by Pharmacia and Upjohn, it is used to treat a variety of growth disorders that result in abnormal shortness. There is also a thriving black market for HGH, in sports medicine and because it has been claimed to be a treatment for ageing (*see* **Sports and biotechnology**).

See also **Drug development pathway**.

Pharmacodynamics

'Pharmacodynamics is the study of what a patient does to the drug: toxicology is the study of what the drug does to the patient.' (Redpath) Pharmacodynamics is what happens to the drug as it passes through the body. The amount of a drug in the body depends on how much is given, how fast it is broken down, and how fast it is excreted. The speed of breakdown is a particularly crucial point for biopharmaceuticals since many recombinant proteins are liable to be mopped up by the body's immune system or by the normal mechanisms that remove 'old' proteins from the blood. Altering the glycosylation patterns of recombinant proteins can alter their pharmacokinetics substantially, which is one reason that glycosylation patterns are important for biotechnological drug producers. Pharmacodynamics are established in ADME studies (standing for absorption, distribution, metabolism, and excretion).

Pharmacokinetics is a more limited term. It means how the pharmacology of the drug (i.e. its effect as a medicine) changes with time.

The biological half-life is a related idea. It means the time taken for half of the drug to disappear from the patient. Often, the drug is cleared in two phases, an initial fast phase and a slower, second phase: this is called two-phase kinetics, and relates to how the drug interacts with different body components.

Pharming

This is the use of transgenic animals to make pharmaceutically useful proteins. The aim is to engineer animals genetically so that the protein we want ends up in the milk. The gene for the biopharmaceutical is spliced on to a promoter and a signal peptide, which makes them express the protein in the mammary gland. Promoters for casein and lactoglobulin are effective, as these proteins are naturally produced in large amounts and only in the milk. Protein levels of up to 35g/litre have been reported. Different groups favour pigs, cows, sheep, goats, and rabbits for this technology. The advantages over fermentation production systems are: avoiding the need for sterile culture; avoiding the need for complex nutrient mixes; and obtaining the protein relatively free of other proteins, and quite free of cell wall materials or potential endotoxins.

Several groups of researchers have made transgenic animals that produce milk containing grams per litre of alpha-1-antitrypsin, a protein with potential for treating emphysema. Pharmaceutical Proteins Ltd have used sheep, and Genzyme and Tufts University have used goats, to make this protein. The original idea of using cows (traditionally associated with making milk) has lost favour. Their longer breeding cycles and small numbers of offspring make breeding them more costly and time consuming, although Gene Pharming in Holland use cattle because of all the sophisticated technology that exists for mass-producing milk from them.

Proteins that have been made by pharming include tissue plasminogen activator (tPA), human growth hormone, urokinase, blood clotting Factor IX, and alpha-1-antitrypsin.

Chickens could also be used to make recombinant proteins. Genetic manipulation of chickens is difficult since the early embryo develops as the egg is being laid down around it. Techniques have been developed for transforming germ cells and of making chimeric chickens. While no commercial programmes have been started, eggs are potentially valuable delivery vehicles for proteins, especially antibodies (which accumulate naturally in the egg white).

Pheromones

Pheromones are chemicals used to signal from one organism to another. They are more specific than scents—the target organism can detect them, in minuscule amounts that are too dilute for other organisms to notice. They are usually complicated organic molecules.

Pheromones are being used as insecticides, because many insects use pheromones to signal their position prior to mating. If you load the air around your crop with pheromone, then pest insects become completely confused: at best they simply stop eating, and at worst they cannot find their mates to continue the pest infestation. Pheromone pesticides for tomato pinworm, the pink bollworm (which attacks cotton), and the coddling moth (which attacks apples) have all been developed. In 1995 the EPA agreed that these were not the same as toxic pesticides, because of the low concentrations they are used at and the fact that they are present in nature anyway.

Biotechnology helps in the development of pheromones by providing research tools for understanding which chemicals affect which species. It can also help to manufacture them, although this can also be done by conventional chemistry. The main value of pheromones, though, is in integrated pest management systems, systems that use knowledge of the biology of a pest to target it with a number of diagnostic and pesticidal strategies. A pheromone, for example, could be very effective during the mating season of an insect, but relatively useless afterwards when another type of pesticide would be needed.

Pheromones can drive moths to fly miles against the wind to find a mate. Despite persistent anecdotes, there is no evidence that a human equivalent exists. Mammals are certainly influenced by smell in their mate choice; however, they retain some mental choice over whether they respond to smells, and this is particularly true of humans.

Photosynthesis

Photosynthesis is the process of building new chemical compounds using the energy of light. Plants build sugars using sunlight, and it is the synthesis of sugars by green plants that is referred to in biotechnology under 'photosynthesis'. The rate and efficiency of photosynthesis is the major limiting factor on how fast plants can grow, which itself limits the productivity of all agriculture and food production. Photosynthesis is carried out in subcellular objects (organelles) called chloroplasts, which look like American footballs filled with green lasagne. The light-capturing part of photosynthesis happens on the membranes of the 'lasagne'.

There are light reactions and dark reactions in photosynthesis. The light reactions capture the energy of light. The dark reactions use that energy to capture carbon dioxide from the air and link it into growing sugar molecules.

The light reactions happen in 'photosystems', called I and II. The light itself is captured by an antenna complex, a large array of protein molecules that is almost 100% effective at capturing photons falling on them. The energy is then transferred to electrochemical reactions that generate ATP, or are used to split water into hydroxide (which forms oxygen gas) and hydrogen (which is used to reduce cytrochrome proteins). The two systems boost the energy of the electrons involved in this process to a high enough level in two steps—one system on its own could not push in enough energy to split the water. The captured energy accumulates as reduced NADH and ATP, which is used in the dark reactions.

The dark reactions centre around the reaction of carbon dioxide with a sugar called ribulose diphosphate, making two three-carbon sugar molecules. It is catalysed by ribulose-1,3-diphosphate carboxylase/oxygenase, always referred to as RUBISCO. A whole series of other reactions then convert the products into other sugars, and thence into plant. RUBISCO is the most common protein on earth. (The most common protein from vertebrates is collagen, which we use to make gelatin, and its total mass on earth is probably less than 1000th that of RUBISCO.) This is the basic photosynthesis process in green plants, and is called the C_3 process, because it centres around three-carbon sugars.

Photosynthesis builds sugars from CO_2 and light, generating oxygen as a side product. Another process is called photorespiration, which does exactly the opposite—it uses light and oxygen to break down sugars into CO_2. Unfortunately, exactly the same enzyme is responsible for photosynthesis and photorespiration, i.e. RUBISCO. So you cannot get rid of the photorespiration. In tropical countries, photorespiration can almost equal photosynthesis, so plants adapted to these conditions use a more complex scheme called C_4. Here, carbon dioxide is carried on a chemical carrier molecule to the chloroplasts, where it is concentrated before it is used in photosynthesis. Thus RUBISCO 'sees' more CO_2. C_4 photosynthesis is not just a matter of different enzymes. The leaves of C_4 plants are arranged so that the CO_2-capturing enzymes are on the outside and the CO_2-releasing enzymes on the inside next to the chloroplasts.

Phylogeny

This is the study of how things are related by descent from a common ancestor. It is useful for gaining clues about what parts of living things do. For example, if we discover a new human gene of unknown function and find that it is related to a gene in yeast that is known to be concerned with controlling the cell cycle, we might guess that the human gene was also concerned with the cell cycle. If a yeast protein was a protein kinase, maybe the corresponding human protein is too. And so on.

Biotechnological phylogeny is usually done by comparing gene sequences, where it is also called phylogenetics (how genes are related by descent). There are quite a lot of technical arguments about how you tell whether genes are really related by descent, and if so, how much.

One of the basic tools of phylogenetic comparisons is multiple sequence alignment. There are lots of subtly different ways of lining up DNA or protein sequences so that the result makes biological sense, and using the 'wrong' algorithm can result in biologically meaningless results. There are also lots of algorithms for compiling phylogenetic 'trees', and they not only give different results, but each can give a family of related but different results. So, using the tools of phylogenetic analysis to decide how two sequences are related needs understanding not only of the DNA or protein involved, but also of the mathematical methods.

Phylogenetic approaches are behind a lot of basic biology, some of which has biotechnological applications. For example, we can look for biodiversity by looking for organisms that do not necessarily look very different, but which our gene studies show are historically very different, i.e. diverged from a common ancestor billions of years ago.

'Classical' phylogeny also uses DNA comparison as a tool, but also uses shape as a guide to how animals are related (particularly bone shape), and ontogeny (how the embryo develops). Thus the relationship between the tiny bones of the human middle ear and the jaw of lizards can be deduced from they way they develop from their respective embryos.

Physical containment

Physical containment is keeping something inside a laboratory by putting physical barriers in the way of its escape. It is the principal way that genetically engineered organisms are kept inside a laboratory and prevented from 'escaping' to the wider world. (The other route is **biological containment**.) There are a range of physical barriers used, many of which are similar to those used in building **clean rooms**: however, in a containment laboratory the idea is to keep the dirt in, not out.

Technologies used include:

Air filtration. Exhaust air from air conditioning is filtered before it is vented to the outside. Often the containment laboratory is kept at a lower pressure than the outside ('negative' pressure) so that any air that leaks leak into the laboratory, not out.

Sterilization lights, usually versions of fluorescent tube lighting that give out a lot of ultraviolet light, are commonly used to sterilize the laboratory's exposed surfaces at night (when it will not give workers sunburn).

Waste disposal. Often, all waste leaving a containment laboratory is autoclaved to sterilize it. This includes apparently innocuous waste like paper towels, as well as obviously contaminated material. An alternative is to incinerate it, but then it has to be sealed to take it to the incinerator.

Personnel protection. Personnel working in a containment laboratory must often wear protective clothing, much as they are required to do in a clean room. This, however, is so that potentially contaminated clothing can be thrown away on leaving the room and not carried into the outside world.

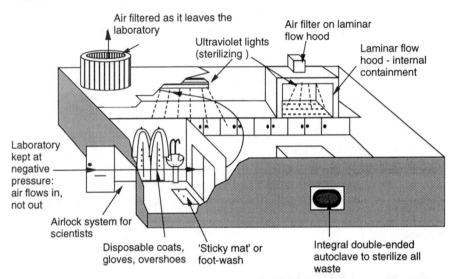

National governments define several levels of containment under which different procedures have to be carried out. Typical levels would be:

- Level 0: any laboratory.

- Level 1: 'good microbiological practice'. This is equivalent to any microbiology laboratory, where normal microbiological techniques are used to ensure that relatively non-hazardous organisms are kept in the lab, and do not cross-contaminate experiments. Typically, such laboratories are used for routine gene cloning involving no expression of a gene that could be hazardous to humans.

- Level 2. The laboratory is kept at negative pressure and air filtered. Any contaminated waste is autoclaved. Initial gene cloning experiments involving high levels of protein expression may be carried out in such laboratories, as well as microbiology involving organisms that have a relatively low hazard risk. As an additional safety precaution, most work would be done inside laminar flow hoods (hoods in which the air is circulated so that any particles generated by the experiment are carried up into the hood's own filter system and not into the laboratory).

- Level 3. The laboratory is only entered through an 'airlock' system, and all waste leaving it is autoclaved. Workers have to wear elementary protective clothing. Work on genetically engineered organisms that are expressing bioactive proteins, and on dangerous but relatively non-infectious organisms such as *Clostridia* would be done in such laboratories.

- Level 4. This is the ultimate containment level in most countries. Air is usually double-filtered on the way out, there is a double-airlock system for personnel with disinfectant bath to wash their shoes/boots in on the way out, and no one is allowed into the laboratory without substantial training (and no one who does not need to be there). Work on 'live' AIDS viruses and genetic engineering of common bacteria to express highly toxic proteins such as ricin could be done in such facilities.

Level 4 facilities are very rare: usually even the most potentially hazardous biotechnology project is adequately contained by Level 3.

Plant cell culture

Like any living organism, plants are composed of cells, which are capable of growing and dividing outside the plant given the right conditions. However, these conditions are rather specialized, since the plant cells themselves are specialized to work most effectively inside a plant. Thus, the conditions for cell culture have to provide the cells with a range of nutrients. In some systems, the plant cells are grown on a callus of other plant cells, called a nurse callus, which provides a 'feeder layer' to provide cells with the hormones and nutrients that they need.

As with animal cell culture, it is essential to keep the cells free from any other contaminating organism like a bacterium or fungus. Although plant cells have a range of defences against infection, the bacteria or fungi can grow very much faster than the plant cells in fermentors, and so outgrow the plant cells, resulting in a large mass of contaminant and either a small mass of plant cells or the death of the plant cells.

Plant cell culture has a wide range of applications in biotechnology, in:

- **Plant cloning**, i.e. the growth of plants from very small pieces of plant tissue, even single plant cells.

- **Plant genetic engineering**.

- Making 'plant' products (like scents or food flavours) from plant cells in culture rather than whole plants. Plants produce a very large number of useful chemicals, but often do so only at certain times of year and in places where growing the plant is difficult or even dangerous. Ideally, if the cells from the plant could be grown in a bioreactor, then some of these inconveniences could be avoided. The problems arise primarily from the way that plant cells in culture produce very little of these **secondary metabolites**. This can be overcome in some cases by growing the cells with suitable elicitors, compounds or mixes of compounds (often from plant of fungal sources) that are observed to increase the rate of production of secondary metabolites in cultured cells.

In this the plant biotechnologist is helped by the plant cell's totipotency. Most plant cells are capable of being grown back into a whole plant: they are totipotent, i.e. they have all the 'potency' of the original plant. This contrasts to animal cells, most of which cannot be grown into anything other than the tissue from which they came.

Plant cell immobilization

As well as the general methods used to immobilize growing cells in a bioreactor (*see* **Animal cell immobilization**), there are several techniques that are relatively specific for immobilizing plant cells.

Entrapment of plant cells in gel matrices is popular: the cells are suspended in small drops of the material, which then set or harden to make little carriers. Materials such as alginates, agar or carageenans (all of which are polysaccharides from seaweed), gelatin, or polyacrylamide have been used. Hollow fibres have been used for plant cells, but are not as popular as for animal cells, in part because hollow fibres are best suited to growing cells that secrete a product, and few plant cells secrete anything in worthwhile amounts. A relatively new method involves immobilizing the cells in polyurethane foam. In these 'foam reactors', small lumps of foam are suspended in culture media and the cells encouraged to grow into the holes inside, where they form minibioreactors.

Unlike animal cells, plant cells are enclosed in a very tough cell wall. This means that plant cells will not spontaneously stick to a substrate as easily as animal cells will. However. you can link them to one chemically without killing them. Plant cells have been chemically linked to nylon and to polyphenylene beads using glutaraldehyde (a standard chemical for linking two biopolymers together).

Plant cloning

One area in which traditional biotechnology has been successful is in plant cloning, based on the techniques of plant cell culture and embryogenesis. The technique is an extension of the idea of taking cuttings to 'duplicate' a particularly valuable plant. With cell culture techniques, the 'cutting' is a single cell.

Cloning from plant cells involves several steps.

Isolating individual cells. If all you want is a number of plants, then the cells need not be rigorously separated from each other, and they could be small chunks of tissue (tissue explants). If you want truly clonal plants (i.e. ones derived from a single cell) then the cells must be separated carefully.

Genetic manipulation of the cells. The isolated cells can then be engineered if we need new genetic characteristics in them (*see* **Plant genetic engineering**).

Callus generation. Culturing the plant cell into a mass of cells, looking like a small piece of chewed paper. The plant cells will grow like this indefinitely, without forming any recognizable plant tissues. Callus-like growths can also be made directly from the meristem, the tip of a plant root or shoot, which contains most of the growing cells. The result is a meristem culture.

Embryogenesis/organogenesis. The callus is encouraged to regenerate roots and leaves (*see* **Embryogenesis**).

Planting. Once the plant cells have generated a recognizable plant, it is safe to put it in the soil since the mechanisms will now be in place to fight off bacteria and fungi, and to get nutrients from simple soil chemicals.

A further step is the use of anther cultures to speed up breeding programmes for obtaining homozygous plant lines. Wild plants (like 'wild' animals, such as people) are heterozygous for many alleles (they will have two different copies of each gene in each cell). This makes breeding more complex, and we would like to have completely homozygous plants. In animals this is not possible, but in plants it often is. Anthers from male plants are cultured, and the haploid cells (i.e. the cells containing only one set of chromosomes, not the normal two) in the anther encouraged to grow clonally into plants. (Formally, these cells are the microspores, and so this is called microspore culture. The pollen can be removed from the anthers first, giving the very similar pollen culture.) Unlike animals, haploid plants are often capable of growing in culture. Since they only have one set of chromosomes, on 'diploidization' (i.e. any technique that will double up their chromosomes to make a normal, diploid plant) both copies of their chromosomes will be the same, i.e. they will be homozygous.

There are two major problems with using plant cloning technology routinely for propagating plants. First, the conditions for getting the callus to grow and then to differentiate are different for each plant. It is largely a matter of trial and error whether you find the right combination for 'your' species. Secondly, the initial plant material must be sterile before you start the culture, which is very hard to achieve for something that spends 24 hours a day sitting in the soil.

The third problem of somaclonal variation arises in some species. If a potato is separated into its component cells and some of these regenerated into potato plants, few of them will turn out to be identical to the original plant. This is a genetic change, a reflection of genetic instability. This is not a feature of the whole plant, which may be bred using normal methods quite well, and so must be an effect of the cell culture system. Why it happens is not understood, but it is one reason why some plants are not cloned in this way (*see* **Somaclonal variation**).

Plant genetic engineering

Plant genetic engineering is a major part of the research effort in biotechnology, because of the potential it holds for improving crop plants. A genetically engineered plant, sometimes called a transgenic plant, is the product of several of the technologies covered in this book. The necessary ingredients to make a transgenic plant are:

- isolating single plant cells,

- getting DNA into those cells,

- regenerating the cells into plants again, and, in some cases

- making homozygous plants from heterozygous transgenics.

The first point is covered under **plant cell culture**, the third and fourth under **plant cloning** and **embryogenesis**.

Getting DNA into plant cells has been difficult, because plant cells are surrounded by a robust cell wall, and, unlike bacterial cells, do not have common mechanisms for acquiring DNA from their surroundings. As with all methods of making truly genetically engineered, multicellular organisms, the key is not only to get DNA into the plant but to get it in in a suitable amount and to have it integrated into the plant's chromosomes.

The common routes discussed are:

- Using *Agrobacterium tumefaciens*.

- Using **electroporation** on plant protoplasts.

- By microinjection. This technique, which has worked so well in creating transgenic animals, has been applied to plants in two ways. Plant cells have been injected with liposomes containing DNA. Providing the liposomes are not injected into a vacuole, this is an effective way of transferring DNA into the cell. The alternative microinjection route is to inject DNA directly into the nucleus of the cell. This is more difficult to do, but gives greater control over the amount of DNA injected.

- By **biolistics** (particle gun) delivery. This is a favoured route, and is efficient at getting DNA into plant cells.

- By transformation of protoplasts. If the plant cell's wall is removed, then the resulting protoplast can sometimes be transformed simply by mixing with DNA (under the right conditions). This has not worked with monocotyledons yet, and seems to have only limited potential.

After a gene has been put into a cell, the one cell among many thousands or millions that has taken up the gene must be identified. This is the selection stage of genetic engineering, and, as with bacterial or yeast genetic engineering, usually relies on a selectable gene that you have transfected into the plant cell together with the gene that you want. This gene may be for resistance to a herbicide (which would kill the plant cell) or for an enzyme that is easy to detect using a simple assay (so you can look through your plant cells for ones that have that enzymatic activity). You can also screen cells for the presence of DNA itself using hybridization. This is more difficult to do with plant cells than with animal cells, because plant cells contain relatively little DNA (compared with yeast or bacterial cells) and it is quite hard to release.

Potential targets for plant genetic engineering fall into a range of types of project.

- Pest resistance: engineering genes into plants that will help them to repel pathogens (*see* **Pest resistance in plants**).

- Herbicide resistance: putting the genes for herbicide resistance into crop plants so that they can be resistant to the herbicides that kill weeds (*see* **Herbicides and resistance**).

- **Nitrogen fixation**: using a variety of routes to make plants 'fix' nitrogen from the air instead of needing fertilizers.

Plant oils

A substantial part of commercial biotechnology is aimed at producing or modifying plant oils. The oils are stored in the plants as triacylglycerols (TAGs), i.e. molecules with one fatty acid linked to each of the three hydroxyls of glycerol.

Common sources of oils include:

- palm and coconut (medium chain oils), used mostly in detergents

- rapeseed (canolla) (long chain oils), used as lubricants, plasticizers, and for making nylon

- castor bean and *Lesquerella* oil (hydroxylipids), used in lubricants and coatings

- jojoba wax used in lubricants and cosmetics

- flax oil (trienoic), used in coating, drying agents, and to a small extent in cosmetics

- cocoa used in chocolate and cosmetics

Enzymatic processes involving the use of plant oils include hydrolysis (to make the fatty acid) and transesterification (to make different esters from the glycerol and fatty acids; *see* **Lipases**).

Plant sterility

An important aspect of plant breeding programmes is obtaining genes that confer sterility. This is in part so that farmers cannot breed from the seeds they are provided with, in part to assist breeding programmes, but mainly so that hybridization breeding methods can work. These produce 'hybrid' grain crops, i.e. crops in which the seed you plant is the result of crossing two other types of grain plant. The two parental strains do not themselves produce high-quality grain. They produce grain that grows into a high-quality crop. This enables characteristics to be combined into one crop plant that could not be maintained by the traditional practice of keeping back a fraction of this year's crop to plant next year. However, it is essential that the grain sold to the farmer is the offspring of the mating of both parental types, and not just one. This requires the breeder to select male plants from one type and female plants from another: as sexing a field of wheat is tedious, this is done by ensuring that the various combinations you do not want are sterile, i.e. set no seed. Usually it is the male plant that is sterilized, and so the genetic effect is often called 'male sterility'.

Biotechnology has provided a range of new ways of making plants sterile, either one sex or both sexes. It has also generated 'restorer genes', which reverse the effect of the male sterility gene. This allows the plants carrying the male sterility gene to be cultivated on their own: without it, the line of plants would die out within one generation because of the lack of males.

Plant storage proteins

Plant storage proteins are proteins accumulated in large amounts in seeds, not because of their enzymatic or structural properties, but simply as a convenient source of amino acids for use when the seed germinates. They are of interest to biotechnologists for two reasons.

Storage proteins as a source of protein. Much of the world's food comes from plant seeds or fruits, and much of the protein in those seeds is storage protein. Any improvement of those proteins could correspondingly improve human diet. Specifically, many storage proteins are poor in some essential amino acids, usually the sulfur-containing ones (cysteine and methionine). They are called Class II proteins, because they cannot provide a good source of protein for humans on their own. A diet that relies on just one storage protein source for nearly all its protein can be deficient in one or two amino acids and lead to a deficiency disease, despite being quite adequate in bulk protein. Improvement of the proteins for food use would seek to engineer them to contain more of the essential amino acids, and so be better balanced, Class I sources of protein.

As protein expression systems. Storage proteins are produced in very large amounts relative to other proteins, and are stored in stable, compact bodies in the plant seed. Several workers are seeking to make the plants produce other proteins in as large amounts (up to 60% of the total seed protein, 15% of the total seed weight) and in as convenient a form. Storage proteins are often glycosylated as well, although often not in the same way as a protein would be glycosylated by mammalian cells.

The favoured route, being tried by Plant Genetic Systems, is to splice the gene for a desired protein into the middle of a plant storage protein gene. This construct will then produce a fusion protein in the seeds, which can be chopped up to yield the desired product afterwards. The favoured protein to do this is the small 2S plant storage protein, and it has been achieved with a model system in *Arabidopsis thaliana* and in *Brassica napus* (oilseed rape). This may not be the ideal protein since, because it is small, splicing a large gene into the middle of it could disrupt its structure.

A more radical approach would be to use the promoters of a storage protein to make a completely synthetic gene. This could be very difficult since, if the protein is not simply to be destroyed, it must also be targeted to the storage vacuoles in the seed. The targeting mechanism for seed storage vacuoles is not known, although proteins have been targeted to the vacuoles of other plant cells successfully.

Plasmid

A plasmid is a small piece of DNA that can exist inside a cell separate from the cell's main DNA. This means that it must be able to replicate itself inside the cell, so plasmids have the correct genetic elements in them to cause the cell's enzymes to replicate them as the cell divides.

Plasmids exist in most microorganisms. Those in bacteria are almost invariably circles of DNA. Some in yeast are linear DNAs, like very small chromosomes.

Plasmids are used extensively in genetic engineering as the basis for vector molecules. Because they are small, they are easy to manipulate. (By contrast the chromosome of *E. coli*, with three million bases, is a molecule 2 nm thick linked into a circle a millimetre round. A test-tube with a billion of those in is too thick to pour, and the shear forces of stirring it will break most of the molecules). Plasmids also have only a few sites for restriction enzymes in them, and so it is relatively easy to cut them open at just one place, then to splice in a piece of 'foreign' DNA and join the ends up again. They can also be manipulated to be present in many copies in the cell, rather than the one copy of normal chromosomes and plasmids (*see* **Vector**).

Plasmids are a specific type of episome, the generic name for any small DNA that can exist as an independent entity inside a cell free of the cell's main chromosomes. Some viruses can also be episomes, existing as DNA within a cell for a long time. (This does not include the retroviruses. These exist as DNA inside a cell, but their DNA is spliced into the chromosomes themselves.) Many plasmids can transfer themselves between cells, which is a very undesirable characteristic for a gene-cloning vector, which we want to stay in the cells we put it in. A first step to turning a plasmid into a usable vector is to delete genes such as '*mob*' (which mobilizes a plasmid) and '*tra*' (which allows it to transfer itself to another bacterium) from the plasmid.

Polysaccharide processing

A common use of industrial enzymes in the food industry is in processing complex polysaccharides such as starch and pectins. Enzymes are involved in several processes:

Liquefaction. The dispersal of starch granules into a gelatinous suspension (i.e. essentially what happens when cornflour is boiled to thicken sauces). The starch is also hydrolysed into shorter molecules by enzymes such as pullulanase and alpha-amylase. Because liquefaction is often carried out in hot solutions, one valuable biotechnological product is heat-stable alpha-amylase and pullulanase, isolated from thermophilic bacteria, which will work at 80 or 90°C.

Saccharification. The formation of low molecular weight sugars, often mainly glucose, from liquefied starch. Acid treatment of starch will break it down into a mixture of glucose and higher sugars, called glucose syrup. The amount of hydrolysis depends on the amount of acid, the time taken, the type of starch involved, and so on. The degree of breakdown is measured as dextrose equivalents (roughly, the fractional extent to which the starch is broken down completely to glucose), and commercially usually falls in the range 40–90. There are also a variety of enzymes that will do this: amylases and pullulanases to break down the starch, invertase to break down sucrose, and glucose isomerase to convert glucose into the sweeter fructose.

Debranching. A chemical rather than a process term, this is the removal of the side branches from the long starch or pectin molecules, leaving long, straight molecules that are easier to break down further. Branched and unbranched polysaccharides also have different gelling properties, and so confer different mechanical properties on food. Enzymes such as pullulanase and isoamylase can perform debranching.

Post-translational modification

This is a blanket term to cover the alterations that happen to a protein after it has been synthesized as a primary polypeptide. They include:

Glycosylation. This is one of the critical post-translational modifications for biopharmaceuticals (*see* **Glycosylation**).

Removal of the N-terminal methionine (or N-formyl methionine). Nearly all proteins are made with a methionine as their first amino acid, and this is usually removed. Sometimes it is removed as part of:

Signal peptide removal. Peptides that are to be inserted into membranes, secreted into special cellular compartments (such as the mitochondrion or into vacuoles or lysosomes), have a short string of amino acids at their front called the signal peptide. This signals to the cell where the protein is to go, and is chopped off as part of the mechanism for getting it there (*see also* **Expression systems**).

Acetylation, formylation. These and a few other modifications put relatively unreactive groups on to more reactive ones. They are often put on the terminal amino group of a protein, producing a 'protected N-terminus'.

Amino acid modification. This is the chemical modification of amino acids after they have been incorporated into the protein chain. There are relatively few examples of such modification, but they can have critical effects on the protein's function. Examples are the modification of glutamate to form gamma-carboxy glutamate by a vitamin K-catalysed reaction in mammalian liver, and the hydroxylation of proline to hydroxyproline in collagen in animals.

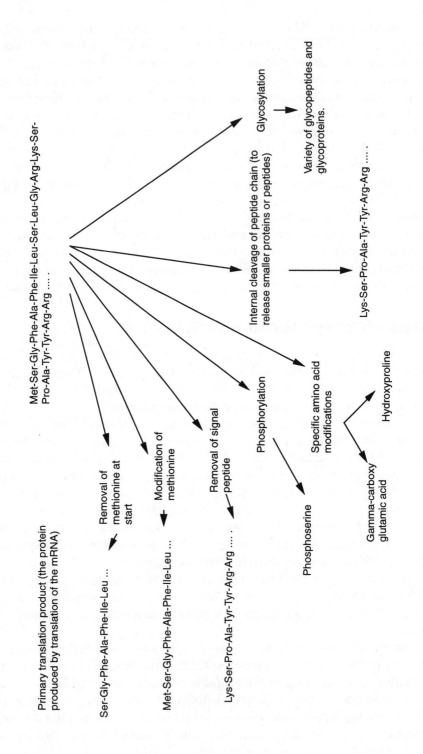

Primary translation product (the protein produced by translation of the mRNA)

Met-Ser-Gly-Phe-Ala-Phe-Ile-Leu-Ser-Leu-Gly-Arg-Lys-Ser-Pro-Ala-Tyr-Tyr-Arg-Arg .....

Removal of methionine at start

Ser-Gly-Phe-Ala-Phe-Ile-Leu ...

Modification of methionine

Met-Ser-Gly-Phe-Ala-Phe-Ile-Leu ...

Removal of signal peptide

Lys-Ser-Pro-Ala-Tyr-Tyr-Arg-Arg ..... .

Phosphorylation

Phosphoserine

Specific amino acid modifications

Gamma-carboxy glutamic acid

Hydroxyproline

Internal cleavage of peptide chain (to release smaller proteins or peptides)

Lys-Ser-Pro-Ala-Tyr-Tyr-Arg-Arg .....

Glycosylation

Variety of glycopeptides and glycoproteins.

Predisposition analysis

This is the analysis of how some people are more likely to get some diseases as a result of their genes. Many diseases have a 'genetic component' and an 'environmental component', and a 'bad' environment or a 'bad' gene can enhance the chances of getting the disease. For some rare diseases of the immune system, such as ankylosing spondylitis there is an 80-fold higher chance that the carriers of some genes will get the disease than the carriers of others. For other diseases the effects are less dramatic. Among those considered as having a genetic component are:

- many immune disorders, including asthma, eczema, Graves disease, allergies

- diabetes

- hypertension

- some types of cancer (but not most cancers)

- hypersensitivities and adverse reactions to drugs and chemicals

And a range of other diseases may have a substantial genetic component, for example:

- schizophrenia

- clinical depression

- cardiovascular disease

These diseases are all influenced by more than one gene, and usually by a large (and unknown) number. They are multigenic (or polygenic) diseases.

The interest to biotechnology of this genetic predisposition is threefold. First, if there is a gene involved, we could hope to use DNA technology to find that gene and then use it as a diagnostic to detect who was predisposed to that disease. Secondly, we could hope to find out what the gene does and hence design a therapy to counteract it. Lastly, we could use our knowledge to identify what aspects of the environment trigger the disease, and hence improve our environment, to everyone's benefit.

A related idea is disease stratification. Here we try to subdivide a disease into more precise groups, depending in this case on the genes involved. Thus specific cancers can be subdivided depending on which oncogenes are mutated in those cancers. This can be important for applying biotechnologically derived therapies, because these are targeted at very precise molecular mechanisms, and so we need to tell whether those mechanisms are the ones relevant to that disease.

Other aspects of disease stratification can actually diagnose a disease more accurately. Doctors hope that genetic diagnostics will enable the accurate diagnosis of psychiatric diseases. For example, is a confused old lady actually suffering from Alzheimer's disease (which is incurable) or from depression or chronic antidepressant use (both of which are curable)? A gene-based test based on understanding what Alzheimer's disease is could tell you.

The principal way of detecting whether a human characteristic is affected by genes is to see whether it 'runs in families'. Family members share more genes with each other than they do with the general population, so if a characteristic is seen to be passed down from parent to child more often than chance would dictate, we can suspect that a gene is involved. However, we must remember that 'inheritable' is not the same thing as 'genetic'. In most Western societies, money is inheritable, but it is certainly not genetic. So the analyses that point to genes as being the cause of a disease have to be quite sophisticated. A common technique is to study identical twins (which have exactly the same genes and usually a very similar environment) and compare them to non-identical twins (fraternal twins), who have different genes and a very similar environment. If the identical twins show much more similarity in some trait (susceptibility to asthma or schizophrenia, tendency for alcohol abuse, or whatever) then we suspect (but have not proven) that a gene is involved. We measure the extent to which a gene is involved as the 'heritability' of that trait, which is roughly the extent to which the variation of that trait between people in that population is a result of genes. A heritability of 100% means that differences in genes are the only factor determining the differences between individuals in the sample studied.

In addition, most of these predispositions will not be caused by 'one' gene, but by a number of genes, all of which will have to be characterized and understood. In addition, the effects of the genes will not be obvious in everybody (they will predispose to a disease, not inevitably cause it). This means that they can only be identified from large statistical surveys. This is a major research undertaking, and this is one of the reasons why, when the genes for many rare genetic diseases have been found, the gene(s) for the much more common diseases, such as hypertension, are still unknown.

Notwithstanding this, several companies have been set up to use DNA techniques to detect predisposition to disease, and one of the aims of the **Human Genome Project** is to provide information about genes that could predispose people to some diseases.

There are also obvious ethical and legal implications of using human genetic information in these ways. They are principally concerned with the person whose information is being used getting appropriate use of that information, not being confused or frightened by it, and ensuring that no one one else gets to use the information without the patient's understanding and consent. (In this, genetic testing is decades ahead of most medical diagnostics) (*see* **Bioethics**).

Prosthetics

Prosthetics are mechanical objects implanted in people, usually to replace a part of them that has been lost to disease. They include joint replacements (especially hips and knees) and intraocular lenses (lenses implanted in the eye). 'Prosthetics' often also cover contact lenses and artificial kidneys, although these are not implanted.

Biotechnology has contributed to the design of new materials for all sorts of implants and prosthetics. Usually these are plastics. They must be mechanically fit for their purpose: plastics must have appropriate flexibility, strength, and resistance to continual bending stresses. They must also be biocompatible, which means that they must both integrate with the body and not cause a reaction from the tissues they abut. Anything placed in the body will become coated with protein, and may cause the cells around it to react. Such reactions must be controlled.

Typical materials used are:

- MMA (polymethylmethacrylate, known by the trade name of Perspex)

- poly-HEMA (2-hydroxyethylmethacrylate), a hydrogel that is converted from a hard perspex-like material to a soft, flexible one in water.

- polyvinylpyrolidone.

Biotechnology is also developing new types of prosthetics, often also called **artificial tissues**.

Proteases

Proteases are enzymes that break up proteins. There are four distinct uses for proteases in biotechnology. Their use depends partly on how cheap they are to make, and partly on how specific they are, i.e. whether they chop up all proteins indiscriminately or only a few proteins at specific points.

Eight thousand tonnes of proteases from microbial and fungal sources are made each year, and most of them are used in detergents. Relatively non-specific proteases are used to digest the protein material in dirt—it is often the denatured protein that makes organic stains hard to wash out. One of the more common is subtilisin carlsberg, made from *Bacillus lichiformis*. Some such detergents are retail products, but more are used in industrial cleaning, as the proteases are powerful enzymes and can strip the protein out of the user's skin if not handled carefully. Subtilisin has been engineered by Novo Nordisk and by Procter and Gamble to improve its resistance to oxidation (oxidizers are another common component of commercial detergents) and 'builders' (calcium chelating compounds). It has also been engineered to make it function better at the pH of washing powder, and to broaden its specificity so that it breaks down a wider range of proteins.

The other main use of proteases is in the food industry, where microbial rennin is used extensively in cheesemaking as an alternative to cows stomach rennin. A rising trend is the use of proteases to tenderize meat and to enhance the flavour of foods by altering the proteins in them. This use requires purer enzymes (since they, or their cooked remains, are going to be eaten), and are usually quite specific, only cleaving a single type of protein at a fairly specific site. An example is collagenase, an enzyme that breaks up collagen, the fibrous protein in connective tissue such as tendon. Collagen contributes substantially to the 'toughness' of lower quality meat: thus soaking low quality meat in collagenase can tenderize it.

The third use of proteases is in biomedical applications. Many of the biopharmaceuticals planned or in development have protease activity (such as thrombolytics), but these are not usually considered to be part of the protease industry. However, proteases with broad activities also have biomedical applications in such areas as wound debridement (removal of the thick coat of protein material that forms on the surface of wounds and can slow healing and encourage scarring) and as aids to digestion. Proteases can be used either as food supplements or in the preparation of predigested foods for people in hospital. Here the enzymes have to be of pharmaceutical purity.

The last use of proteases is in biotransformation reactions. Although the normal reaction of a protease is to cut peptides up, if they are used in conditions where there is very little free water (for example in non-aqueous solvents), or where the amino acids are freely available but one of the peptides made from them is removed as soon as it is formed, then proteases can be used to make short peptides. Thus the dipeptide artificial sweetener 'Aspartame' can be manufactured from a derivative of aspartic acid and methylphenylalanine using a protease to join them together.

Protein crystallization

A key part of most ways of finding out a protein's 3D structure, and hence being able to use that structure to design drugs, is making crystals of the protein. This is difficult, since protein molecules are not as well behaved as simple salts, and the larger they are the worse behaved they are. The trick is usually to perform the crystallization very slowly and in exactly the right solutions—finding the right solutions can take a lot of expertise and time.

Novel approaches to protein crystallization include crystallizing under high pressure or in free fall (i.e. in orbit). High pressure reduces the amount of movement in the protein molecule, enabling much faster crystallization in some cases. Crystallizing in free fall (also called microgravity) means that the crystals do not have to touch the side of the container they are in, and so their growth is not affected by that container. Eight companies and 10 research institutes had protein crystallization experiments on the Space Shuttle Columbia's mission in January 1990.

The study of the protein crystals so formed is called protein crystallography. It is usually done with X-rays: the pattern of X-rays that are diffracted from a crystal of a protein is extremely complex, and depends on the way all the atoms are arranged inside the crystal. From a good pattern the atom distribution (or more exactly the distribution of electric charge, i.e. the electron density) can be deduced. X-rays can come from a conventional X-ray tube, but a more popular source nowadays is synchrotron radiation, because it is highly monochromatic (it has only one wavelength) and is very intense. It is from an electron density map of this sort that the computer graphics pictures of how a protein folds are drawn (*see* **Computational chemistry**).

Protein engineering

Protein engineering is the design, production, analysis, and use of altered, non-natural proteins. This can be a Herculean task if we do not use a natural protein as a starting point, so usually protein engineering involves modifying existing proteins.

Protein engineering has a number of aims:

Improving protein stability. Protease enzymes that have been genetically altered for greater stability are in widespread use in detergents (*see* **Protein stability**).

Altering antibody properties. There are a range of specialist technologies concerned with antibody engineering, and these are discussed in the entries for **chimeric/humanized antibodies** and **Dabs and other engineered antibodies**.

Altering the substrate specificity of an enzyme. Most enzymes catalyse only a very narrow range of reactions, and it would be helpful to be able to alter that range so that they acted on other, more commercially useful products. Protein engineering can seek to do this by altering the amino acids around the active site of the enzyme, the bit of the molecule that actually binds on to the substrate and catalyses the reaction. By altering the amino acids the forces holding the substrate in place are altered, and hence the molecules that the enzyme best recognizes are altered. A spectacular example of this was the conversion of malate dehydrogenase (MDH) to lactate dehydrogenase (LDH), two enzymes that catalyse similar types of reactions on different substrates. Unfortunately, neither MDH nor LDH are particularly useful enzymes, and this approach has not been successful on any commercial enzyme. Engineering of proteases such as subtilisin for use in detergents has been a more practical example, in this case broadening the substrate specificity (that is, making it less specific).

Altering pharmacological action. Much protein engineering is aimed at biopharmaceuticals (*see* **Pharmaceutical proteins**). In this field protein engineeering seeks to alter the biological activity of proteins that have effects that can be harnessed as drugs, by making the effects more potent, more specific, coupling them to targeting mechanisms so that they only effect a few cells or cell types, improving their survival time in the patient, or reducing side-effects. This can include altering the way that proteins aggregate. Therapeutic insulin tends to form multimolecule complexes, which slows its rate of absorption by tissues. Replacing proline-28 in the B chain of insulin with aspartate stops it aggregating in the vial, and allows it to be absorbed much more rapidly.

More ambitious aims of protein engineering involve the production of completely new protein structures or substructures. These are often built around known structures, such as alpha helices or beta sheets, or around 'super-secondary structure'—assemblies of several alpha helices for example (helix bundles), which are known to be particularly stable bits of protein. Even larger 'modules' or 'domains' can be assembled—many proteins are structured from such domains in nature, each domain having a specific function (*see* **Protein structure**).

Combinatorial approaches can also be used to optimize protein structure (*see* **Combinatorial chemistry**).

Protein sequencing

Determining the sequence of amino acids in a protein is done chemically via a cycle of reactions that chop one amino acid off at a time. There are several machines that perform these quite complex sequences of reactions automatically. The number of amino acids that can be determined depends on the amount of protein available and the nature of the amino acids. None of the reactions in the cycle is 100% efficient, and the efficiency varies to some extent depending on which amino acid is being removed for analysis. Thus, after a while, the amount of each amino acid being released by the reaction cycle becomes too small to detect against the 'noise' of other amino acids released from those proteins that were not broken in previous cycles. Clearly, also, the protein must be reasonably pure, otherwise the result is a mixture of amino acids at each step.

The standard chemical method is called the Edman degradation. The process starts from the amino end of the protein (the N-terminus). In some proteins the N-terminal amino acid has another small chemical group attached to it—usually a methyl, acetyl, or formyl group. The presence of this group makes it impossible to start the reaction cycle. Therefore, some pre-preparation of the protein is needed before a sequence can be determined.

Other methods including mass spectrometry (MS), especially fast atom bombardment (FAB) mass spectrometry, are gaining popularity (*see* **Mass spectrometry**).

Because of the difficulties in sequencing proteins, and the limit of around 40 amino acids from any one peptide that can be sequenced in one experiment, many workers prefer to clone the gene for the protein (if they can) and sequence the DNA, using the genetic code to deduce the amino acid sequence of the protein. There are potential problems with this approach, however (*see* **Genetic code and protein synthesis**).

Protein stability

Proteins are not very stable in chemical terms: they are easily denatured (i.e. converted to inactive forms). Denaturation occurs when the protein chain of amino acids, usually folded into a specific, tight-coiled conformation, unfolds: the carefully arranged 3D structure of its surface is lost, and usually whatever its function was is lost as well. This change can be brought about by heat, acids, alkali, and by some chemicals such as urea and guanidine, which are known as chaotropic agents because they are extremely effective at opening up the structure of water-soluble macromolecules into a floppy, 'random coil', which has no real fixed 3D structure.

If enzyme reactions can be carried out at higher temperatures, or antibodies made more stable so that they can last longer, biotechnologists would be very pleased. So there is a lot of work in trying to improve protein stability. The lines of work are:

- Using another, more stable enzyme, especially from a thermophilic bacterium (*see* **Thermophile**).

- Increasing the number of disulfide bonds within the protein. These bonds, formed between cysteine residues in the protein once it has folded up into its proper shape, help to lock it into that shape (*see* **Disulfide bond**).

- Increase internal hydrophobicity. Often, the amino acids that end up inside the correctly folded protein are water-hating (hydrophobic) amino acids: if the protein is unfolded, that exposes them to water, which requires energy so does not tend to happen.

- Add other stabilizing interactions. A wide range of other interactions of amino acids with each other help to hold a protein in its correct state. These include hydrogen bonds and ion (or salt) bridges.

In all these latter three cases, the protein engineer aims to add or alter amino acids to increase the number of stabilizing interactions in the protein. This needs a detailed understanding of the 3D structure of the protein, information that can be very difficult to obtain (*see* **Protein crystallization**).

Proteins can also be stabilized by adding specific stabilizing agents to their preparation. Very few enzymes are sold as pure protein—most have many other materials in their 'formulation' to stabilize them. Some of these can have a dramatic effect, extending life times from hours to weeks. Exactly what is in each stabilizer depends on the enzyme concerned.

Folding and stability is also important when a protein is to be made by recombinant DNA technology. Frequently, a protein made at high levels in a bacterium is not made in its native (i.e. normal) conformation. This may be because the protein precipitates inside the cell as an inclusion body, or it may be because the protein is synthesized or modified in different ways in a bacterial cell. Thus, part of the purification procedures for many recombinant proteins involve steps that partially unfold the protein and then refold it again, this time under conditions that allow it to fold up properly. (This can also help purification, by selectively unfolding and refolding the desired product: contaminant proteins fail to unfold, or fail to fold up again, and so can be distinguished from the product.) Clearly, it must relatively easy to fold up the protein if this strategy is to work—some proteins cannot be refolded into their native structure once they have been unfolded.

Protein structure

Proteins are highly complex structures. For convenience, and partly to reflect biological reality, their structure is divided into several different levels.

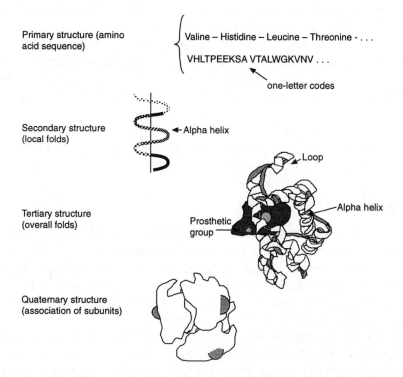

Primary structure (amino acid sequence)

Valine – Histidine – Leucine – Threonine - . . .

VHLTPEEKSA VTALWGKVNV . . .

one-letter codes

Secondary structure (local folds)

←Alpha helix

Loop

Tertiary structure (overall folds)

Alpha helix

Prosthetic group

Quaternary structure (association of subunits)

Primary structure. This is the chemical sequence of amino acids in the protein. It is unique and definitive for that protein. Implicit in it is that the primary structure is a string of the 20 'biological' amino acids joined by peptide bonds.

Secondary structure. This is how the primary structure folds into small, distinct shapes, usually consisting of a 3–10 amino acids. These units are very stable, and occur in many different proteins. Common ones are alpha helices (short rods), beta sheets (flat, twisted plates), and a variety of turns, which join them. Alpha helices are often shown in graphics of protein structure as solid cylinders, beta sheets as parallel arrays of flat bands or arrows. Anything from 30 to 90% of the amino acids in a protein can be part of a secondary structure feature.

Super-secondary structure. Some of the secondary structures assemble into groups of three or four element which again are quite stable and common amongst many proteins. The most common example is the helix bundle, a tight bundle of four alpha helices.

Tertiary structure. This is how the whole protein (secondary structure features and other amino acids) folds together in space to make a molecule. Quite often the tertiary structure is made up of structural domains, regions of the protein that are quite distinct and have a distinct function. Thus there are recognized domain 'shapes' for binding nucleoside triphosphates, which turn up in many different metabolic enzymes that use ATP as a substrate. Some scientists believe this is because the enzymes we have today have been assembled out of more primitive domains that were once individual proteins. The tertiary structure of the domains is usually much more similar than the primary structure—their amino acid sequence can be very different, but they fold into similarly shaped proteins.

Quaternary structure. Many functional proteins are assembled from several amino acid chains. How the different chains (polypeptides) assemble together is their quaternary structure.

Proteome

The 'proteome' programme, also called the 'phenome', is the name given to some of the investigations that must follow the genome programme if we are to make sense of what the genes we have discovered actually do. Specifically, proteome seeks to perform protein biochemistry using the same philosophy of high throughput, automated analysis that has been applied so successfully to the genome projects.

Proteome programmes seek to characterize all the proteins in a cell, identifying at least part of their amino acid sequence. This can then be linked by computer to the gene that codes for that protein—thus we know not only the whole gene and protein sequence, but also that that gene is really working in that cell. This is a version of functional genomics.

To make the chemistry work at high throughput, peptide or protein sequencing must be done by mass spectrometry; conventional Edman degradation approaches are too slow. The proteins must be separated efficiently before they go into the MS system, and 2D gels or HPLC are the methods usually used for doing this.

At the start of 1996, 'the proteome programme' was only a phrase, although Large Scale Biology Inc. has a smaller version of it. Certainly, if we are to understand what the genome is doing (as opposed to what its sequence is), then something like a proteome programme will be needed. However, it is also likely that the current enthusiasm is at least in part because protein biochemists have at last seen a way of getting back at those DNA people, who have upstaged them for the last 20 years.

Protoplasts

Plant, fungal, and most bacterial cells are surrounded by a tough, thick, cell wall. A protoplast is such a cell from which we have removed the cell wall, leaving the cell naked and surrounded only by its plasma membrane.

There are a number of reasons for wanting to do this, but all involve the strength of the cell wall itself. Often, plant breeders wish to fuse the cells of two quite different plants that cannot be cross-bred by conventional methods. However, the cell wall gets in the way. Again, getting DNA into plant or yeast cells for genetic engineering is extremely difficult, with the cell wall essentially impervious to any large molecules. (Getting DNA into bacteria is an exception because bacteria have mechanisms for absorbing DNA from the medium surrounding them.) Thus, many manipulations using these types of cells require starting with protoplasts.

Plant and yeast protoplasts are generated by dissolving their cell walls with appropriate enzymes, which will digest the carbohydrate (plant) and chitin (yeast) in the cell wall without affecting the lipid and protein cell membrane.

The cells of yeasts and some plants can be regenerated from protoplasts, providing the cells have not been shocked too much when they were turned into protoplasts in the first place. Thus, protoplasts that have been genetically manipulated can be turned back into normal cells. This is desirable because protoplasts are very fragile (even more fragile to chemical and physical attack than animal cells in culture) so they are extremely difficult to use in a commercial biotechnological process. Plant cells that have been regenerated in this way can then be used to regenerate whole plants; so using protoplasts of plant cells can be a step in plant genetic engineering.

Purification methods: large scale

One of the central parts of the downstream processing of a fermentation product is purification. Large-scale purification methods are used to take a crude fermentation supernatant or cell homogenate and isolate the product from it in a fairly pure form. Industrial enzymes are often sold in this semi-pure form as a bulk product. If they need to be really pure, then they have to go through a second purification step (*see* **Purification methods: small scale**). Purification of the cells from a culture is usually called **harvesting**, and relies on rather different methods.

The economics of producing almost anything are usually governed by the cost of purification and distribution. For expensive biotechnology products, purification is the largest of these costs. Growing *E. coli* in a tube of broth is very cheap. Purifying a recombinant drug from it needs time, skill, and materials, all of which cost money. This is especially relevant in biotechnology, where the products are also governed by strict regulations that demand not merely purity but proof of purity (*see* **GLP/ GMP, QC/QA**). In general, then, the less purifying you need to do the better. This usually means that the more concentrated the material is to start with the cheaper it will be, as illustrated in the graph. For very dilute materials, only a very high selling price can justify the extreme cost of extracting the pure material from the starting bulk. Pharmaceutical proteins are an example of such materials.

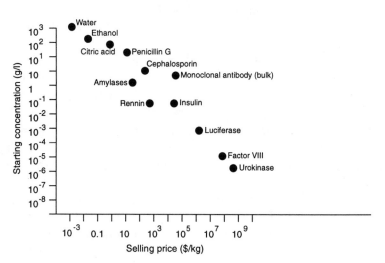

There are a range of purification methods that are cheap enough to use on large volumes of material. These include:

Salt precipitation. Adding salt so that a particular group of proteins precipitates from solution. Also sometimes called 'salting out'. Simply adding water to the precipitate usually makes them dissolve again. Ammonium sulfate precipitation (adding concentrated ammonium sulfate to a protein solution) is often used in preparing proteins at small and medium scale. A low concentration of ammonium sulfate is added to precipitate some unwanted proteins, the precipitate removed, then a higher concentration added to precipitate a partially purified mixture containing the protein that is wanted. Such a preparation is sometimes called an 'ammonium sulfate cut'.

Liquid–liquid separation. Also called two-phase separation, this uses the fact that the material you want will dissolve well in one solvent while

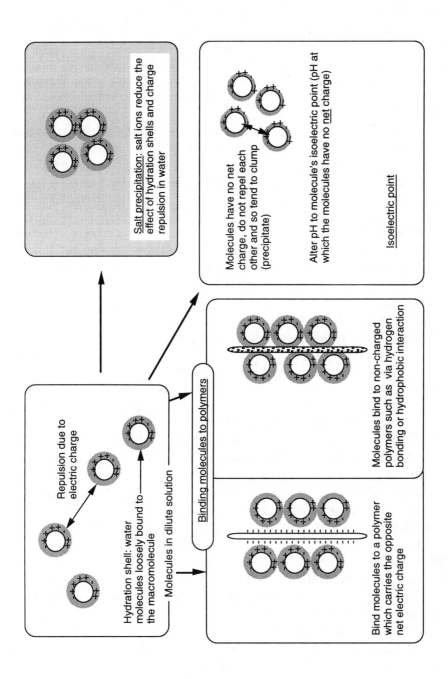

Salt precipitation: salt ions reduce the effect of hydration shells and charge repulsion in water

Molecules have no net charge, do not repel each other and so tend to clump (precipitate)

Alter pH to molecule's isoelectric point (pH at which the molecules have no net charge)

Isoelectric point

Repulsion due to electric charge

Hydration shell: water molecules loosely bound to the macromolecule

Molecules in dilute solution

Binding molecules to polymers

Molecules bind to non-charged polymers such as via hydrogen bonding or hydrophobic interaction

Bind molecules to a polymer which carries the opposite net electric charge most of the impurities will not. The two are intimately mixed, and then separated (by allowing to stand, by filtration systems, or by gentle centrifugation). This only works, of course, if the two liquids are immiscible. This can be performed several times, reducing the amount of contaminant in the sample phase each time. For large-scale preparations it is essential that the two phases are cheap since it rare that they can be recycled efficiently. One is usually water (as that is the basis of the culture medium) and so the other is a material like benzene, ether, or oil.

Two-phase aqueous extraction. Here the protein is shaken up with a polymer-based mixture which, on being left to stand, separates into two distinct layers [polyethylene glycol (PEG) and salt will do this trick, for example]. The two layers contain different amounts of salt and polymer, but both are water based. The conditions are arranged so that the product ends up in one layer, and most of the contaminants in the other.

Polymer precipitation. Some polymers, particularly PEG (polyethylene glycol) can bind gently to proteins and make them precipitate of their own accord. This is in fact a variation of the aqueous phase partitioning (one of the phases in this case is the precipitated protein).

Heat denaturation. This is simple and effective if your protein is heat stable (thermostable): you just heat up the mixture and most of the proteins denature, and so 'coagulate' and settle out of solution. The one you want remains soluble. This only works with some proteins. It can also be used under some conditions to separate proteins from non-protein products (e.g. metabolites).

Isoelectric point separations. Most proteins are quite insoluble at a particular pH (their isoelectric point or pK_i). If you add acid or alkali until the pH (the 'acidity') of the solution is at this isoelectric point, then those proteins will precipitate. Again, adding water usually redissolves the precipitate.

Dialysis. Here the protein is put on one side of a semi-permeable membrane, and water or salt solution on the other. A semi-permeable membrane is one that lets through small molecules, but not large ones. So salts, impurities, etc. in a protein solution are let out, and the protein remains in. The process can be accelerated by an electric current in electrodialysis, which drives the ions in one direction. Electrodialysis is used to deionize (i.e. remove ions from) several industrial-scale protein materials, including whey.

Supercritical fluids can also be used to extract biomolecules. Supercritical fluids are gasses that have been compressed to the density of liquids above the temperature at which they can form liquids. Their ability to dissolve chemicals depends strongly on their temperature and pressure, and so they can be adjusted to dissolve what you want. Supercritical carbon dioxide has been used to extract the caffeine from coffee for many years: its properties can be 'tuned' to dissolve the caffeine and leave virtually all the other components of the coffee in place. It is also being explored as a method of extracting and partially purifying many other biomolecules, including proteins; some proteins can even function as enzymes in supercritical fluids (*see* **Supercritical fluid enzymology**).

Purification methods: small scale

Many biotechnological products have to be extremely pure, for use as drugs or in producing fine chemicals, so the relatively crude purification methods that isolate them from large-scale culture are not good enough. A further purification step is needed. There are many such methods, but since they are complicated and expensive they are usually only used on a small scale (micrograms to grams rather than grams to tonnes). Most are chromatography methods. Here the mixture is passed down a tube that is packed with some material to which some components in the mixture stick and others do not. It does not matter whether the product you want sticks or not, providing the contaminants do the opposite.

- **Affinity chromatography**.

- Gel filtration. This is a chromatographic method in which the molecules are separated by size (*see* **Chromatography**).

- Ion exchange. This separates molecules according to their charge. Since the charge of a molecule depends on the pH, a combination of varying pH and ion exchange chromatography can prove very effective at purifying proteins.

- Hydrophobic chromatography. This type of chromatography uses the different affinity that different molecules have for hydrophobic materials, i.e. for materials that are hydrophobic ('water haters'), like plastics (as opposed to hydrophilic ('water lover') materials, like paper).

Popular versions of chromatographic separation methods are FPLC and HPLC, which have been scaled up from laboratory tools to production methods in some cases. HPLC (high performance liquid chromatography) pumps the mixture through the chromatography 'column' at very high pressures, ensuring very precise separation in a short time. FPLC (fast protein liquid chromatography) is a more specialized technique for separating proteins, which, because many biotechnological products are proteins, has found widespread use. The pressures used in FPLC are much lower than in HPLC, so the apparatus can be substantially cheaper.

QC/QA

QC is quality control, testing a product to see that it meets the required quality. QA is quality assurance, making sure that the process you are using to make the product is appropriate for making a product of the required quality. Both are general concerns in all industries, and cannot be covered in depth in this book.

Both are very important to a wide range of biotechnological processes, especially ones producing pharmaceuticals or foods. Some of the QA/QC procedures that are specific to biologically based products are:

- sterility (are there any living organisms in the product?) and bioburden (how many organisms are there?), both performed using conventional microbiology.

- endotoxin (or pyrogen) testing, usually done using the Limulus amoebocyte lysate (LAL) test. Endotoxins are lipopolysaccharides from bacterial cell walls, which cause extreme immune system responses, sometimes fatal.

GLP and GMP processes are meant to assure quality in product development and product production, respectively. In fact, they assure that the product is always the same, not that it is always of high quality. Products produced by biotechnology are not usually attacked for their technical quality, but for their deliberate performance: for example, no one disputes that BST is pure and highly effective, but people argue strenuously about whether it should be used on cattle at all. One well-publicized exception was the Showa Denko tryptophan case (*see* **Amino acids**).

Quantitative trait loci (QTL)

A quantitative trait is a trait or characteristic of an organism that varies continuously across the population, and hence is almost always controlled by a combination of several genes and environmental factors. Human height is one such trait: we clearly inherit a general tendency to be tall or short from our parents, but many genes contribute to this, as well as nutrition. A quantitative trait locus (QTL) is a gene that contributes towards a quantitative trait. A synonym is metric trait locus.

QTLs are important in agricultural breeding programmes. Traits such as plant productivity, protein quality, growth rates in plants and animals, fat content in animals, etc. are all under the control of QTLs, and so breeding an animal or plant with the right collection of alleles at their QTLs is a major aim of the breeding industry.

Mapping QTLs is therefore a major concern. A critical aspect to be discovered is the variance caused by the QTL. How much of the variability of the trait is owing to genes as a whole, and to that QTL in particular? Is the variance additive with other genes (e.g. if you find two gene alleles that will add six inches to your height, will someone possessing both 'tall' alleles of both genes be 12 inches taller than average)? A related concept is penetrance, which is a measure of how often in the population the gene has any effect: sometimes an allele for a trait will be present, but not show its effects because of interference from other genes or from environmental effects. (This is not the same as it being a recessive allele, when we expect it to be 'hidden'.)

Many human genes could be called QTLs, but generally these are referred to as 'predisposition genes' instead, predisposing us to develop a specific disease, or to grow a certain way.

Rational drug design

This approach to drug discovery seeks to model the molecular structure of the target of the drug, and then design a drug molecule that will fit it. This is therefore also called structure-based drug design (SBDD). This contrasts to the alternative, which is to screen a large number of compounds for drug activity, choose the most promising and make a whole lot of variants, choose the most promising of them, and repeat until a suitable drug is found.

Rational drug design (RDD) involves knowing the chemical structure of the drug's target, usually a protein. Protein structures are difficult to obtain: it is relatively easy to obtain the amino acid sequence of a protein if you can purify it, but determining how the peptide chain folds up in space is difficult. Finding out the structure usually involves cloning the genes for the proteins to which drugs bind and making them in large amounts in an expression system. The protein must then be crystallized, and the structure of the crystals deduced using X-ray analysis (*see* **Protein crystallization**). Increasingly, protein structures are extrapolated from the structures of related proteins, rather than being discovered *de novo*.

The Roche drug Saquinavir, an inhibitor of the HIV protease, approved in 1996, is a successful example of a drug 'designed' almost entirely by RDD methods. However, this is unusual, and most drug discovery programmes use RDD methods to complement more traditional biology and chemical screening approaches. The disproportionate level of attention paid to RDD arises in part because it offers an alternative to the exhausting extensive screening programmes by which drugs used to be discovered, and in part because it is done on a computer and produces coloured pictures.

Receptors

Cells signal to each other by chemical means—only nerve cells, as far as we know, use direct electrical signals. The proteins that receive the chemical messages are generically called receptors. They are favoured targets for drug discovery since they are usually very specific to one cell type, but they are also important in fields as diverse as nitrogen fixation biology and cow rumination. Cell receptors and other cell surface molecules are responsible for cell–cell recognition—the mechanisms by which a cell 'knows' which type of cell it is next to. ICAMs are examples of such cell–cell recognition molecules (*see* **Cell adhesion molecules**).

Receptors generally are on the outside of cells, where they can 'see' the chemical signals. (Exceptions are receptors for steroid and thyroid hormones, *inter alia*, which are in the cell nucleus because these fat-soluble hormones can penetrate the cell membrane.) They therefore need a 'signal transduction' system to carry the signal from the outside, where the chemical binds to the part of the cell that receives the message.

Two types of signal transduction systems are common. Second messengers are small molecules that act as 'hormones' inside the cell, triggering other proteins to act. One common such second messenger is cyclic-AMP (cAMP). Protein kinases (serine kinases or tyrosine kinases) put phosphate groups on target proteins and alter their activity. Often, the resulting proteins then become kinases, and phosphorylate other proteins, and so on down a 'kinase cascade'. Often second messenger systems and kinase systems interconnect.

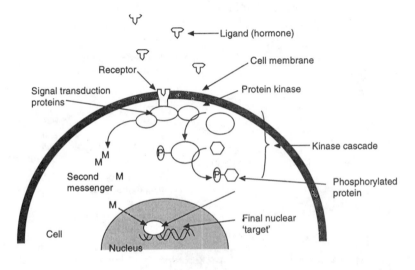

Many receptors are members of the G protein-coupled receptor gene family, also called seven-helix membrane spanning (SMS) receptors. These form a tube across the cell membrane and interact with a protein called G protein on the inside of the cell. When the message molecule binds to the receptor, it signals the G protein to catalyse further actions inside the cell, such as synthesis of cAMP. There are hundreds of SMSs: the family includes receptors for hormones, proteins, neurotransmitters, and (in the eye) light.

There are a range of other signal transduction mechanisms as well. Several biotechnology companies, including Pharmagenics and Sugen, are working on drug discovery by specifically investigating the signal transduction mechanisms inside cells, because altering these mechanisms should allow one to alter whether a cell proliferates or differentiates or not.

Receptor binding screening

This is one of the biotech-based methods for discovering conventional (i.e. 'chemical') drugs. The method relies on the fact that many drugs act by binding to specific proteins (receptors) on or in cells. Normally, these proteins bind to hormones or to other cells, and control the cell's behaviour, although they may be enzymes or structural elements of the cell. The drug interferes with the normal role of the protein.

Finding a drug that has a particular effect on a cell or animal involves exposing the cell or animal to the drug and then looking for the often subtle effect. Receptor binding assays isolate the receptor protein, and then search for chemicals that latch on to that receptor. The ones that do may be good drugs, but the ones that do not are pretty certain not to be, so you have narrowed the field.

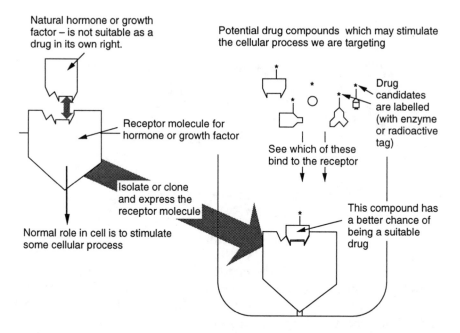

The problems are twofold. First, you have to know what the relevant receptor is. (Indeed, for many drugs there may not be any receptor that is sufficiently specific, or localized on sufficiently few cells. Anticancer drugs suffer from the problem that cancer cells often do not have any unique proteins that the drug can 'target'.) Secondly, even when you have identified the receptor, there are usually only a few thousand molecules per cell, so you have to process several kilograms of mouse to get a few milligrams of receptor. So the receptors are often isolated from cloned cell lines that have been selected to over-express them, or from cloned genes that express the receptors in yeast or mammalian cells.

There are several companies involved in using receptor screens, including most of the major pharmaceutical companies and several small companies such as Protos and Receptortech, which are dedicated to '**rational drug design**'. The most spectacular example is Affymax, a company that has developed chemical methods for depositing huge numbers of peptides and oligonucleotides on to small silica 'chips', and is using them to screen those peptides and other compounds for their ability to bind to receptors.

Recombinant DNA technology

This is the blanket term for the technologies that have made the recent boom in biotechnology possible. It is also called biomolecular engineering, especially in France ('ingenieur biomoleculaire'). Recombinant DNA techniques allow the biotechnologist to isolate and amplify a single gene out of all the genes in an organism, so that it can be studied, altered, and put into another organism. The technique is also known as gene cloning (because you produce a lot of genetically identical genes), and the result sometimes called a gene clone, or simply a clone. An organism manipulated using recombinant DNA techniques is called a genetically manipulated organism (GMO).

Recombinant DNA technology covers the following areas:

Isolating genes. This involves splicing the gene of interest with a **vector** and putting the result into a suitable organism, usually a bacterium or yeast. This new DNA is made of at least two bits of DNA (the target gene and the vector) and is called a recombinant DNA. The organism then grows, multiplying the gene–vector combination as it does so, to produce a clone of cells. The DNA is then said to have been 'cloned into' the vector. The piece of DNA we have inserted into the vector is called, with uncharacteristic obviousness, the 'insert'.

Identifying and characterizing genes. This involves finding which clone contains the gene you want. This is done using biochemical methods of increasing power to discriminate one gene from another, culminating in **DNA sequencing**.

Sub-cloning is a related technique. This takes a large gene clone and breaks it up into smaller pieces, making a new clone from each. This means that what was originally a large piece of DNA is now in smaller, more convenient pieces. This is often used to take a large piece of DNA with many genes on it and separate the genes into one-gene clones.

Modifying genes. This involves replacing anything from a single base to a whole chunk of the gene with other DNA, using (*inter alia*) site-directed mutagenesis.

Putting the genes into another organism. In some cases this is not necessary because you are looking for information about the gene. However, for biotechnologists putting the gene to use it is usually essential, and so they put the gene into another organism using either transfection, transduction, transformation, biolistics, electroporation, or microinjection.

Other key technologies mentioned in this book are the polymerase chain reaction (**PCR**), and **homologous recombination**.

Recombinant DNA: bits and kits

There are a number of pieces of the technology of DNA cloning that are usually referred to without further explanation. The more common ones are:

- Adaptor/linker. These are short oligonucleotides that are used to join disparate DNA molecules together. To do the actual joining, DNA ligase is needed.

- DNA polymerase. An enzyme that makes DNA. To do so, it must have a DNA molecule to copy (the template) and a short DNA molecule to start with (the primer). It then adds bases on to the primer, copying the template until it gets to the end.

- DNA ligase. Also sometimes T4 DNA ligase. This enzyme joins two double-helical DNA molecules together to make one longer one.

- Klenow enzyme. A version of DNA polymerase.

- Methylation. This is a process (again carried out by specific enzymes, methylases) that puts methyl groups on specific bases on DNA. The presence of these methyl groups can stop some restriction enzymes cutting at that site, and in mammalian cells it is important in controlling genes.

- Restriction enzymes. These are enzymes that cut double-stranded DNA at very specific base sequences, and nowhere else. Thus, they cut cloned DNA into a few pieces only. The place at which they cut is called a restriction site, and the map of all such sites on a clone is called a restriction map.

- Reverse transcriptase. An enzyme that makes DNA, but uses an RNA template to do so, not a DNA template.

- RNA polymerase. There are several of these around, notably SP4 RNA polymerase. These are used to make an RNA copy of a DNA. It needs a template but does not need a primer.

- Taq polymerase. Another DNA polymerase made from *Thermus aquaticus*, and an enzyme that is stable when heated to 95°C, used for PCR.

There are a lot of 'kits' on the market, collections of reagents, enzymes, DNAs, and even organisms that have been developed as a package that works together to process the purchasers' samples. Among the more common are packaging kits (which are used in bacteriophage cloning), *in vitro* transcription and translation kits (which carry out transcription and translation in a 'test-tube'; *see* **Genetic code and protein synthesis**), kits for site-directed mutagenesis, labelling DNA with radioactive, fluorescent or chemical labels, and so on. There is a school of thought that there are so many kits around that molecular biology has been reduced to a game of putting together the right kits and writing up the result. Having done it both with and without kits, I think that kits have a lot going for them in allowing the scientist to concentrate on doing creative experiments rather than making all the reagents needed.

Regulation

Biotechnology sometimes complains that it is a heavily over-regulated industry, but in practice it is no more regulated than many other industries, and especially those that rely on relatively new technology. Several aspects to the regulation of biotechnology are covered in this book.

- Patents and intellectual property rights (*see* **Patents**).

- Safety of the microorganisms and genetically engineered constructs (*see* **Microorganism safety classification**).

- Safety of genetically engineered organisms to be released into the outside world (*see* **Regulation of organism release**).

There is also the area of the ethical right that individuals or humans, as a species, have to do any specific part of biotechnology. This is a contentious area of debate, in which sincerely held views vary, from the belief that biotechnology is one of the most inherently beneficial technologies ever invented, to the belief that the technology threatens the whole structure of life on earth (*see* **Bioethics, Mythogenesis, Yuk factor**).

Regulation of organism release

Regulations about the deliberate release of organisms, and particularly genetically manipulated organisms, vary widely. The USA has a fairly consistent set of regulations controlled by the EPA. European regulations vary enormously from the extremely restrictive (Denmark) to the extremely liberal (Italy, Greece) as judged by the American yardstick. By the end of 1989 there had been 140 deliberate release experiments in the USA, about half that number in Europe.

Deliberate release experiments in the USA are the subject of intense and often very public debate about their safety. In Europe, where public access to private data is more limited, laws such as Britain's Environmental Protection Law allow for public access to private data about potential deliberate release experiments so as to allow for the same level of public involvement in deliberate release experiments as the US experience has taught Europeans to expect. By the end of 1992 all countries in the European Community had to abide by European directive 91/220 regarding notification and control of deliberate release.

Deliberate release debates took a new turn in late 1996 when there was heated debate in Europe over importation of American grain that had been genetically engineered to carry, among other genes, an antibiotic resistance marker gene. This raised the possibility that such a gene could spread to the 'wild' bacterial population when the grain was eaten by humans or animals. The scientific implausibility of this was a minor consideration since the debate soon became highly politicized, with the same standards of safety being applied to food products as would be applied to live recombinant organisms.

Regulatory authorities (USA)

The USA has the most advanced biotechnology industry in the world, and is seen as the leader in regulating that industry. In particular, the Food and Drug Administration (FDA) rules of drug development are regarded as the 'gold standard' that any biotechnology company must reach, even if they are based outside the USA. This is in part because where the US leads, other countries follow, and in part because the US is the biggest single market for almost all biotechnology products.

There is a wealth of regulatory bodies in the USA which oversee the biotechnology industry. Some important regulatory agencies are:

- National Biotechnology Policy Board (NBPB). Provides an advisory scientific board for the Secretary of Health and Human Services on the scientific issues behind biotechnology regulation.

- President's Office of Science and Technology Policy (ASTP), which replaced the previous Biotechnology Science Coordinating Committee (BSCC). Has a broad remit to evaluate the scientific basis of biotechnology regulation, and advise the federal government on regulatory issues. The board's remit and membership overlaps substantially with the NBPB.

- Food and Drug Administration (FDA). Oversees and regulates all medical drugs and devices, and new food and cosmetic products, ensuring that they work, and are not harmful. An autonomous agency, it is the principal regulatory agency that any company must appease to launch a new drug or medical device on to the market. In general, FDA regulations set the pace for other countries in biotechnology, because the US market dominates biotechnology products and so all companies want to make sure that their products and processes fit FDA regulations. FDA regulations cover what the efficacy of a drug is (and hence how its trials are done), how it is manufactured (*see* **GLP/GMP**), and how it is formulated. It is notable that, since 1958, the burden of proof that a drug or food additive is safe falls on the producer, and the FDA is not responsible for proving that it is not safe.

- Environmental Protection Agency (EPA). Has responsibility over deliberate release of organisms into the environment.

- Health Care Financing Administration. Developing a biopharmaceutical is time consuming and expensive, and the number of patients that can benefit from it are usually small compared with many conventional drugs. The HCFA has a substantial role in determining the acceptable price for a new drug, and hence whether the company that developed the drug can make enough to recoup its investment and generate funds for future research. This has affected biopharmaceuticals particularly: streptokinase, an established 'clot-buster' drug, costs $186/dose; tPA, a genetically engineered alternative, which some studies says is no more effective, costs $2200/dose. The HCFA's comments are particularly relevant since most biopharmaceuticals (indeed, most drugs) are aimed at the elderly, many of whom are covered by the federal Medicare programme (which has 34 million elderly and disabled clients) in the US.

Replica plate

This is a simple technique for reproducing and selecting bacteria. A number of bacteria are grown on a petri dish. A pad (traditionally sterilized felt) is lowered carefully on to the plate, and when it is taken off some of the bacteria stick to it. It is then lowered on to another plate, where some of the bacteria stick. This second plate then carries a replica of the pattern of organisms on the first plate. The replica plate may now be incubated, and the bacteria on it tested quite destructively for some property. The ones that come out as having the best result are then identified, and the corresponding group of organisms on the original plate can be identified because they are in the equivalent place.

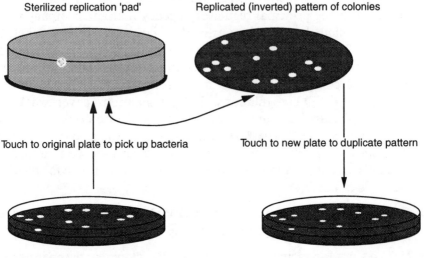

Sterilized replication 'pad'

Replicated (inverted) pattern of colonies

Touch to original plate to pick up bacteria

Touch to new plate to duplicate pattern

Original petri dish with colonies

Replica plate

A related technique is the plaque lift and colony blot. Here the 'pad' is a filter membrane, and after picking off some microorganisms the filter is treated to liberate their internal DNA and proteins and treated to detect specific proteins or genes. Again, the bacteria or bacteriophage that contained those proteins or genes can be identified by their position on the original plate.

Reporter gene

This is a gene that we introduce into a cell to signal when some cellular event is happening. Often reporter genes are used to signal the activation of a whole lot of other genes. The reporter gene makes an enzyme that is easy to detect: often luciferase, the enzyme that catalyses luminol-based bioluminescence, or green fluorescent protein (GFP). The gene for the 'marker' protein is joined with a controller gene, which responds to what we want to measure. Thus, if we want to measure the effect of steroids on a cell, we might tie the luciferase gene on to a steroid-sensitive promoter—when steroids get into the cell, the cells will make luciferase and will glow. When we succeed in blocking the effect of steroids, the glow will die.

Examples of the use of reporter genes are: studying how signal mechanisms affect gene control in cells and tissues (for example, kinase cascades; *see* **Receptors**), detecting viral infection of cells, showing what effects a new gene or mutant has, and studying the different effects of hormones or protein factors like cytokines on different types of cells. Reporter genes are responsible for pictures of glowing plants and bacteria, which make good press.

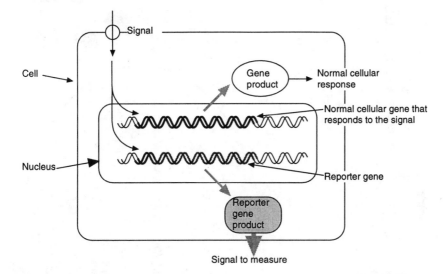

Retroviruses

Retroviruses are viruses whose RNA genes are copied on to DNA as part of their life cycle. The DNA is then usually inserted into the DNA of the host cell, and it can remain there for many cell divisions as a 'provirus', until some signal triggers it to be transcribed on to RNA and so be translated into viral proteins, making more virus. The only thing distinguishing the provirus from any other DNA in the cell is its base sequence.

Retroviruses are of interest to biotechnology for two reasons.

1. Several retroviruses are of medical importance. The AIDS virus, HIV, is a retrovirus, as are several other immune system-targeting viruses (the HTLV family), and some viruses that can cause cancer in laboratory models (the oncogenic retroviruses). Thus retrovirus biology is very important to the search for treatments and cures for AIDS, among others.

2. The ability of retroviruses to infect a cell and then insert their DNA copies into the chromosomes of that cell has also been harnessed to make DNA cloning vectors that can get foreign DNA reliably integrated into mammalian chromosomes. These have been used to

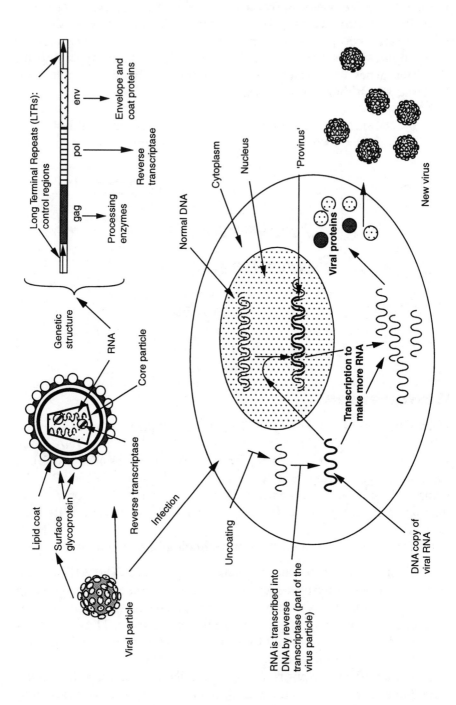

Long Terminal Repeats (LTRs): control regions

gag → Processing enzymes

pol → Reverse transcriptase

env → Envelope and coat proteins

Genetic structure

RNA

Core particle

Lipid coat

Surface glycoprotein

Reverse transcriptase

Viral particle

Infection

Uncoating

RNA is transcribed into DNA by reverse transcriptase (part of the virus particle)

Normal DNA

Cytoplasm

Nucleus

'Provirus'

Viral proteins

Transcription to make more RNA

DNA copy of viral RNA

New virus

transfect mammalian cells, and to create transgenic animals by infecting EC (Embryonic carcinoma) cells with retroviral vectors (*see* **Chimera**). The vectors have to have only part of the viral DNA in them, otherwise they would simply produce infectious virus. Thus, the ideal retrovirus-based vector has those genes that are needed to insert the DNA into the chromosomes but no others. Sometimes this requires that the engineered vector is infected into the cell along with a 'helper virus' which provides some of the genetic functions necessary, but which does not itself get into the cells.

Reversed phase biocatalysis

Some enzymes work on reactants or products that are almost completely insoluble in water. Others work using water as a substrate, and it would be useful to be able to remove the water from the reaction to make it 'run backwards'. In both cases, it is useful to be able to operate an enzyme reaction in a solvent other than water.

Organic phase catalysis and **supercritical fluid** catalysis offers ways of doing this, but an alternative, which is not so radical, is reversed phase biocatalysis (also biphasic biocatalysis), in which an enzyme dissolved in microscopic droplets of water is suspended in an organic solvent containing the reaction substrate and/or product. The enzyme's substrate diffuses out of the solvent in very small amounts, is acted on by the enzyme, and diffuses back into the solvent. Because the droplets are very small, the rate of diffusion is very fast and so the reaction proceeds at a useful rate.

A variation is to use a solid support to hold the enzyme in a totally organic solvent. The solid support has a single-molecule layer of water adsorbed on to its surface: the enzyme sticks to that, and is simultaneously immobilized (so that it is easy to remove as part of a particulate solid once the reaction is done), activated by the water, and stabilized by immobilization. Inorganic materials such as silica or celite are usually used.

These systems have the advantage that you do not have to dehydrate the enzyme so thoroughly before the reaction (organic phase catalysis needs a thoroughly dehydrated enzyme to work properly), and so can be much easier to get working.

Reverse genetics

Reverse genetics is the type of genetic analysis that starts with a piece of DNA and proceeds to work out what it does. By contrast, normal genetics ('forward genetics') starts with the phenotype (what the organism looks like) and proceeds to work out what the genetic structure is, ultimately decoding the DNA itself.

Such feats of gene cloning as the isolation and characterization of the cystic fibrosis gene are often called reverse genetics; however, although these use an impressive panoply of recombinant DNA techniques they still start with an observed phenotype (the disease) and work via ever more detailed genetic techniques to a genetic explanation of what is going on. Reverse genetics has been used, for example, in understanding the genetic structure of a range of viruses, including the AIDS virus. Here, the DNA structure is known in detail, but what it does is not known. So mutations are found or made in the DNA, and then their effect on the phenotype is discovered. In this way the function of those bits of the gene is worked out.

RFLP

This abbreviation stands for restriction fragment length polymorphism and is usually pronounced 'riflip'. It means a piece of DNA that varies between two individuals: whether the DNA has a function or not, or whether the variation is important is irrelevant. The term refers only to the way of detecting the variant, which is by the use of the very specific DNA-cutting enzymes called restriction enzymes. The essence of an RFLP is that one variant DNA is cut by a particular enzyme at one site, the other is not. This means that the fragments produced by that enzyme on those DNAs have different lengths.

RFLPs have found wide use as 'marker' genes for genetic studies. Here, the RFLP is used to detect when a piece of DNA has been inherited by an individual from one parent (rather than the other). If the RFLP is near to a gene that we wish to track but cannot detect directly, then there is a fair chance that the target gene has been inherited along with the RFLP. The RFLP is termed a linked marker since it is physically and genetically linked to the gene we are actually interested in.

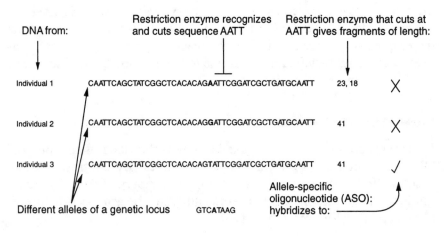

A related term is the allele-specific oligonucleotide (ASO). This is an oligonucleotide that will hybridize to the DNA from one individual but not to that of another, because the DNAs differ by one or two bases. The variant forms of the DNA are called alleles. Both RFLPs and ASOs have found substantial use in human genetics, and in animal and plant breeding programmes.

Ribozyme

Also called catalytic RNA, these are RNA molecules that catalyse chemical reactions, often the breakdown of other RNAs. Their discovery in the mid-1980s overturned the idea that only proteins could be biological catalysts.

Ribozymes have potential in two areas. They are widely touted as potential pharmaceutical agents since their action against other RNAs can be extremely specific. They could, for example, attack a viral RNA without affecting the normal RNAs in a cell. Thus they have potential as antiviral agents and, through their potential ability to attack RNAs from oncogenes, anticancer agents. Ribozymes as therapeutics are still a research technology, however. Although very specific in the test-tube, like antisense RNA they can have unexpected effects when introduced into cells. There is also the problem of how to get them into cells. RNA is destroyed very easily by chemical or enzymatic attack, so has to be protected by encapsulating in (for example) liposomes to get it to the cells it must affect (*see* **Antisense**).

The other area is to use ribozymes as industrial catalysts, selecting suitable catalytic activities through **Darwinian cloning**.

Scale up

Scale up is the process of taking a biotechnological production from a laboratory scale to a scale at which it is commercially useful. A few biotechnological processes can be run on laboratory-scale systems (for example, the production of reagents for research use, such as monoclonal antibodies). All others have to be done in much larger installations than a research laboratory can handle.

The difficulty with scale up is that a tonne of fermenting bacteria seldom behaves in the same way as a gram of the same bacteria, unless it is divided into a million separate tubes. In general, it is not possible to take the conditions that have worked well in the laboratory and apply them to an industrial process. Instead, they are gradually adapted to ever larger scales of production, each step usually being between a four- and a 10-fold increase in size over the previous one. At each stage, the optimum amount of various chemical parameters, and mechanical parameters (such as stirring rate, air-supply method, and rate) must be determined, based on the biotechnologist's experience with previous production systems and a general knowledge of scale-up procedures. There is some mathematical modelling available to help, but even so experimentation is essential.

The problems of scale up were not well understood by the early genetic engineers, and so in the mid-1980s there was a severe shortage of scientists skilled in this field. However, it is now understood that a marvellous laboratory result does not automatically translate into money in the bank, as scale up might be prohibitively expensive.

Scanning tunnelling microscopy (STM)

This new type of microscope has been promised to be the ultimate way of discovering the structure of biomolecules (among other things). A related technique is the atomic force microscope. In essence, an ultra-sharp needle point is scanned slowly over something, and either the force on the needle or the electric potential of the needle tip is monitored. When the tip encounters an atom sticking out above the general surface, the extra force/current is measured. By scanning back and forth across a surface, a picture of the hills and valleys can be built up on an atomic scale.

There are two areas of application in biotechnology, neither advanced beyond a laboratory curiosity stage. The first is directly detecting the physical shape of complex molecules, getting round the need for the pure crystals that X-ray methods need. Arscott and Bloomfield at Minnesota University have produced pictures of the double-helical structure of a synthetic DNA using STM. By hitting molecules under the STM with light (and so altering their shape), something can be deduced about the chemical nature of individual bits of a new molecule, as well as their size and shape.

The other, even more radical, idea is to use STM as a way of actually moving atoms around, creating new chemical entities. So far this has been confined to drawing letters with individual atoms on crystal surfaces, the atoms being xenon (at IBM in San Jose) or sulfur (at Hitachi in Tokyo). In principle, this could lead to the direct fabrication of new biomolecules which would be enormously hard to make by any conventional method: however, this is definitely 'Buck Rogers' stuff at the moment.

SCP (single-cell protein)

Coined in 1966 at Massachusetts Institute of Technology (MIT), the term single-cell protein refers to protein biomass used as a food additive for animals or people. Either isolated protein or whole bacterial cells (suitably processed) may be called SCP.

The drive to develop SCP came from the realization that the 'food shortages' seen in many Third World famines were primarily shortages of protein, not of food bulk *per se*. Similarly, the limiting factor in many animal feeding systems is how much protein is available for animal growth, not the total calorific content. The idea behind SCP technology was to use bacteria, growing on a cheap carbon substrate and with a cheap nitrogen source such as ammonia, to make protein fit for human, or at least animal, consumption.

As with many large-scale fermentations, the key to making SCP economic is to find a carbon source cheap enough. Oil and natural gas have been tried, but are only marginally economic even when the oil and gas prices are low. Gas oil, the oil left after petrol and paraffin has been distilled out of crude oil, has also been tried (using the organism *Candida lipolytica*), but again is only marginally worth it. Methanol made from natural gas is a good potential substrate since bacteria find it easier to use (they need less oxygen to grow on methanol than on methane, and it is very soluble in water). ICI developed a large-scale biomass process based on a methanol-using bacterium, *Methylophilus methylotrophus*, to produce a partially purified protein product ('Pruteen'). The production-scale plant had a volume of 1000 cubic metres and a capacity of 70 000 tonnes of SCP per annum; despite the economies of scale, however, this was at best only marginally economic, despite ICI's use of genetic engineering to improve the effectiveness of the bacterium's metabolism in using ammonia to make the protein.

Cellulose, wood, waste starch, paper processing effluent, and other complex sources of carbon have all been suggested as potential SCP substrates, using *Cellulomonas* and *Alcaligenes* organisms; however, none of these are efficient enough to be economic. Algal products such as *Chlorella* and *Spirulina* use sunlight and CO_2 as raw materials, which are cheap enough; however, it is rarely economic to isolate protein from their tough cells, and so these are used as products in their own right (*see* **Food**).

A problem with SCP is that most microorganisms have a much higher nucleic acid (DNA and RNA) content than animals or plants, which can cause health problems if eaten raw. So the nucleic acid must be removed. In addition, microbial cells can absorb or make toxic materials during fermentation, and the cells themselves may be extremely indigestible or allergenic. This has limited the use of SCP in human food, and meant that most effort has gone into using it as an animal feed supplement. In this it competes directly with soybean meal and fish meal, which are extremely cheap

One SCP-like food has gained general acceptance, the meat substitute Quorn (*see* **Food**).

Sea water

There have been many and varied plans for extracting metals from sea water, often lured on by the idea that a cubic mile of sea water contains over a thousand tonnes of gold. In fact it does not (on average it contains 40 kilograms) and that is spread over a very large volume of water. No process yet devised is cheap enough to get the gold (or anything else other than salt and a few other chemicals) out of it.

Biosorption and bioaccumulation is a biotechnological route to obtaining value from sea water. The idea is to use bacterial cells to accumulate a specific metal from the water: all you have to do is pass the water over the cells, and then afterwards 'wring them out' into a much smaller volume, resulting in concentrated gold solution. Despite this attractive-sounding proposition, it is never economic to do this when the actual costs are worked out, including (for example) the cost of pumping 4 billion tonnes of sea water through your extraction apparatus, and replacing the apparatus regularly as it is eroded by the salt water.

The sea is a source of tremendous biotechnological wealth, but mainly algae, microbes, and other living things.

Secondary metabolites

Primary metabolites are the chemicals commonly found in most living things, and which are essential for them to live. Compounds such as glucose or glycine would fit into this category. Secondary metabolites are chemicals that are usually unique to one organism or class of organisms, and which are not essential for cell survival. They perform more specialist functions, like being involved in specific stages in the organism's life cycle, degrading unusual food sources, or (usually) fighting off other organisms. Many of the chemicals that plants or microorganisms produce that are of biochemical interest, including antibiotics, are secondary metabolites.

Unlike primary metabolites, which most organisms contain most of the time, the production of secondary metabolites is very dependent on the environment of the organism. Thus, small changes in culture conditions of an actinomycete (actinomycetes are the most commonly used sources of new secondary metabolites) will dramatically alter how much of a particular chemical they produce.

Plants often produce secondary metabolites as defences against infection or being eaten: caffeine in coffee, atropine in nightshade, and the vinca alkaloids in Madagascan periwinkle are examples of quite poisonous compounds made to ward off attack. These secondary metabolites are usually not produced efficiently in isolated, cultured cells. However, their production can sometimes be stimulated by elicitor compounds or preparations, which are often fungal or plant extracts.

Secondary metabolites are used for many purposes. The two most common are:

As drugs. Many drugs were discovered when a plant or fungal extract was found to have pharmacological activity. Almost invariably this activity is caused by a secondary metabolite. Often the chemical structure of the metabolite is so complicated that it is still extracted from its natural source since making it by chemical synthesis would be too expensive. **Antibiotics** are often secondary metabolites, as are alkaloids.

Flavours and fragrance compounds, other than sweet flavours and salt, are usually secondary metabolites, often from plants. (Meat flavours arise rather differently, from chemical reactions between fats, protein breakdown products, and sugars in the meat.) Several companies, such as Universal Foods and Universal Flavours and Fragrances, are working on using plant cell culture and cloning methods to produce flavour or fragrance chemicals by fermentation.

In general, secondary metabolites are the product of specialist anabolic pathways. Like many such pathways, they are end-product inhibited, which means that it is very difficult to get the organism concerned to make large amounts of the metabolite.

Secretion

Secretion is the active export of a material from a cell or organism. The secretion of proteins by bacterial and mammalian cells is very important in their production by biotechnology. If a foreign protein being produced by a cell can be secreted, then it is usually much easier to purify it away from all the other proteins that the cell is making, as these mostly stay inside the cell.

Proteins that are to be secreted from a cell have a short peptide on their front end (the signal peptide) which acts as an export label. The signal peptide is chopped off the protein as it is exported (during a step called 'processing'), so the final protein does not have this extra peptide on it. Genes for naturally secreted proteins code for this peptide. Genes for proteins that are not normally secreted do not, and so this signal peptide must be engineered on the front end of the 'new' gene. Secretion vectors are expression vectors that allow this. They have a promoter and then a short section of a gene that codes for this signal peptide. Any gene spliced in exactly next to the signal peptide gene will produce a fusion protein (one with the signal peptide joined on to the front of the protein) which should then be exported from the cell.

Sewage treatment

Sewage treatment is one of the most widespread biotechnological processes in urban Western societies, which produce huge amounts of human and animal waste. Sewage treatment methods vary widely, but all have a biological basis to break down the organic material in sewage and convert it into something that can be safely discharged into rivers or seas.

All sewage treatment falls into several stages.

- Filtering—to remove large solid objects (paper, sticks, sand, etc.)

- Settling—to allow the particulate material to settle out. This sludge is usually then composted to decompose any organic matter, and then used for landfill or as fertilizer.

- Biological treatment—the resulting liquid is treated using micro-organisms to remove the remaining organic matter. This treatment can be by:

- (i) trickling bed systems (also called trickle filter) in which the liquid is sprayed over porous mineral or plastic beds with a film of organisms growing on it.

- (ii) activated sludge process, in which the sewage is incubated with organisms derived from the sewage sludge, with air or oxygen pumped through the mixture. The oxygen concentration and microorganism concentrations are kept high, so that the microorganisms can grow at the same rate as solid waste is supplied to them. This is an aerobic digestor process.

- Further settling—the microbial biomass produced during biological treatment is allowed to settle out, resulting in reasonably clean water. The sludge is either recycled to the fermentation system or further incubated in a digestor to make fertilizer.

In countries with lots of space and/or a high tolerance of smell, sewage is usually treated in open, shallow trickle beds where the bacteria get the oxygen they need from air by diffusion. In many European countries this is being replaced by a version of the deep shaft system, and activated sludge process. This is more akin to a fermentor. Oxygen or air is pumped into a column of waste, and absorption is enhanced by the high pressure at the bottom. The result is a more compact and faster-acting digestor system; however, it is more complex (and hence more expensive) to build, and to run. Activated sludge digestors usually have the cells in them in free suspension, mixed with the liquid waste. A variant is the upflow sludge blanket system, where the cells form flocs that are kept in suspension by the upflow, but remain together.

An important aspect of sewage processing is the reduction of the amount of organic carbon compounds in the sewage, expressed as biological oxygen demand (BOD). BOD is the amount of oxygen that is needed for microorganisms to use up all the nutrient sources in the water. BOD value are the amount of oxygen, in milligrams per litre of water, that is consumed by microorganisms' respiration in five days. Typical BOD values are:

- Outfall from a sewage treatment plant into rivers <45

- Farm 'dirty water' 300–1000

- Farm manure slurries, raw sewage >3000

In conventional sewage the organic material is metabolized by the microorganisms in the treatment plant, ending up as carbon dioxide and biomass. The biomass material (sewage sludge) is usually burned, composted, or used as fertilizer. Alternative methods generate methane (biogas) from this material, but this is not a common use. If carbon is not removed from waste material before it is dumped into the environment, the result is eutrophication, making the environment over-rich in nutrients or minerals at the expense of oxygen. This leads to rapid growth of anaerobic bacteria, depletion of oxygen in water, and death of oxygen-requiring organisms like fish. Eutrophication was a common fate of rivers in urban areas in the mid-twentieth century.

Site-directed mutagenesis

This is the introduction of specific base changes (mutations) into a piece of DNA using recombinant DNA methods. There are several ways of doing this, but all the ones in common use involve using a synthetic DNA that has the mutation you want built into it to replace the equivalent piece of DNA in the original gene. This is also called oligonucleotide-directed mutagenesis. This can be done by copying a new version of the gene from the old version, using an enzyme [usually acting on a single-stranded DNA, such an M13 clone (*see* **Bacteriophage**)], or by cutting out the old version of the section of the gene you want to mutate and splicing in a new, mutated version.

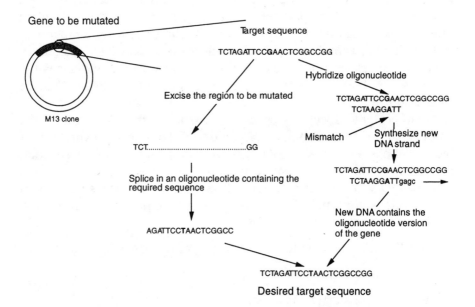

The alternative to site-directed mutagenesis is some version of random mutagenesis, where the DNA is mutated at random by chemical treatment and then the mutant you want selected out from the resultant mixture.

Soil amelioration

The improvement of poor soils, usually using bacteria or fungi. (This contrasts with bioremediation, which is the cleaning up of toxins, usually in soils). Amelioration includes breaking down organic matter, forming humus (i.e. good soil structure), making minerals such as phosphates in the soil available to plants by solubilizing them, fixing nitrogen, and sometimes an element of bioremediation as well.

Soil improvement methods were touted as being the way to 'green the deserts' in the 1960s; however, they have not worked all that well, mainly because the deserts are not very promising material to start on for climatic as well as chemical reasons. Most of what was then called soil amelioration is now included in bioremediation.

Solar energy

There has been quite a lot of interest in using biotechnology to generate fuels or power from sunlight. This is what plants do all the time, of course, but getting them to do it for humans has proven difficult.

The simplest method is to grow plants, and then turn them into fuel: this can be by extremely traditional routes (burning wood) or by growing organisms with high oil contents to make fuel oil. Attempts to make fuel oil using algae have generally proven uneconomic, as have the use of photosynthetic bacteria to make hydrogen. (Bacteria that generate hydrogen or methane from biomass are more successful, and indeed are the basis of **biogas** technology).

More speculative schemes plan to use the electrochemistry of photosynthesis to generate electricity directly. This can be done either by using intact cells (analogous to a bacterial biosensor) or by isolating protein complexes from the photosynthetic apparatus and using them as chemical reagents. The protein complexes considered have included the photosystems (I or II), which transduce light energy into electrochemical potential in the chloroplast, and more specific parts of the photosynthetic apparatus such as the antenna complex, which actually captures photons and passes them to a reactive centre. Power outputs to date have been vastly exceeded by the effort and energy needed to make the materials for the experiment, and the complexity of the photosynthetic apparatus inside a cell makes it an outside chance that this will be a workable system.

An alternative approach is to use a synthetic chemical system. One example is a chemical reaction series based on ruthenium. A ruthenium compound [ruthenium(II) tris-(2,2'-bipyridine)] is a reducing agent in its normal state, but a powerful oxidizing agent when excited by blue light. With a metal oxide catalyst and methyl viologen (MV) as an electron acceptor, this compound can transfer electrons from water to MV. The reduced MV can then be used (in theory) to reduce other compounds. Yields are not good enough to make this more than of research interest, however.

Solid phases

Many biotechnological processes rely on immobilizing something (a cell, an enzyme, an antibody) on a solid phase. This means anything that is not a liquid or a gas, but in practice usually means silica (glass) or plastic.

There are many reasons for doing this, discussed elsewhere. In general, the purpose is to allow easy separation of two molecules that would otherwise both be in a solution together. Thus enzymes are immobilized on to solid phases in bioreactors so that they can be separated from their substrates and products, which remain in solution. The principle is the same as making coffee from ground beans—the solid coffee grounds are easy to separate from the liquid coffee because they are large lumps, and so can be separated by settling or filtering.

Immobilizing enzymes can also sometimes stabilize them, so that they are active for longer. This is thought to be because the solid material, which is never completely uniform, provides microscopic 'pockets' that have an environment that is unusually conducive to the enzyme or antibody. However, no one really knows why this works.

The rate of reaction of a molecule that is immobilized on a surface with others in solution can be quite different from that when they are both in solution. A critical factor is the boundary layer or diffusion layer. The enzyme can convert all of its substrate within range very fast, but then has to wait for some more material to move to its molecular neighbourhood before it can act on that. Such a reaction is said to be diffusion limited, because, no matter how good the enzyme is, the rate of reaction is only as fast as the rate at which the substrate can diffuse to the enzyme from bulk solution. Diffusion layers can be reduced by stirring or shaking, but can never be removed entirely.

Somaclonal variation

This is the variation seen between the individuals in a clone. It applies particularly to plant clones, i.e. plants that have been grown up from the cells of a single 'parent' plant. If you separate a plant into its component cells and culture them under the right conditions, then you can get each cell to grow into a new plant. For some plants this is relatively easy: gardeners will be aware how easy it is to grow a whole new potato or dandelion plant from a tiny bit of root left in the ground. In theory all such plants should be identical, and for a few species this is so. However, for most (particularly for potato) there is substantial variation between the offspring, often more variation than usually arises between plants of that species. The variations are heritable, so some at least are caused by mutation. This is called somaclonal variation.

This can be a problem or an opportunity for the plant breeder. It is a problem if you want to use plant cloning technology to grow a lot of your prize plant: the offspring of most cloning methods will not be the same as the parents. The opportunity is that this is another way of generating new plant types which would be harder or impossible to generate by conventional plant breeding.

Sports and biotechnology

Despite the fact that recreation, and specifically sports, is as big a business as medicine, and approaches the chemical and agricultural industries in size, biotechnology has consistently ignored the lighter side of life in preference for healthcare or process industry products. The only major exception seems to be in the discussion of the potential abuse of biotechnology products for sporting advantage. Two specific cases are widely discussed: these may or may not be real rather than potential abuses since 'authoritative' rumours abound, but there is almost no factual support.

Growth hormone. The market for growth hormone for medical treatment is quite small; however, the market perceived for the drug is substantially larger, and must include some 'indications' not thought to require growth hormone treatment absolutely when the protein was first expressed in bacteria. Two areas of 'new' application are for short stature and for sport. Kabi Pharmacia placed advertisements in the medical literature in late 1991 suggesting that growth hormone could be a 'cure' for the childhood 'condition' of being short (not pathologically short, but just in the lower few per cent of the normal human range for children of that age). This could be defensible on psychological grounds. An application that is indefensible on medical grounds is the use of growth hormone to try to make people unusually tall, so that they have an advantage in some sports such as basketball. For this to work, growth hormone would have to be administered during early adolescence. The abuse of growth hormone by adults who try to use it to increase muscle mass is fairly well established. Rumours that people have tried to acquire growth hormone to give to their children are widespread—whether this is an urban myth, along with the women who put poodles in microwaves and the people who discover rats in hamburgers, or is based on any true event is not clear.

EPO (erythropoietin). This biopharmaceutical was developed to boost blood production in a number of diseases such as anaemia and kidney failure, where patients lack red blood cells, and where other therapies, especially for leukaemias, have depleted the bone marrow and so the patients develop an iatrogenic anaemia. There is a suggestion that athletes have used, or have tried to use, EPO to boost their levels of red blood cells above the normal level to give their blood greater oxygen-carrying ability. This may give them greater endurance in long-distance running. It would almost certainly be dangerous, increasing the viscosity of the blood and hence the risk of heart attack, stroke, or haemorrhage. Dutch cyclist Johannes Draaijer died of a heart attack at 27 years of age in 1990 under the suspicion of using EPO.

Standard laboratory equipment

There are a few pieces of kit that all biotechnologists seem to use and to refer to, called by trade names analogous to 'Hoover' or 'Kleenex'. Among the more common are:

- Multiwell plate. Also 96-well plate or microtitre plate. A postcard-sized plastic dish with eight rows of 12 small round wells in it. Used extensively in cell culture and molecular biology where you need to perform the same reaction on up to 96 samples at once. Machines for automatically washing and detecting colour in 96-well plates are common.

- Gilson. Any type of micropipette, a device that will measure small volumes (i.e. 1 μl–1ml) of liquid routinely. Goes with Gilson tip, the replaceable tips that go on the end of the pipette unit and are the only part in contact with the liquid.

- Eppendorf. A centrifuge the size of a 'mini' hi-fi deck that sits on the bench: also the disposable 1.5 ml plastic tubes that fit inside the centrifuge.

- Universal. A cylindrical tube with a screw cap holding about 20 ml and now usually made of plastic.

- Kit. A set of reagents and materials that allow a scientist to perform a research step without the need for any other materials other than his research subject. There are, for example, DNA labelling kits, PCR kits, *in vitro* transcription/translation kits, cloning kits, etc., all of which contain everything you need to perform that particular protocol (including instructions) except for the DNA you are going to do it on. Older molecular biologists mumble about how good it was in the old days when they had to make their own reagents, forgetting that the process of getting a specialist to make things for you is what got us down from the trees.

Stem cell growth factors

These are compounds, almost always proteins, that act to make stem cells grow faster. Stem cells are cells that, while not themselves critical parts of muscle or blood, grow into cells that make those tissues. Thus they are the 'stem' from which the 'leaves' of tissue arise. Stem cells therefore have two roles: to make more stem cells, and to make their differentiated 'progeny' cells.

The best characterized stem cells are in the bone marrow. These stem cells (about 1 in 100 000 bone marrow cells) make the precursors for all the cells in the blood. These stem cells are called 'totipotent' because they can make any of the various types of blood cells. As their offspring develop they become fixed ('determined') into the track of making one type of cell or another, and then eventually develop the final character-istics of the cells concerned ('differentiation') and are released into the blood. The same sort of thing happens in muscle, skin, and developing nerves (including the brain).

Clearly, if the stem cells are to play their role there must be a balance between how fast they make new stem cells and how fast they turn into their differentiated daughter cells. Too much differentiation and there will not be enough stem cells left for the future. Too much division and you will end up with a cancer. A battery of controls affect how this balance is regulated: aberrations in these controls can lead to cancer (*see* **Oncogene**). The controls can also be altered artificially to correct disease states.

The most discussed stem cells are the blood (haematopoietic) stem cells. True stem cell factor (SCF) was isolated in 1990, but a range of other factors, which affect cells in various stages of determination and differentiation, have been found and their corresponding genes cloned, usually with the aim of developing them for drug use. Some of these are listed in the entry on **growth factors**.

Sterilization

There are a number of established ways of sterilizing equipment and materials for biological use. Clearly, if a microorganism or cultured cell is to be grown, either for research or for production, it is vital that no other microorganism be present to compete with it, possibly killing it and certainly producing unwanted contamination. Thus, sterilization is a vital part of any biotechnological process.

Note that sterilization is not necessarily the same as cleaning. Cleaning means removing 'all' the dirt from something, usually all the contami-nants that you can conveniently detect. One bacterium is not usually detectable, but if a sample contains one live bacterium, then it is not sterile. Thus a plate taken out of a conventional dishwasher will be clean, but almost certainly not sterile. A sterile material is something that has no living organisms in it. However, it can be very 'dirty', as long as the dirt is sterile too. Thus, a recently run car engine block will be sterile (because of heat) but definitely not clean. Often, biotechnological 'cleaning' systems aim to sterilize as well.

There are four generally used approaches to sterilizing things.

Heat. All organisms are susceptible to heat, although some are more susceptible than others. Heat can be dry or wet. Wet heating to 121°C in an autoclave (essentially a large pressure cooker) is a very popular way of sterilizing equipment and reagents since it is cheap and easy to do. Lesser levels of sterility can be obtained by boiling, or by heating to 75°C for a short time (pasteurization). The most heat-resistant biological entity known is the BSE agent, which is reputed to be able to stand heating to 300°C (*see* **Transmissible encephalopathies**).

Chemicals. Many chemicals are inimical to life. Extremely corrosive materials such as chromic acid are used to strip all biological residues off glassware, including all living organisms. However, more benign biocides (chemicals that kill microorganisms but leave most other things intact) are more commonly used. Many are used as cleaning agents, since, unless swallowed, they are relatively harmless to humans. A variation on chemical treatment is treatment with a biocidal gas, usually ethylene oxide. This has the advantage that equipment being sterilized does not then have to be dried out. Usually biocides are not suitable for sterilizing liquids, because there is no way of getting them out again afterwards. Exceptions are ozone and hydrogen peroxide.

Irradiation. Gamma-rays will sterilize anything, but are dangerous and relatively expensive to produce. Ultraviolet (UV) light is an effective sterilizing agent, and somewhat safer. However, UV does not penetrate very far into most liquids or solids, so it is usually only useful for sterilizing surfaces. UV of wavelength 245 nm is the most effective (it is absorbed efficiently by DNA), but sterilizing lamps usually have a broad spectrum output between 220 and 300 nm. Recently, very intense, white, visible light has been demonstrated to be a surface-sterilizing agent.

Filtration. This is only suitable for liquids or gases, but is extremely effective: usually a 0.2 μm filter (i.e. a filter with holes in it about 0.2 μm across) will remove all living things except viruses from a fluid.

Different sterilization methods must be chosen for different applications—there is no 'correct' method for all applications. The key problem to be overcome is materials compatibility. Thus many plastics are discoloured and rendered brittle by gamma-rays, and melted by excess heat. Many fermentation and cell culture media cannot be autoclaved, as this would destroy some of the essential nutrient in them.

The effect of sterilization is measured by measuring viable cell counts or bioburden. These are both measures of the number of living cells in the sample. Process industries have developed very sensitive methods for detecting just a few live bacteria in large samples of drugs or sterilized foodstuffs. Traditionally, this has involved culturing a sample of the food in bacterial culture medium ('broth') and seeing if anything grows. Some of this can be replaced by ATP luminescence tests. ATP (adenosine triphosphate) is a component of all living cells, but is very rapidly broken down in dead ones. The enzyme luciferase uses ATP to generate light (*see* **Luminescence**), and so the enzyme can be used to detect living bacteria in a sample through detecting their ATP, and hence the light generated. A sensitive photomultiplier tube can detect the light generated from the ATP from a handful of bacteria. The drawback to this method is that many other things in the food can also contain trace amounts of ATP, or can stop the enzyme working, giving false positive or false negative results, respectively.

Stirred tank bioreactor

This is the simplest type of continuous bioreactor. A reaction vessel has liquid medium pumped into it, gas pumped in (usually through a sparger at the base), and a stirring system to keep the ferment uniform. Material is taken off continuously without separating the components. The stirred tank bioreactor is the most common type of system used for fermentations, and is used in every size from a few hundred millilitres in a laboratory system, to hundreds of cubic metres in a large production plant. It is suited to many microbial growth fermentations. In contrast, enzyme reactions tend to be carried out in plug flow reactors, where the material passes through the reactor in a 'plug' that is processed all at once, rather than as a continuous process.

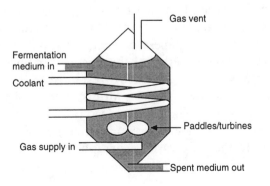

The variations on the stirred tank bioreactor include the size and nature of the stirring systems (paddles, turbines), whether there are baffles inside the reactor to break up the fluid flow, and how the gas and liquid are introduced. More complex modifications tend to result in fundamentally different types of bioreactor.

See also **Tank bioreactors**.

Strain (cultivar)

A strain of an organism is a type that is genetically distinct from other representatives of the species to which the organism belongs, but which is not different enough to be called a new species. Members of a strain are much more genetically similar to each other than to members of other strains. The word 'strain' is normally used for microorganisms to describe a particular organism that has been isolated or engineered to have some property, like growing well or making a lot of a product. Isolating and improving strains of microorganisms is a major part of the process of making them suitable for an economic biotechnological process (*see* **Strain development**).

For animals, the term 'breed' or, sometimes, 'race' means much the same thing—a genetically homogeneous collection of animals, usually derived from one pair of parents, which is significantly distinct from other animals of the same species. Breeds or races can interbreed with each other, where animals of different species almost never can. Thus there are a large number of different 'breeds' of dogs (huskies, labradors, poodles, and so on) which can interbred to form a 'generic' mongrel dog. For plants, the term 'cultivar' and 'variety' have a similar meaning. 'Strain' is sometimes used for plants, rarely for animals.

Human 'races' are not really equivalent to animal 'races' or 'breeds' since humans of different 'races' are in fact genetically almost identical. The human idea of 'race' is as much to do with culture and social training as with biology. However, there are a few objective differences between humans from different parts of the world (apart from cosmetic ones like colour). Most people of Japanese descent, for example, have a much lower ability to metabolize ethanol than people of European descent, and do not generally have the ability to digest large amounts of lactose as adults, with the result that they get drunk more easily and are less tolerant of milk products. These differences mean that it is important to carry out clinical trials of new drugs in different places in the world, just in case reaction to the drug varies geographically as well.

Strain development

Also strain improvement. This is the general term for improving the genetics of an organism so that is carries out a biotechnological process more effectively. The aim is to create a strain of the organism that makes what you want, makes it in large amounts, does not make much of anything else (so that you can purify your product easily), uses cheap and easily obtainable things to grow on, and does not require excessively careful control of culture conditions. The idea of an 'improved strain' is exemplified by coniferous trees used for wood pulp production: they grow almost anywhere on soil, air, and water, and you can make most of their mass into the product simply by pulping it. (This is why wood pulp is cheaper than interferon.)

There are many routes to strain development.

Incremental selection. This takes the current strain, treats it with mutation-causing chemicals (mutagens) and looks at a large number of the descendant strains to see if any have acquired a mutation that makes them more productive. This is a time- and labour-intensive operation, but frequently the most useful route to improving the production of chemicals such as antibiotics or amino acids from fermentations. This is a random screening approach, in which large numbers of random variants of an organism must be screened. Often the key to success lies in how these numbers can be screened rapidly and automatically, i.e. the system's 'screening capacity'.

The other methods are more directed.

Hybridization. This is taking two strains and combining them genetically. It has been used extensively in agriculture, but, because the organisms used are so diverse, often cannot be used so successfully in biotechnology. A variant, which is more applicable to bacterial systems, is:

Conjugation. Here only a few 'desirable' genes are transferred between one strain and another.

Genetic engineering. This seeks to alter the genetic make-up of an organism by directly introducing genes into it. These could code for a more efficient enzyme, or block the action of an enzyme that destroys the product you require. This is a more complex and costly path, but one which may be the only route open if 'traditional genetics' has failed.

Often, the key to successful strain improvement via any route is finding a selection procedure. This is a set of conditions under which the strain you want has an advantage over all others. For finding strain that makes an enzyme that breaks down one particular compound or group of compounds, this can be straightforward. For example, an oil-eating bacterium can be selected by growing a population of bacteria in a medium where the only carbon source is oil. Thus, the only bacterium to flourish will be one that can metabolize the oil, and the faster it can metabolize it the faster it can grow. Such relatively straightforward selection procedures are rarely available.

Strain isolation

This is the isolation of any bacterium, yeast, or indeed animal or plant, from the outside world. In general there are two approaches to strain isolation for microorganisms.

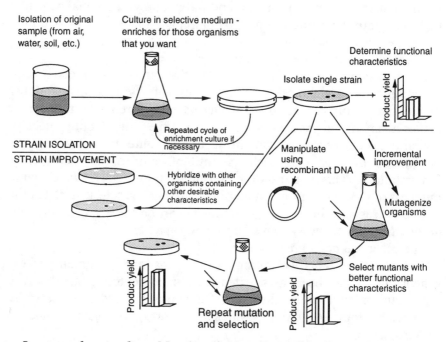

Large-scale sampling. Nearly all biotechnologically useful microorganisms are isolated from the soil, which holds between 1000 and 1 000 000 000 microorganisms per gram. The microorganisms that exist in a particular place depend on the local soil ecology, and clearly this varies greatly. Thus, one approach to finding the ideal organism is to sample as many soils as possible. Many major chemical and pharmaceutical companies have (or had) programmes under which any staff member travelling to exotic parts of the world brought home a small sample of soil to be used in the in-house screening programmes. The key to success here is not the number of samples you can collect, but the variety of environments from which they can be collected. Hence the biotechnological interest in 'black smoker' vents, hot springs, arctic sea water organisms, and so on.

The other approach is to find an environment in which organisms will have had to develop the characteristic you want in order to survive. Favoured sites are the effluent paths or waste tips for chemical plants, which tend to accumulate microorganisms that can break down all the chemicals that are present in their local environment. But many other possibilities exist. The organisms that break down methane, for example, were originally isolated from the soil around a cracked gas main.

An important part of this isolation is to develop a method of selection or 'enrichment' of the organism you want. This involves making an enrichment medium that will select for the particular set of characteristics you are looking for. So, if you want an organism that breaks down oil, you dig up some soil samples around leaking oil pipes, and grow the bacteria concerned on a medium in which the only source of carbon is oil. Then, anything that can degrade oil will grow, and everything else will not. In practice, there is more to the design of enrichment media (or selection media) than this since many other aspects of the organism's likely chemistry must be accounted for.

The terms enrichment media and selection media are also used when talking about the selection of mutant organisms, or even mutant cultured cells. Thus the HAT medium used to select for hybridomas (the cells that produce monoclonal antibodies) is a selection medium, because it allows hybridomas to grow but prevents the growth of other cells.

Despite all the efforts of biotechnologists to develope recombinant DNA methods for optimizing bacteria for biotechnological uses, as often as not it is the original selection method that has the greatest effect on whether the organism is going to be the basis of a commercial process or not.

Strategic alliance

Not unique to biotechnology, this term means an alliance between two companies formalized as a legal agreement and usually aimed at developing some common area of business. Because setting up a complete biotechnology R&D department can be time consuming and expensive, biotechnology companies and pharmaceutical companies frequently set up strategic alliances with each other to get access to know-how they would otherwise have to develop in-house. The partners might be after a stable source of funds and market support, a particular R&D technique, production facilities, product formulation and packaging ability, experience with regulatory authorities, or marketing and sales expertise. The value gained depends on which side of the alliance you fall; however, the essence of an alliance is that both sides benefit while remaining independent.

Strategic alliances are different from specific research contracts and from mergers or acquisitions. Research contracts are often called alliances, but actually are normal contracts to employ one party to do something for the other party. The only thing going from the contractor to the researcher is money. Mergers and acquisitions are the more intimate fusing of the business interests of two companies, and one partner usually loses their autonomy in the process. Probably the bestknown biotechnology acquisition/alliance of all was the acquisition of 60% of Genentech by Hoffman LaRoche in 1991. Genentech is probably big and energetic enough to retain its own identity, so making this a strategic alliance rather than the takeover the balance sheet makes it appear.

Alliances are often organized around a major development programme, e.g. to develop a new product or technology. Such agreements are usually funded by milestone payments or royalties. The former are made when the partner doing the research achieves specific goals, e.g. getting a prototype to work or getting a good Phase I clinical trials result. Royalties are a percentage on the sale price of the final product, and so do not start to roll in until the joint development is completed.

Substrate channelling

This is a neat idea, which has been shown to work in research but not in large-scale application yet. The idea is to link together two enzymes that perform a series of reactions. The first enzyme takes substrate 1 and turns it into product 1. The second takes product 1 and turns it into product 2. If both enzymes are added to a solution of substrate 1, then product 2 will accumulate. However, quite a bit of product 1 will have to accumulate before there is enough for the second enzyme to work on. A faster and more efficient way of doing this is to link the two enzymes together physically, by making a fusion protein of them or by linking them together chemically. Then, as soon as product 1 is made by the first enzyme, it is handed over to the second enzyme (which is right next door) and turned into product 2.

This has potential advantages in situations where product 1 is very unstable, or is liable to be acted on by other enzymes to turn it into an undesirable side product. It is called substrate channelling because the process works as if there is a channel sending product 1 from enzyme to enzyme without it ever being released into the solution.

A related idea is to link a co factor on to the enzyme. This has been done with the NADH cofactor of glucose dehydrogenase. As most dehydrogenases need NADH (or the related NADPH), if it is chemically linked to one enzyme then any other enzyme that wants to use that molecule has to come very close to the first to get its NADH. This, in effect, links the two enzymes together, although they are not physically bonded most of the time.

Supercritical fluid enzymology

All materials have a critical temperature (T_c) above which their gases cannot be turned into a liquid by compressing them. At this critical temperature, liquid and gas can coexist if the pressure is at the critical pressure (P_c). For carbon dioxide, for example, the critical temperature is 31°C. Thus at room temperature, if you compress carbon dioxide enough (as in a gas cylinder) it will turn into a liquid. Above 31°C, no matter how hard you compress it it will not liquefy—the gas will just become more and more dense.

This highly compressed gas behaves in part like a gas and in part like a liquid. It is called a supercritical fluid (SCF), and has some useful properties for chemical and biotechnological processes.

- Diffusion in supercritical fluids is usually much faster than in liquids, so diffusion-limited reactions (which covers quite a number of enzymatic reactions) can occur faster.

- The solubility of chemicals in SCFs depends very sensitively on pressure. Thus reagents can be dissolved or products removed by precipitation when you alter the pressure. Some chemicals that are only sparingly soluble in water can be made extremely soluble in SCFs by choosing the right pressure and temperature.

- The pressures and temperatures involved are not damaging to many biopolymers.

SCFs have been used for several model enzyme reactions. In general it helps to include a small amount of water (which also dissolves in some SCFs) to assist enzyme stabilization; it is also essential if the enzyme uses water as a substrate.

Against the advantages, of course, is the disadvantage that SCFs must be kept at high pressure. One of the much-advertised advantages of enzymes is that they work at mild temperatures and pressures. Working at 100 bar pressure in SCF removes one of these advantages. Thus SCFs are only useful for enzymatic catalysis if some other feature of using SCFs clearly compensates for the additional complexity of working with pressurized gas.

Supercritical fluids are also used to purify materials, using supercritical fluid extraction (SFE). SCFs can have very specific abilities to dissolve chemicals, which can be 'tuned' by altering pressure and temperature. Thus SFE is used commercially to extract caffeine from coffee, leaving virtually all the other chemical components of the coffee behind. The solvent used (supercritical CO_2) is removed by reducing the pressure, whereupon it all turns back into normal gas.

Tank bioreactors

Bioreactors, also called fermentors, are the vessels in which fermentation takes place. Tank bioreactors are vessels in which the microorganism is grown in a large volume of liquid. This contrasts to fibre/membrane bioreactors (*see* **Hollow Fibre**) and **immobilized cell bioreactors**. The large majority of bioreactors used in biotechnology are tank bioreactors, and most tank bioreactors are **stirred tank bioreactors**, because stirring helps to distribute gas and nutrient to the growing organisms effectively.

The bioreactor must provide a mechanism for introducing reagents and the microorganism into the reactor vessel, providing substrate (i.e. food) to the microorganism (including oxygen for aerobic fermentation), stirring it, and keeping it at the right temperature, pH, etc. Temperature control is especially critical for large volume fermentations since metabolizing microorganisms produce a great deal of heat. Variations in layout include different sizing and spacing of the baffles (which ensure that the volume is mixed thoroughly by the stirring) and different types of impellers (stirrers). These can be a wide range of shapes and sizes: disk (Rushton) turbine, open turbine, marine impeller (i.e. like a ship's screw).

The other main variation between reactors is the gas injection mechanism. These are almost always via a sparger or a diffuser (a pipe or plate with holes in it, which shoots bubbles into the base of the reactor over a wide area at the base of the tank), but a wide range of shapes of pipes or plates have been used. The shapes, which include rings, crosses ('spiders') or dead-end pipes, must be selected for the particular shape and size of the reactor, and the amount of gas that has to be injected.

There is a great deal of expertise in designing a suitable bioreactor for culturing an organism or cell type. As a consequence, there are more companies specializing in bioreactor design, control, and engineering than in recombinant DNA techniques and reagents, despite the much higher profile of 'gene cloning'.

Targeted drug delivery

This is using any method to deliver a drug to the site in the body where it is needed, rather than allowing it to diffuse into many sites. There are three approaches to such targeted drug delivery.

The first is to encapsulate the drug in something, usually a lipid coat (i.e. a **liposome**). The outside of the coat is itself coated with a material that binds to the target cells, either an antibody specific for those cells, or a glycoprotein, or a receptor molecule or ligand. The liposome travels round the blood until it finds its target, where it sticks to the cell and delivers the drug. In the case of liposomes, the lipid membrane fuses with that of the cell, releasing its contents into the cell.

The second approach links the targeting mechanism directly to the drug. Here the drug must either operate outside the cell, or be able to get itself into the cell on its own. A much talked about application is linking toxin proteins to antibodies: the proteins can get inside cells and there destroy the cellular machinery, but only when carried near enough to the cell by the antibody. Such conjugates are called **immunotoxins**. The application here, clearly, is to destroy cancer cells or, conceivably, cells infected with long-term viruses such as AIDS or HPV.

The problem with both these approaches is how to get the drug–carrier complex from the bloodstream into the target tissue: unless the target is the endothelial cells of the blood vessels or a few cell types in liver, lung, or kidney, nothing as large as a liposome is going to be able to escape from the blood vessels to get to them.

The third approach is to make the drug as a 'prodrug' which goes to all tissues of the body, but which is metabolized to the active drug only by one tissue because that tissue has a high level of an enzyme that cuts the prodrug into inert 'carrier' and active drug. This is easier to do for tissues such as liver and kidney, which have a battery of rather specific enzymes. Several drugs, such as the antiviral drug gancyclovir, are actually prodrugs that are activated by their target cells but remain less active in other cells.

Textiles

Biotechnological products are used in some areas of the textile industry to process cloth or cloth precursors. The majority of products are enzymes. The applications include:

Amylases to desize cotton. 'Size' is a coating of starch put on many fibres to stop them breaking during weaving. It has to be removed from the final cloth, which is usually done with acid or high temperatures. Amylase enzymes can be used instead. (Sizing was also used to make the cloth on the wings of early aeroplanes more rigid and airtight.)

'Prewashed' jeans. 'Stonewashed' jeans are actually washed with pumice stone to make them look worn. However, this process does actually wear them. Cellulases can be used to erode a small fraction of the cotton fibres more uniformly than pumice stone, so producing chemically aged jeans.

'Polishing' fibres with cellulases. Cotton fibres are covered with microscopic protruding fibre ends, which can catch and cause the cloth to go 'bobbly' with wear. Novo Nordisk has developed a cellulase product that polishes these fibres off, a process sometimes called biopolishing.

Bleaching and bleach removal. Enzyme bleaching systems similar to those being tried for wood pulp bleaching have been applied to cloth. Generally, the systems are oxidase or peroxidase enzymes that generate peroxide radicals, which react with coloured materials and oxidize them to colourless products.

Enzymes are also used extensively in detergent manufacture (*see* **Proteases**).

Biotechnology has also provided some limited new materials, such as the 'chitopoly' polymer made of chitosan and another polymer. The chitosan is meant to have antifungal properties, useful for underwear. Agracetus is working to fill the hollow centres of cotton fibres with polyhydroxybutyrate (PHB) using enzymes, so as to produce an intimate, natural cotton/polyester mixed fibre.

Thermal sensors

Thermal sensors, sensors that detect tiny changes in heat or temperature, are well known in many applications. Such sensors are often used in gas chromatography systems to detect molecules emerging from the GC column. There have been some attempts to harness thermal sensors as biosensors. Here, a probe detects the heat given out when an enzymatic reaction occurs. This could be much more flexible than enzyme electrodes since, while relatively few enzyme reactions involve the transfer of electrons that could be picked up by an electrode, nearly all result in the release of heat. The problem is that, for small samples of dilute material, the amount of heat released is tiny, so very small and very sensitive heat sensors are needed.

Thermophile

A thermophile is an organism that grows at a higher temperature than most other organisms. Generally, since a wide range of bacteria, fungi, and simple plants and animals can grow at temperatures up to 50°C, 'thermophiles' are considered to be organism that can grow above 50°C. They can be classified fairly arbitrarily depending on their optimal growth temperature into 'slight thermophiles' (50–65°C), 'thermophiles' (65–85°C), and 'extreme thermophiles' (> 85°C). Thermophiles and extreme thermophiles are usually found growing in very hot places: hot springs and geysers, and domestic hot water pipes, for example. Deep sea exploration submarines have brought back microorganisms from the 'black smoker' volcanic vents on the mid-ocean ridges which can apparently grow at over 100°C, under pressure.

Thermophiles are of interest to biotechnology because of the economics of fermentation and biotransformation. Many industrial processes could be catalysed by enzymes, but the enzymes are too slow. The reactions may be speeded up by heating, but this rapidly destroys the enzyme. If you can use an enzyme that works at high temperatures, then you get the benefit of heat and catalysis. Heating up the reaction is also desirable because it reduces viscosity and increases the rate of diffusion of reagents, and so reduces the amount of stirring and pumping energy needed, and the heat prevents other enzymes from working or (usually) contaminating organisms from growing in your reactor.

Enzymes from thermophiles can, of necessity, withstand such high temperatures. They also frequently exhibit increased stability to organic solvents. Thus there is substantial interest in isolating these enzymes and using them in industrial processes. Because the bacteria themselves are usually tricky to grow (and must be grown at high temperatures), once a suitable enzyme is identified it is common to seek to 'clone' its gene into a bacterium that grows at a more moderate temperature. This also means that the enzyme can be purified from all the other proteins in the bacterial cell simply by heating it up: all the other, non-thermostable proteins will precipitate, leaving a reasonably pure preparation of the target enzyme.

A range of thermostable enzymes are used in industrial processes. As with all research into isolating enzymes from bacteria, one key feature is to have a large number of diverse sources of candidate organisms to screen. This is why Iceland, with one of the world's densest concentration of different types of hot spring and geyser, has been the source of the majority of the publicly available thermophilic organisms in use.

Tissue culture

This is sometimes used interchangeably with '**cell culture**'. Strictly, it means the cultivation of tissues, i.e. multicell assemblies, outside the body. It uses very similar techniques and materials to cell culture, which are discussed in the entry on cell culture.

Whole tissues are cultured when scientists want to examine the cells' interactions with their extracellular matrix (i.e. the stuff that supports the cells in the body), when they wants to examine how cells interact, or when there is no known way of culturing those cells in isolation so that they maintain the properties they have in the body. Thus liver slices are used quite widely to study the metabolism of drugs since the technology to culture liver cells in isolation and have them behave like liver cells is still in development. Brain slices are also used to examine how nerves interact as a network.

The technology of tissue culture and cell culture overlaps with that necessary to build **artificial tissues** from isolated cells.

Toxins

Living things make some of the most dangerous compounds known, like ricin (castor bean toxin) and pertussis (whooping cough) toxin. One molecule of *Botulinus* toxin, delivered to the inside of a nerve cell a billion times more massive than it is, will kill the cell, or less than a tenth of a microgram will kill an adult human. By comparison, the lethal dose of plutonium is guessed to be half a microgram. Such powerful poisons have potential uses, and biotechnology has the potential to make them relatively safely.

Toxins can be used on their own as therapeutics. *Botulinus* toxin is being developed as a way of blocking unwanted muscle spasm. Clearly it cannot be injected generally, as other drugs would be (it would kill the patient). However, if minuscule doses are injected into the muscle they can paralyse that muscle. The amount of protein is so small that it escapes the notice of the immune system, and so the body does not make antibodies that could neutralize future doses. Allergan and Porton International are producing commercial version of *Botulinus* toxin for this application.

Toxins can also be coupled to other things to give them a lethal 'sting'. **Immunoconjugates** are probably the best example.

Making such toxins is difficult, even with all the panoply of micro-biological containment available (*see* **Physical containment**). People have tried cloning the genes for these protein toxins into bacteria to express them more efficiently (they are normally present in extremely small amounts). Such scientists tend to find themselves in the middle of empty rooms whenever they talk about their ambitions at conferences.

Transfection, transduction, transformation

These terms are all used to mean ways of getting DNA into cells, usually animal or bacterial cells. The meanings are different depending on the type of cell being discussed.

Transfection. Strictly, this means carrying a piece of DNA into a cell as part of a virus particle. For mammalian and plant cells, it is used more generally to mean almost any way of getting DNA into a cell.

Transduction. A relatively specialist genetic technique, this means transferring a piece of DNA from one organism to another via natural DNA exchange processes. It occurs almost exclusively in bacteria, and is a method of genetically engineering large pieces of DNA such as the *Agrobacterium tumefaciens* Ti plasmid.

Transformation. For bacteria, this means getting the bacterium to take up DNA that the experimenter has added to its medium. The bacteria that are able to do this are called 'competent'. Demonstrating transformation was one of the key proofs that DNA was the genetic material. For mammalian cells, transformation means changing the cell from one whose growth is limited by its neighbouring cells into one whose growth is limited only by the media available to it. Transformation is a step in the development of cancer cells, and is also a crucial part of generating an 'immortalized' cell line. Because these two meanings of 'transformation' grew up beside each other, genetic engineers who are manipulating mammalian cells often say that they 'transfect' the cells with DNA, rather than 'transform' them, even if what they actually do is add naked DNA to the cells.

There are several common methods used to place 'naked' DNA (i.e. DNA that is not encapsulated into a virus particle, a liposome, or some other carrier system) into cells.

- Bacterial cells. Bacterial cells that are 'competent' (i.e. in a suitable physiological state, which may be achieved by growing them in the right way and suspending them in the right buffer) will spontaneously take up DNA from the solution around them. The common factor involved is usually the need for magnesium salts in their medium.

- Bacterial protoplasts can also be transformed by fusing them together in the presence of DNA. This can be done using polyethylene glycol (PEG). The cells' membranes join up in the presence of PEG forming many-cell masses, and some of the external solution, containing the DNA, gets trapped inside the cell in the process.

- Mammalian cells can be transfected by adding DNA to them as a calcium phosphate precipitate.

Transgenic

A transgenic organism is one that has been altered to contain a gene from another organism, usually from another species. While this would suggest that any genetically engineered organism could be called 'transgenic', the term is usually only applied to animals. Bacteria and yeasts are always called 'genetically engineered', and with plants it is an even choice which term to use. Creating transgenic plants is a relatively mature science, and is discussed under **plant genetic engineering**.

Creating transgenic animals is more complex. The germ cells (i.e. the egg, sperm, or newly fertilized zygote) must be altered—altering some of the cells in the adult (the somatic cells) is no help at all (although it may be valuable for other reasons; *see* **Gene therapy**). Thus, unlike plant genetic engineers, who can regenerate a new plant from almost any cell in the old plant, animal genetic engineers must develop methods for getting DNA into the germ cells. There are several ways of doing this.

Microinjection. This, the first successful method, simply injects the DNA into the egg's nucleus (diameter about one hundredth of a millimetre) with a very fine needle. Microinjection needs considerable skill. This is the only method that works for cows, sheep, goats, and pigs.

Transfection. This is chemical treatment of the egg with the DNA. While this works well for somatic cells, it is a bit dodgy for eggs. An Italian group claimed that they had found a simple way of making sperm absorb DNA from solution; however, no one else has been able to repeat their results.

Electroporation. Not very successful with animal cells, and not at all successful with eggs (*see* **Electroporation**).

Using *embryonic carcinoma* cells (EC cells) (*see* **Chimera**).

Retroviral vectors. Some viruses, notably the retroviruses, can carry DNA into a cell and splice it into the cell's own DNA. There is a lot of interest in harnessing this ability to engineer genetically all sorts of animal cells.

Transomics. This is an injection technique, but instead of injecting pure DNA its practitioners dissect out sections of chromosomes under the microscope and inject them. As chromosomes are only a thousandth of a millimetre or so long, (and much thinner) this is not a profession for myopics or those with shaky hands.

Foreign genes introduced into transgenics are usually called exogenous (in animals) or ectopic (in plants) genes.

Transgenic animals: applications

Trangenic animal technology is used mostly to produce research results. The favoured tool here is the 'knock-out' mouse (*see* **Knock-outs and mutants**). There are three areas in which transgenic animal technology has been used to create biotechnological products, as opposed to research results.

The first is creating animal models for disease. This is probably the most successful application to date (*see* **Transgenic Disease Models**). The second is as production systems for foreign proteins, usually therapeutics. This is known as **pharming**. The third application area is in farm animal improvement. Sixty per cent of the fixed cost of producing a pig is the cost of feed, so if a pig could be engineered to turn that feed into meat more efficiently, that would represent a substantial saving for the farmer. In principle, expression of a transgenic growth hormone (GH) gene in the pig should do this; however, in experiments to date the side-effects of engineering GH genes into pigs or cattle have outweighed the potential benefits. In addition, the debate over the use of injected BST suggests that, even if the genetic engineering is successful, it will be controversial on regulatory and social grounds.

Other engineering of farm animals has looked at improving **wool** quality, and milk quality by introducing more milk protein into cows' milk.

Transgenic disease models

One application of transgenic animals is to model human diseases. Having an animal model of a disease can be very valuable when human patients for a disease are rare, when it is impossible to find them before the disease is well established and so the early stages cannot be studied, or when it is not ethical or practical to do studies on humans. However, quite a few human diseases are not mimicked accurately by any animal model. Transgenic technology seeks to create animals, especially mice, that get diseases that are in some specific way characteristic of a human disease. These can then be used for screening for potential new therapies or drugs. Among the models used are:

1. Humanized ('humanoid') mice for AIDS research. True transgenic mice with the human CD4 gene can be infected with the AIDS virus. Another model (SCID-Hu mice) have no functional immune system of their own, but have human immune cells inserted into them to make an immune system that is susceptible to AIDS. (These are properly called chimeric animals, because they are a mixture of cells or tissues from several different animals.) SCID mice can be made in several ways which knock out their immune system, including by massive whole-body irradiation and genetic engineering of them to include a toxin gene that is expressed at high levels in their lymphocytes, so destroying them.

2. Models for diabetes, and several other diseases in which specific cells are absent or do not work properly. A toxin gene is placed in the animals, spliced on to a promoter sequence that only expresses that toxin gene in one particular tissue (in the case of diabetes, in the beta cells of the pancreas). The toxin then kills those cells off, leaving the rest of the animal healthy. Such genetic constructs are called toxigenics.

3. Models for cancer. Models for cancer usually have an oncogene inserted into them, so that they develop a specific cancer at an anomalously high rate (*see* **Oncomouse**).

4. Models of immune function. A key aspect of the healthy immune system is its ability to tell the normal constituents of the body from

other, potentially invading materials. A wide range of diseases arise from a failure of this mechanism. Transgenics are being used to find out how the immune system 'learns' to discriminate self from non-self, both by introducing foreign protein genes into mice and by creating toxigenics that lack certain sets of lymphocytes. These studies have implications for many diseases, such as diabetes (which has an autoimmune component), arthritis, allergies, and multiple sclerosis

Another approach is to use **homologous recombination** to disrupt a gene in the animal, and so model directly a human disease in which the gene is faulty. Several of the collagen diseases, such as osteogenesis imperfecta, have been modelled in this way.

Transgenic plants: applications

Transgenic plants have been tested widely for agricultural use, and some have come into economic production. There were early concerns that the transgenic trait would reduce productivity, thus rendering the crop uneconomic, and the only way to test this is to grow the plant under real farming conditions in large amounts to see just what the yield is. In some cases it has been shown that the genetic engineering of the plant does make it unprofitable to grow compared with its unengineered counterpart, despite its theoretical superiority, but this is not always the case. For example, genetically engineered flax, containing the acetolactate synthase gene from *Arabidopsis*, which allows it to be resistant to sulfonylurea herbicide, grows as well and as productively as the original strain, despite carrying the biochemical burden of the extra gene.

Typical applications of genetically engineered (transgenic) plants include:

Improved disease resistance in crops. Some virus resistance has been engineered into plants by giving them genes for virus coat proteins: the excess of the virus proteins in the plant prevents the live virus from replicating effectively. This is called pathogen-derived resistance (PDR), and was first shown with tobacco mosaic virus. Pest resistance has been very successfully engineered into plants by giving them the genes for chitinase (an enzyme that attacks insects' skeletons), or *Bacillus thuringiensis* toxin (B.t.k; *see* **Biopesticides**).

Protein production in plants. There has been a lot of work, and some progress, on using plants to produce recombinant proteins, particularly antibodies. The attraction is that plant seeds can contain 30% of their mass as one protein, and plants only need air and water to grow.

However, such production rates are never approached in practice for recombinant proteins.

Herbicide resistance in crops. A common target is to engineer crop plants to be resistant to herbicides, so that fields can be sprayed with herbicides that will kill off weeds but not affect the crop plants. Concerns that this will lead to increased herbicide use have not been realized in practice.

Transmissible encephalopathies

This is a generic term for the diseases bovine spongiform encephalopathy (BSE) (also 'mad cow disease'), scrapie (affects sheep), Creutzfeldt–Jacob disease (CJD) and kuru (affect humans), and transmissible mink encephalopathy. These are slow, degenerative disease of the brain where the brain ends up with many spherical lesions in it that appear almost empty under the microscope, looking like a sponge, hence the name. The diseases are generally believed to be caused by a protein called a prion; this is a variant of a normal body protein that can somehow trigger the body to turn its own prion precursor protein (PrP) into prion. The causative agent is very hard to destroy: boiling it, digesting it in acid, or leaving it in the sun for a week seem to have little effect.

The encephalopathies started to cause concern in the biotechnology industry because of the possibility that the agent that causes the disease will get into biotechnological products produced from cultured cells. Many cell culture systems use fetal calf serum as part of the medium in which the cells grow. The fear is that the scrapie/BSE agent could get into the cells, and from there into the biotechnological products. The Dutch health board refused approval of the Ares–Serono growth hormone on this basis in 1990. Many European countries banned British beef imports in 1996 because of similar concerns about BSE getting into food. A 'new variant' of CJD was identified in Britain in 1996 which may be linked to BSE, and 13 people have died of it since (and another three have it but have not yet died, as of March 1997). The more common form of CJD causes 40–50 deaths a year in Britain.

Transplant

Transplantation is putting a new organ, tissue, or (more recently) cell into a patient. Traditionally the new organ or tissue has come from another human being. Biotechnological approaches are allowing animal sources and synthetic tissues to be used.

The fundamental problem with putting an organ or tissue from one human into another (an allograft) is that of tissue rejection. Tissues are rejected when the recipient immune system recognizes them as carrying foreign proteins on their cell surfaces. In particular, most cells carry specific proteins called major histocompatability complex (MHC) proteins [human leucocyte antigens (HLA) is an older synonym] which vary hugely between people, so the chances that they are the same between any two individuals is tiny. The MHC proteins are concerned with the way the immune system detects foreign proteins, so if they themselves are foreign the receiving body has a strong reaction to them.

Some cells 'present' such antigens much better than others, and so cause much stronger tissue rejection. Immune cells, especially dendritic cells, are probably the 'worst', and removing them from a transplanted tissue reduces the rate of rejection. Other approaches to reduce rejection include masking the donor's cell surface antigens with antibodies, irradiating the cells or tissues with ultraviolet light, or giving the recipient large doses of immunosuppressive drugs. This last course is the standard method, but has severe drawbacks, so the search is continuously on for better methods. Amongst these are genetic engineering approaches to removing the rejection antigens: transgenic animal technology has been used to create replacement organs that could, in principle, be transplanted into humans (*see* **Xenograft**).

Biotechnology also uses transplant technology as a part of cell therapy, removing cells from a patient, engineering them, and then reintroducing them into the same person or into someone else. These are usually haematopoietic cells (cells that make up the blood and their precursors and especially cells from the bone marrow). Bone marrow transplants (BMTs) are a major component both of some gene therapy regimes and of cancer therapies. Many cancer treatments destroy the bone marrow, which consequently has to be replaced.

A version of somatic cell therapy that does not involve genetic engineering at all is autolymphocyte cell therapy (ALT), or autologous gene therapy. This removes a cancer patient's lymphocytes (as opposed to their bone marrow) and uses a combination of cytokine treatments in the laboratory ('*in vitro*'), which stimulates them to reject the patient's own cancer.

Transposon

A transposon is a genetic element that can move around the genome. Most genes stay where they are with respect to other genes, unless mutation rearranges the genome in which they reside. Transposons break this rule. They are able to create copies of themselves elsewhere in the genome, or even in other genomes if they are present in the same cell. Thus, for example, a transposon might copy itself out of a bacterial genome and into the genome of a bacteriophage as the phage infects the bacterium. Some transposons splice themselves out of their original site to do this, but most simply copy themselves, so the end result of the copying process is two copies of the transposon where one existed before.

The process of a transposon moving is called transposition. It has been harnessed in a variety of ways by geneticists and genetic engineers to move genes around in bacteria and, to a lesser degree, in plants.

Many transposons carry useful genes as well as being 'selfish' DNA that propagates itself around the genome. Most antibiotic resistance genes are carried on transposons in some bacteria, as are genes for things like heavy metal resistance.

The way that many transposons move is reminiscent of the way that retroviruses reproduce, in that the transposon is transcribed on to an RNA, which is then copied back into the genome as a DNA. Because of this similarity, such transposons and retroviruses are sometimes grouped together and called retrotransposons.

Treatment Protocol Program

An FDA initiative to allow terminally ill patients to be given experimental drugs before they have cleared all the hurdles of full regulatory approval. This concept was brought to public notice by AIDS patients, who objected that the rate of approval of new drugs for AIDS was so slow that they would be dead before anyone got a drug on the market to cure their disease.

Triple DNA

Most introductory textbooks will tell you that RNA is 'single stranded' and DNA is 'double stranded', i.e. that DNA consists of a helix of two strands wrapped round each other. However, it has been known for a while that RNA can be triple stranded, and recently triple-stranded DNA has been demonstrated too (also sometimes called triplex DNA). It has several potential applications.

The 'third strand' of triple DNA binds to the other two by specific base pairing, and so it can be used as a reagent that recognizes a specific DNA sequence. (The base pairing 'rules' are rather different from those in normal double-stranded DNA.) This can be used to detect a specific sequence in a double-stranded target (rather like a conventional DNA probe detects a sequence in a single-stranded target DNA) or to block gene activity. In this second role, a short oligonucleotide is made that binds to a gene, preventing the RNA polymerase from opening it and copying it on to RNA. This has the same effect as an antisense agent made against that gene product, but acts at the DNA level. So far, this has worked to a limited extent in the 'test-tube' only. This is an application of **aptamer** technology.

Triplex DNAs can also be used to build very specific DNA-cutting reagents. If one is linked to a molecule that cuts DNA, the third strand can be shown to act as a sequence-specific nuclease, i.e. a reagent that will cut the DNA at (or just next to) a very specific site. Several 'artificial nucleases' of this sort have been made.

DNA can also be forced into a left-handed helical form, called Z-DNA. This structure occurs naturally in gene promoters, and so making it appear (or disappear) may be another route to blocking gene function.

There are a number of other, related complex structures that have been made from DNA, for various purposes. Chiron made branched polymers of DNA as an aid to increasing the sensitivity of hybridization assays. Nadrian Seeman has used oligonucleotides to make cage-like structures, opening the intriguing possibility of using DNA as a biomaterial.

Tumour marker

A tumour marker is any molecule that shows that a cancer is present. Usually each type of marker is produced by only a few types of cancer, so as well as showing that a cancer is present it helps to identify the type of cancer, and hence the most suitable therapy. Tumour markers can be used for diagnosis or, potentially, as targets for biopharmaceutical drugs such as immunotoxins.

Tumour markers fall into two categories. The first type are the products of **oncogenes**, and hence their presence represents part of the reason that the cell is a cancer cell to start with. The others are apparently incidental to the cause of the cancer, but are always found associated with a particular type of cancer. Such proteins are usually made in only a few cell types in the healthy body, but cancer cells make them in much larger amounts [such as prostate-specific antigen (PSA)] or in inappropriate places (such as the hCG, human chorionic gonado-trophin, usually only made by the placenta in early pregnancy but also produced by some cancers). Quite a few markers, like CEA (Carcino-Embryonic Antigen), are proteins normally found only in the fetus (sometimes called oncofetal proteins).

In addition there are a wide range of antigens (i.e. proteins to which antibodies bind) that have been identified by a monoclonal antibody as being associated with a type of cancer, but whose normal function is obscure. A number of them are glycoproteins or carbohydrates: cancer cells often add sugar units to proteins in a slightly different order from normal cells, so creating different glycoforms of those proteins. It is these differences between the glycoforms that is detected as a 'marker' by the antibody. Polymorphic endothelial mucin (PEM) is an example—the protein is covered in sugar residues in normal tissue, but exposes some of its polypepide chain in cancers.

Although tumour markers can be used as a diagnostic for cancer, they are not infallible. What doctors want to know is whether a cancer is likely to be life threatening. What the tumour marker tells them is that a particular cell type is present. These two things may not be the same: 50% of American men aged 50 have PSA-positive 'tumour' cells in their prostate, but in most cases the cancer poses no threat; only 3% of American men die of prostate cancer. Thirty per cent of those with positive PSA tests apparently have no cancer cells at all. So tumour markers have to be used carefully, especially for large-scale screening programmes.

Tumour markers are also targets for tumour vaccines (also called cancer vaccines). If the marker is linked to another protein or injected with an adjuvant, it could stimulate the immune system to make antibodies that recognize the marker on the cancer cells, and so destroy the cancer. This neat idea has not been very successful, because the body becomes tolerant of the cancer during its development. If it did not, it would have destroyed it before. A compromise approach is to use a monoclonal antibody against the marker as an **immunotherapy**.

Vaccines

Vaccines are preparations that, when given to a patient, elicit an immune response that subsequently protects the patient from a disease. Usually the vaccine consists of the organism that causes the disease (suitably attenuated or killed) or some part of it. Attenuation of a virus (or bacterium) is manipulating it so that it does not loose its ability to grow in culture, but looses some or all of its ability to cause disease in animals. Usually bacteria, and to a degree viruses, slowly loose their ability to colonize living things (and hence to cause disease) as they are cultured outside the body. There are a range of biotechnological approaches to producing vaccines.

Viral vaccines. These are vaccines that consist of genetically altered viruses (*see* **Viral vaccines**).

Enhanced bacterial vaccines. Bacteria have been genetically engineered to enhance their value as vaccines (when dead). Intervet markets a genetically engineered *E. coli* as a vaccine for pigs. BioResearch (Ireland) has produced a bacterial vaccine for furunculosis in salmon by deleting some genes that are critical to causing the disease from the bacterium. The genetic engineering aims to build a bacterium that has the antigenic properties of the original (i.e. stimulates the immune system), but is not a pathogen.

Recombinant DNA techniques can also be used to generate attenuated strains, by deleting pathogenesis-causing genes, or by engineering the protective epitope from a pathogen into a safe bacterium. Thus vaccination with the safe bacterium causes the body to make protective antibodies against the pathogen. This is a research programme so far.

Biopharmaceutical vaccines. Proteins, or sections of proteins, that are the same as the proteins in a virus' or bacterium's wall, can be made by recombinant DNA methods as vaccines. This is a standard biotechnological route, and has the advantage that there is no possibility that the resulting vaccine will contain any live virus particles. Peptide vaccines are often incorporated by genetic engineering into a larger carrier protein to improve their immunogenicity (i.e. how well they cause the body to become immune), or their stability.

Multiple antigen peptides (MAPs). Developed by J. T. Tam, these are peptide vaccines that are chemically stitched together (usually on to a polylysine 'backbone'). This means that several vaccines can be delivered in one shot.

Polyprotein vaccines. This is a similar idea to MAPs, but here a single protein is made by genetic engineering in which the different peptides form part of a continuous polypeptide chain.

Vaccines have also been produced by a range of non-standard

methods. The conventional method is to produce the vaccine from an attenuated version of the disease-making organism, grown in cultured cells or hens eggs. Alternatives include:

- Producing single protein vaccines in recombinant microorganisms.

- Producing vaccines in transgenic plants. Plants will not have any immune response to a human virus, and will not get ill from them. They can produce lots of the protein and could, in principle, be used as food to provide an oral vaccine. Some preliminary experiments suggest that this could work.

Vaccinia virus

Vaccinia viruses are DNA viruses from the same family as cowpox and smallpox. However, they are safe viruses to work with, and so have been used for several biotechnological applications.

Vaccinia has been used as the basis of an expression vector system. It can infect a wide range of cells, and has a lot of DNA, quite a bit of which can be removed using suitable genetics. Thus quite large foreign genes can be spliced into it, and then the recombinant virus can be used to infect a wide range of cells, allowing the biotechnologists to choose the cell most suitable for the process. Vaccinia virus vectors have been used quite widely in research since they can be used to express proteins in mammalian cells. Because they can hold quite a lot of DNA, they can be used to make more than one protein at once in a cell, which can be useful for making proteins with more than one polypeptide chain (multi-subunit proteins).

Vaccinia has also been used as the basis for 'live virus' vaccines. It does not itself cause severe disease, and because it can infect a wide range of species it can be used to produce a range of animal vaccines, which are the first targets of this sort of technology. Provisional approval for a field trial of a live vaccinia virus vaccine was granted in the USA in 1990.

Foreign genes are usually introduced into vaccinia by recombination, rather than by isolating the vaccinia DNA and manipulating it *in vitro*. This is because the vaccinia virus is too large to be manipulated conveniently.

Cowpox and raccoon pox viruses, which share some of the useful characteristics of vaccinia, are being looked at as alternative vector systems.

Vector

A vector in biotechnology is usually a DNA segment that allows another piece of DNA to be 'cloned' using recombinant DNA techniques.

DNA does not replicate itself: it needs a battery of enzymes to replicate inside a cell. The enzymes coordinate DNA synthesis with cell growth by only starting the synthesis of a DNA molecule at a specific time in the cell's growth cycle. To allow this to happen the DNA must contain a 'start here' signal, called the origin of replication. Thus any DNA that is to be cloned must contain an origin. A unit of DNA that has an origin of replication (and a signal to stop replication at the other end, if that is necessary) is called a replicon. Since most bits of DNA do not contain an origin, they must be given one: this is done by splicing them together with an origin-containing bit of DNA, called a vector. Vectors can be thought of as little replicons into which we can add other DNA.

This is the basic function of vectors. To make them convenient to use, they have a range of other properties.

Most cloning vectors are episomes, i.e. they are genetic elements that can be replicated separately from the host cell's chromosome (i.e. the rest of its DNA). Episomes can be plasmids (small loops of DNA with no function that is deleterious to the cell) or persistent viruses (bits of DNA with the potential of coding for virus particles) (see **Plasmid**). 'Traditional' vectors, such as the pBR series and the 2μm vectors for yeast, are plasmids, while the lambda series of DNA-sequencing vectors are based on a bacteriophage (a bacterium-eating virus). Other viruses, such as T7, are also used, and bits of them have been used in the construction of more exotic beasts such as cosmids. (Cosmids, much used in large-scale gene cloning, are plasmids that can be packaged up into lambda virus particles, but only when you have put 40 000 bases of foreign DNA into them. Thus the packaging process is an excellent way of ensuring that you have a plasmid with a large stretch of DNA inserted into it.)

A few vector types integrate their DNA into the chromosomes of their targets. Examples are YIps (yeast-integrating plasmid), a plasmid that inserts itself into the chromosomes of yeast, and retrovirus vectors for mammalian cells. [Other yeast-cloning vectors, such as the YEps (yeast episomal plasmids) based on the '2-μm plasmid' remain separate from the chromosomes like a bacterial plasmid does.]

The DNA replication parts of a vector are specific to the organism it is intended for; a bacterial plasmid will not (usually) replicate in mammalian cells. You can construct a vector such that it has the DNA replication origins for two organisms, for example, for *E. coli* and mammalian cells, so that the vector can be manipulated in *E. coli* (which is easy to grow) and then transferred directly to mammalian cells (where we want to do the experiment). Such two-host vectors are called binary vectors or shuttle vectors.

Vectors contain a range of genetic elements to make cloning with them easier. These can include:

Selectable genes. These code for something that allows the cell to survive an otherwise hostile environment. A common one is a gene for resistance to an antibiotic: growing the engineered organism in the presence of antibiotic will select those organisms that contain the vector (and hence whatever genes we have spliced into the vector).

Polylinker. This is a piece of DNA made to contain many restriction enzyme sites, so the vector can be cut at that one place for splicing in other genes.

Specialist origins. Other variations on origins of replication are:

- High copy number plasmids, which exist as many copies inside a cell, not the usual one or two.

- Runaway replication plasmids, where, on some trigger signal (usually a change in temperature) the normal controls on how much plasmid DNA there is in a cell break down and the cell fills up with the plasmid.

Promoters, enhancers, leader peptides . These are elements to help you express a gene that is cloned in the vector (*see* **Expression systems**).

Because there are so many possible vectors that can be assembled out of these components, some vector systems are made not as complete vectors but as cassette systems, where different selectable genes, origins, etc. can be slotted together to make the vector of your choice.

A critical feature of a vector is the amount of foreign DNA that it can hold. The practical limits (in thousands of base pairs, kilobases) are:

'Conventional plasmid vectors'	15
Lambda phage vectors	20
Cosmids	45
P1 bacteriophage vectors	85
BACs (bacterial artificial chromosomes)	300
YACs (yeast artificial chromosomes)	2000
MACs (mammalian artificial chromosomes)	at least 1000

Vectors for gene therapy

A major barrier to getting gene therapy into the clinic is the lack of effective ways of delivering genes to cells in living patients. Several viruses have been engineered to act as gene therapy vectors, with varying success. The most commonly discussed are:

Retroviruses. Retroviruses efficiently transport their RNA into cells, copy their RNA into DNA, and then insert that DNA into the cell's chromosome. In principle this ability could be harnessed to carry other DNAs into a patient's cells, and the viruses' ability to grow removed (producing a replication-deficient retroviral vector, RDRV). However, such vectors are hard to engineer to be safe. Manufacturing gene therapeutics based on retroviruses is also difficult since the genetics of the cells you grow the virus on can readily generate replication-competent viruses (RCVs), i.e. variant viruses that can grow in their target cell like a virus, and so start an infection *see* **Retroviruses**). A practical difficulty is that retroviruses cannot infect non-dividing cells, like muscle or nerves.

Adenoviruses. Adenoviruses can incorporate their DNA into the cell's nucleus without inserting it into the chromosomes. They can infect cells that are not dividing, which is a major advantage over retroviruses. However, the genes that are engineered into them are not expressed permanently in the receiving cells.

Adeno-associated virus (AAV). These are vectors based around the virus particle from adenoviruses, but with no viral genes. The foreign DNA has the ends of the adenovirus DNA spliced on to its ends, and is placed in mammalian cells that also contain some adenoviral genes in their chromosomes. The viral genes make adenovirus coats, which are filled with the foreign DNA. AAV virus-based therapeutics for cystic fibrosis, which take the 'correct' CF gene directly to the lungs of affected patients in a spray, are in the clinic.

Alphaviruses. Viruses such as semliki forest virus (SMV) and sindbis have been research subjects for decades, and have been developed into powerful expression vectors in recent years. They could be converted into gene therapy vectors by engineering replication-defective versions of the vectors, which would then express large amounts of a therapeutic protein inside the target cells. However, this is still at research stage.

Herpesviruses. The herpesviruses carry DNA into specific types of cell, where they lie low for months or years. Engineered versions have been made to carry foreign DNA into cells either in replication-deficient viruses or in otherwise 'empty' virus coats. Again, this is largely a research-stage idea.

The alternative approach to viral vectors is a chemical system, such as wrapping the DNA into liposomes or complexing it with lipids, such as lipofectin, that speed the DNA into cells. This gets round the safety problems of viruses, but is very much less efficient—less than 0.1% of cells affected, compared with over 10% for some adenovirus systems.

Vertical integration

A must term for management consultants, this means a company that can perform all the parts of development, production, and sale of something. In the pharmaceutical industry, a vertically integrated company would be one that researched, developed, manufactured, marketed, and sold a drug.

There is a major difference in philosophy between companies that seek vertical integration and those that do not. The latter see themselves as providing a service for larger, 'mainline' pharmaceutical companies: they discover or invent drugs, develop new ways of delivering them, or provide research or development capabilities for making the drugs. By contrast, some biotechnology companies feel that it is their destiny to become large drug companies, doing everything from drug discovery to knocking on GP's doors. They want to become FIPCOs—fully integrated pharmaceutical companies. A less ambitious goal is to become a FIDDO (fully integrated drug discovery and development organization), leaving the manufacture, marketing, etc. of the product to established pharmaceutical companies.

Outside healthcare, in areas such as environmental clean-up or specialist chemicals synthesis, the same criteria do not apply, since the biotechnology companies act as service suppliers for other companies and individuals in many industries: no biotechnology company making food products wants to run high-street supermarkets, for example.

A related idea is the virtual company. This is a company that assembles its capabilities by alliance or contract with other organizations, rather than building them 'in-house. Thus, for example, a pharmaceutical company could licence the world-class research of a biotechnology start-up, get preclinical development and clinical trials performed by clinical research organizations (CROs), have manufacturing done by a contract chemical manufacturing company, and even hire an off-the-shelf sales force. The company's value lies in its ability to put this consortium together, make it all work, and fill in the gaps. This is no mean feat, and so very few truly virtual companies exist in biotechnology.

Viral vaccines

Also called live virus vaccines, these are vaccines that consist of live viruses rather than dead ones or separated parts of viruses. Clearly, the virus itself cannot be used, as that would simply give you the disease, so, instead, one of two genetic engineering methods is used to produce a virus that will elicit an immune response to a viral pathogen but will not cause the disease itself.

The first method is to engineer the disease virus genetically so that it is harmless, but can still replicate (albeit inefficiently, sometimes) in cultured animal cells. This has the same effect as producing an 'attenuated' virus, i.e. one that has been grown in the laboratory until it loses its ability to cause disease. However, the genetic engineering route seeks to make sure that the attenuated virus has no chance of mutating back to a 'wild type', pathogenic virus, by deleting whole genes or replacing key regions of genes with completely different genetic material.

The other approach is to clone the gene for a protein from the pathogenic virus into another, harmless virus, so that the result 'looks' like the pathogenic virus but causes no disease. The foreign protein must not itself be dangerous, of course. Vaccinia and adeno-viruses have been used in this way, notably to make rabies vaccines for distribution with meat bait: a trial of such a vaccine was carried out in summer 1990 in the USA. A similar vaccinia virus engineered to produce the haemagglutinin and fusion proteins of rindepest virus (a disease of cattle) has been developed successfully by the University of California at Davis.

Walking

There are several techniques known as 'gene walking' or 'chromosome walking'. They are all methods for cloning large regions of a chromosome. The diagram illustrates the basic idea. Starting from a known site, a gene library is screened for clones that hybridize to DNA probes taken from the ends of the first clone. These clones are then isolated, and their ends used to screen the library again. These clones are then isolated and their ends used . . ., and so on. This can go on as long as is needed to get from where you are (usually at a 'linked marker'—an RFLP site known to be near the gene of interest) to where you want to be.

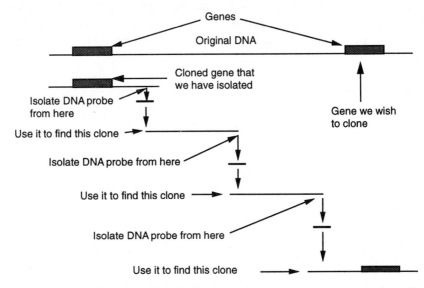

There are variants called 'gene jumping' or 'chromosome jumping' that allow some of the intermediate steps to be cut out: these rely on rearranging the original chromosomal DNA during cloning.

To 'walk' a chromosome quickly, it is useful for the clones to each cover an extremely large amount of DNA, otherwise each 'step' will only cover a small amount of the genome. Thus cosmid vectors (which hold 40 000 bases of foreign DNA per clone) and YAC vectors (which can hold up to a million bases) are preferred.

Waste disposal

Waste disposal is a major concern of environmental biotechnology (the other major concern being bioremediation—the removal of environmental poisons). Biotechnology addresses waste disposal in many ways.

Sewage disposal. All sewage disposal is through biotechnological routes to degrade the organic matter. In crowded areas of the world like Holland and Japan, sewage disposal is a major problem (*see* **Sewage treatment**).

Digestor design. Related to sewage disposal is the design and operation of digestors, bioreactor systems designed specifically to destroy organic material rather than convert it into something else (*see* **Digestor**).

Composting and landfill. Much of Western culture's waste is thrown into holes in the ground, where the organic matter slowly decomposes. This is especially difficult for the mixture of organic materials present in common household waste, (called municipal solid waste, or MSW). Crowded parts of the world are running out of suitable holes, so finding more efficient methods are of substantial concern for biotechnologists. This is especially true of the composting stage, which applies to any material with organic content. Such landfill material generates a lot of gas as it decomposes, a gas mixture called landfill gas which is made up of 50–65% methane, 30–40% CO_2, and small amounts of nitrogen, hydrogen, and volatile organic chemicals. Large sites can generate 1000 m^3 of landfill gas per day, which represents a substantial explosion risk if it is not released. It can be used as a source of biogas fuel, but must be scrubbed first to remove hydrogen sulfide and other smelly and poisonous trace gases.

Wood

Wood processing has attracted increasing attention from biotechnology, in part because traditional methods produce a lot of effluent that is considered to be environmentally unfriendly, and in part because wood is a biological material that is well suited to processing by biological means. Nearly all the bioprocessing of wood is aimed at paper production, which takes wood chips and turns it, via wood pulp, into clean, white cellulose for making paper.

The six areas on which biotechnology has focused are:

Tree genetics. Breeding trees is very hard; conventionally, people have found a tree that happens to be suitable, and cloned it by micropropagation. However, modern genetic mapping technology allows you to test a seedling for the desired combination of genes, long before they make themselves known in the adult tree. However, few of the traits we want are caused by single genes (*see* **Quantitative trait loci**).

Preprocessing. This is the removal of pitch and resins from wood. The wood from most trees contains a substantial amount of complex, oily chemicals which preserve the wood from being attacked by insects and bacteria. This has to be removed, and can be by 'fermenting' wood pulp with microorganisms that grow on pitch or digesting it with lipases that break the pitch down into water-soluble materials.

Pulping. Usually, wood chips are converted to pulp mechanically or using chemicals. Enzyme methods are being investigated. The object here is to break down the lignin and other non-cellulose materials that hold the cellulose fibres together. Several fungi that make ligninases are known, and these can contribute to breaking the wood down. At the moment such methods are used in conjunction with mechanical mashing. Enzyme or fungal treatment softens the wood, reducing the power needed in the mechanical pulpers.

Fibre modification. The nature of paper depends strongly on the sort of fibres that it is made of. The cellulose fibres can be modified by trimming off surface irregularities. Generally, biotechnological methods are too expensive for use in the paper industry (although the same process is used in fibre polishing in textile manufacture).

Biobleaching. The colour of paper is extremely important. Wood is coloured because of a large number of compounds that permeate the fibres, primarily materials that fall under the general heading of 'lignin'. Ligninases have been used to bleach pulp without resort to the chlorine and chlorine oxides usually used in the paper industry. Xylanases are also used here: these break down polysaccharides other than cellulose and so release coloured materials trapped in the pulp. (It is important that the xylanases be free of any contaminating cellulase since this would break down the cellulose as well.) Novo Nordisk market such a xylanase preparation, called Pulpzyme.

Waste disposal. Producing new paper and recycling old paper generates a lot of waste water that contains a lot of carbon compounds. These can be a substantial pollution problem, raising the BOD of the waste water to unacceptable levels (*see* **Sewage treatment**). Biological treatment of wood pulp waste is a way of removing or reducing these effluent problems. The same technologies have been tried for processing other agriculturally derived waste, such as straw. They are mainly made of the same cellulose and hemicellulose as wood, with small amounts of lignin, and can be broken down by a combination of grinding, steam or hot water processing, enzyme treatment, or direct bacterial breakdown. Fungal breakdown is the normal mode of breakdown of woody materials in European countries, and many bioprocessing approaches to both woody waste treatment and to bleaching use direct fungal fermentation to break down lignin.

Wool

One of the of the targets of the application of genetic engineering to farm animals is to improve the quality or quantity of wool that sheep produce. This is a complex problem, but one that a number of research groups are working on, especially in Australia, which produces a substantial fraction of the two billion kilograms of wool produced annually world-wide.

Improved wool production focuses on:

1. Inserting the gene for growth hormone in sheep. This has been tried, and seems to produce an increase in wool production, although no one is sure why.

2. Inserting new genes for keratins into sheep. There are several types of keratins in wool, and altering the ratio of them may improve the quality of the wool. This is an empirical approach, because it is not clear what effect any particular gene insertion will have on the wool, even if it makes protein in the right cells and at the right time.

3. Inserting the genes for improved synthesis of cysteine into trans-genic sheep. Keratin, the protein in wool, has a lot of cysteine, which is a limiting factor in the rate of growth of wool. Sheep cannot usually make cysteine themselves since they lack the relevant enzymes, so the engineering aims to give the sheep enzymes from bacteria that can make cysteine from sulfides generated in the rumen.

4. Engineering feed plants. An alternative method to getting more cysteine into sheep is to engineer the plants they eat to contain more cysteine. The problem here is that the rumen bacteria break down a lot of the cysteine in the food, and so improving the forage plants may not improve wool output. Some storage proteins from peas are proof against rumen breakdown, and may be suitable.

5. Engineering rumen bacteria. An alternative route is to manipulate the rumen bacteria to convert cellulose in the feed into chemicals that the sheep can use more efficiently, or to make more essential amino acids, and especially cysteine, available to the sheep. This research is still in a very early stage, in part because, to model accurately what the bacteria do, you need something very like a sheep's stomach as an incubator.

Xenobiotics

A xenobiotic is a chemical that would not normally be found in a given environment, and usually means a toxic chemical that is entirely artificial, such as a chlorinated aromatic compound or an organomercury compound. Biotechnology brushes with xenobiotics in three areas. First, in determining their toxicity and effects on living systems. Secondly, biotechnology has developed methods for removing them through bioremediation or enzyme-based degradation. Lastly, a range of biotechnological products aim to replace the compounds, which, if they get out of their target site, are classified as xenobiotics. Among these are the chemical herbicides and pesticides that biocontrol agents and biopesticides hope to replace.

Nearly all drugs are xenobiotics, and so much of healthcare biotechnology is concerned with them as drugs rather than as chemicals. Proteins are not xenobiotics, of course. A substantial element of how xenobiotics interact with mammals is their metabolism, particularly that mediated by a group of enzymes called cytochrome P450s, which catalyse the first stage of attack on a wide range of different xenobiotics.

Xenograft

A xenograft is a grafted organ or tissue that comes from another species. Because our immune system would violently reject such a graft (especially because of the alien major histocompatability complex antigens on them, discussed in the entry on **transplants**), strategies have to be developed to prevent immune rejection. These include:

- creating transgenic animals that show the human transplantation antigens, not the animal ones.

- creating animals that show no transplantation antigens at all, by gene knock-out.

- encapsulating the organ or tissue so that the human immune system does not 'see' it.

- inducing tolerance (also called 'anergy') in the human, so that they come to accept the foreign tissue as their own.

So far, none of these have been successful.

The most developed practical human xenografts come from pigs, although most of the work is done in other species for convenience. For organs such as the heart, the donating animal has to be similar in size to humans, and with a similar physiology: pigs and baboons are favourites, and since pigs are much easier to breed they are the focus of much of the recombinant DNA work.

There is substantial ethical debate about xenografts, which offend many religious groups. There are also general practical objections centring around the concern that the xenografted tissue will carry new infections, particularly new viruses, into the human population. The historical example behind this case is probably AIDS, which some scientists believe spread from being a relatively contained disease in monkeys to a pandemic in humans. It is ironic that one of the groups pushing for early clinical trials of xenografting are AIDS activists, who want to try transplanting baboon lymphocytes into AIDS patients.

YACs

YACs are yeast artificial chromosomes, and are cloning vectors that are gaining a lot of use in the human genome project. They consist of those pieces of DNA that define the ends (telomeres) and the middle (centromere) of a chromosome in yeast. Both these elements are needed to allow a chromosome to be replicated in yeast cells: if there is no telomere then the ends of the chromosome are liable to be broken off, or to join on to other chromosomes. If there is no centromere, then newly made chromosomes will not be pulled into the new cells during cell division. In addition, there is an origin of replication so that the DNA is replicated.

These elements are placed in a single DNA fragment, which can be used as a vector to clone foreign DNA into yeast. The advantage of a YAC is that there is no effective limit to how big the piece of DNA can be. Thus, while conventional bacterial cloning using bacteriophage or plasmids is usually limited to cloning foreign DNA fragments of a few tens of thousands of bases long, YACs can and have cloned fragments millions of bases long. This makes mapping whole genomes of DNA much easier since the whole genome map has to be assembled from far fewer YAC maps, and also makes cloning very large genes such as the gene for muscular dystrophy (which is at least two million bases long) more straightforward.

YACS are technically difficult to use, and a common problem is that DNA from two different regions of the original genome can end up in one YAC, giving the erroneous appearance that they were originally linked together, a result called a chimeric clone.

BACS (bacterial artificial chromosomes) and MACS (mammalian artificial chromosomes) use the same concept with the genetic elements from their respective species. Other vectors capable of cloning large pieces of DNA are the bacterial vectors based on cosmids and P1 (*see* **Vector**).

Yuk factor

A flippant term for the very real observation that the public, and indeed many scientists, judge the ethical acceptability of experimental procedures and biological manipulations in accordance with a scale of personal distaste. Thus the creation of the first cloned carrot in the 1960s was greeted with amusement in the press, while the creation of the first cloned frog in the early 1970s was treated with interest and some caution, and the creation of Dolly the cloned sheep in 1997 resulted in widespread alarm. Similarly, tests that rely on newts have less negative impact on public relations than those on rats, and rats are considered more acceptable than rabbits or dogs.

In general this reflects a concern for animals that look or behave more like human beings. The ultimate public condemnation is therefore reserved for the potential scientific interference with human fetuses or children. This is a very real scale of values, and one that many scientists do not take seriously enough (hence their calling it a 'yuk factor' rather than a 'value scale'). In public debate the yuk factor is sometimes a deciding one: much of the opposition to Monsanto's promotion of BST as a biopharmaceutical to boost the milk production of dairy cattle was based not on arguments about farm economics but on the feeling that it must be horrible for the cow to be turned into a milk-producing machine.

Zoonosis

Zoonosis is infection by an organism that usually infects another species, for example, infection of humans by hanta viruses (which usually only infect rodents). It is an increasing concern for biotechnology for three reasons

New diseases. Many of the 'new' diseases which are claimed to be a major heath threat to Western cultures, such as Ebola, Marburg, and Lassa viruses, are probably zoonoses. They are 'natural' diseases of tropical primates, but can also infect humans. With increasing travel between tropical and temperate zones, it is likely that more such cross-species infections will turn up.

AIDS. The reason that this may be important is that AIDS, and possibly Lyme disease, may originally have been diseases of other species, which jumped the species barrier to infect humans as zoonoses, but which subsequently became 'naturalized' human illnesses. AIDS in particular is similar to a range of illnesses that are endemic to old world monkeys. Understanding how this jump happened could both give insight into AIDS biology and into understanding whether diseases such as Ebola, which are now very rare, could become common amongst humans.

Xenografts. One way that AIDS was transmitted was by the most common transplant procedure, blood transfusion. There is concern that grafting animal organs (xenografting) could transmit animal infections to humans through a route that would otherwise be impossible for the bacterium or virus to traverse. Viruses are the major concern here since, once introduced into a host, they might mutate very rapidly to give rise to a human version of a pig or goat disease. The possibility that a version of Creuztfeldt–Jacob disease has been passed on to humans from cows enhances people's concern about this type of zoonosis.

Further reading

As this books seeks to cover the whole of several academic disciplines and at least three complete industries, no simple collection of books will give you detailed information on everything touched on here. I have not included any sources on the ethics and social implications of biotechnology: the bookshops are full of such books, but please, dear reader, read the facts first and their authors opinions afterwards.

There are two journals which are excellent sources of information of all sorts on biotechnology:

Nature Biotechnology, pub. Macmillan Magazines Ltd (London). ISSN 1087 0156. Monthly	Industry and technical overviews, news, and some research articles.
Trends in Biotechnology, pb. Elsevier Trends Journals, (Cambridge, UK) ISSN 0167 7799. Monthly	Scientific reviews, reviews of scientific meetings and analysis of technical trends.

There are any number of books on biotechnology and related topics. The following are widely regarded as useful introductory (i.e. usually undergraduate level) texts.

General science

Bruce Alberts, Dennis Bray, Julian Lewis, Martin Raff, Keith Roberts, and James D. Watson, (1997). *Molecular Biology of the Cell*, (3rd edn). Garland Publishing. ISBN 0 8153 1620 8.	Cells and how they work.
Benjamin Lewin, (1997). *Genes VI*. Oxford University Press. ISBN 0 19 857778 8.	How genes work
Wesley A. Volk and Jay C. Brown, (1997). *Basic Microbiology*, (8th edn). Addison Wesley Longman ISBN 0 673 99560 7	How bacteria work

General biotechnology

Robert A. Meyers (ed.), (1995). *Molecular biology and biotechnology. A comprehensive desk reference*. VCH ISBN 1 56081 925 1	Large collection of articles, very focused on molecular biology
Alexander N. Glazer and Hiroshi Nikaido, (1995). *Microbial biotechnology: fundamental of applied microbiology*. W.H. Freeman ISBN 0 7167 2608 4	Counterbalancing text on microbial biotechnology

FURTHER READING

S.M.Roberts, N.J.Turner, A.J.Willetts, M.K.Turner, (1995). *Introduction to biocatalysis using enzymes and microorganisms*. Cambridge University Press IBSN 0 521 43685 0

Uses of biological processes in chemistry

Pauline M. Doran, (1995). *Bioprocess engineering principles*. Academic Press ISBN 0 12 220856 0

Fermentation, bioreactors and related topics

The Internet is also a great source of information, and also of mis-information and downright lies, about biotechnology. See the site http://www.oup.co.uk/molsci/biotech linked to this book for a list of links to some more informative sites.

Index